Application of the Systems Approach to the Management of Complex Water Systems

Application of the Systems Approach to the Management of Complex Water Systems

Editor

Slobodan P. Simonovic

MDPI • Basel • Beijing • Wuhan • Barcelona • Belgrade • Manchester • Tokyo • Cluj • Tianjin

Editor
Slobodan P. Simonovic
Department of Civil and Environmental Engineering University of Western Ontario London
Canada

Editorial Office
MDPI
St. Alban-Anlage 66
4052 Basel, Switzerland

This is a reprint of articles from the Special Issue published online in the open access journal *Water* (ISSN 2073-4441) (available at: https://www.mdpi.com/journal/water/special_issues/syst_approach_appl).

For citation purposes, cite each article independently as indicated on the article page online and as indicated below:

LastName, A.A.; LastName, B.B.; LastName, C.C. Article Title. *Journal Name* **Year**, *Volume Number*, Page Range.

ISBN 978-3-03943-769-6 (Hbk)
ISBN 978-3-03943-770-2 (PDF)

Cover image courtesy of Slobodan P. Simonovic.

Contents

About the Editor

Slobodan P. Simonovic has made seminal contributions to the development of systems engineering approaches to the planning, designing, and managing of complex water resource systems in the search for sustainable and robust physical and societal solutions, based on stakeholder value systems and ethical principles. He has utilized probabilistic and fuzzy simulation and optimization to address subjective and objective uncertainties in managing water resources systems. Moreover, Dr. Simonovic has contributed to the solution of complex reservoir operations problems; developed effective flood management measures; improved assessment of climate change impacts on local scales; and developed decision support for integrated water resource management.

Editorial

Application of the Systems Approach to the Management of Complex Water Systems

Slobodan P. Simonovic

Department of Civil and Environmental Engineering, The University of Western Ontario, London, ON N6A 3K7, Canada; simonovic@uwo.ca

Received: 13 October 2020; Accepted: 16 October 2020; Published: 19 October 2020

Abstract: This paper provides an introduction to, and an overview of, the Special Issue on the application of systems approach to the management of complex water systems. The main motivation in proposing this Special Issue was that today, more than ever, we need a systems approach to assist in dealing with the difficulties introduced by the increase in the complexity of water resource problems, consideration of environmental impacts, and the introduction of the principles of sustainability. This issue offers an opportunity to review applications of the systems approach to water resource management and draw lessons from worldwide experience relevant to future water problems. The Special Issue includes 15 contributions that offer an interesting view into contemporary problems, approaches, and issues related to management of complex water resources systems. It will be presumptuous to say that these 15 contributions characterize the success or failure of the systems approach to support water resources decision-making. However, these contributions offer some interesting lessons from the current experience and trace possible future work directions.

Keywords: systems; complexity; water resources; management

---

1. Introduction

During the past five decades, we have witnessed a tremendous evolution in water resource systems management. From the early days and the introduction of the approach by [1] and some of the most significant texts [2–4] up to today's practice, it is very clear that the approach matured and became essential to support water resources decision making. Three of the characteristics of this evolution should be noted in particular: (1) the application of the systems approach to complex water management problems has been established as one of the most important advances in the field of water resource management; (2) the past five decades have brought a remarkable transformation of attitude in the water resource management community towards environmental concerns, and action to address these concerns; and (3) applying the principles of sustainability to water resource decision-making requires major changes in the objectives on which decisions are based, and an understanding of the complicated inter-relationships between existing ecological, economic and social factors.

Today, more than ever, we need appropriate tools that can assist in dealing with the challenges introduced by the increase in the complexity of water resource problems, consideration of environmental impacts, and the introduction of principles of sustainability. The systems approach is one such tool. This Special Issue offers an opportunity to review some applications of the systems approach to water resource management and draw lessons from worldwide experience relevant to the solution of future water problems.

Let me repeat the basic definition of a system here. Simonovic [4] defines "a system as a collection of various structural and non-structural elements that are connected and organized in such a way as to achieve some specific objective through the control and distribution of material resources, energy, and information". The systems approach is characterized by emergence (the whole is different than the sum of its parts), self-organization (cooperation, interdependence and competition yield stabilizing homeostasis), nonlinearity (small changes in part of the system can have excessively significant effects across the whole), and feedback loops (the outputs of the system affect its inputs).

The experience presented through contributions of the Special Issue and [5] offer the following summary of the current state of the water resources systems approach: (i) the water resources systems approach today offers a scientific interdisciplinary context for dealing with the complex practical issues of water management and prediction of the water resources future; (ii) the systems approach is helping all those who are responsible for water resources management to organize water related information and improve the decision-making; (iii) the implementation of the systems approach allows us to address complex problems in close collaboration with the general public; (iv) the systems approach, allows through clear articulation of assumptions, use of models, identification of feedback relationships, and monitoring system behavior, and helps decision-makers better anticipate future conditions and make smarter management decisions; (v) the tools of systems analysis (simulation, optimization and multi-objective analysis) provide decision-makers with the information for full understanding of the dynamics that direct the interactions between the social (people and economy), natural (water, land and air) and constructed systems (buildings, roads, bridges etc.); (vi) the systems approach is contributing to the improvement in human behavior by using systems thinking; and (vii) the systems approach leads to greater practical and safer uncertainty management policies for increasing the resilience of water systems to changing conditions.

In the review presented in this Special Issue [5,6] it is pointed out that "a success reached today must contribute to further evolution of the water resources systems approach to successfully address the serious water challenges faced by society. The future activities must continue: to deal with the most difficult complex water problems; to conduct further practice-based as well as fundamental research; and provide further capacity building".

2. Contributions

The Special Issue includes 15 contributions that offer an interesting view into contemporary problems, approaches, and issues related to the management of complex water resources systems. It is not easy to classify the contributions published in the Special Issue. Their order of presentation in the Issue reflects my understanding of the contributions.

The Special Issue Organization of Contributions

The Issue opens with the paper by Prof. D.P. Loucks [7], who is one of the leaders in this field and has provided invaluable contributions that influenced academia, industry, and governments. His message focuses on the transition in the water resources systems approach from preoccupation with methodological issues to implementation experiences and innovation. Prof. Loucks sends a message that a crisis in water is no longer an abstraction for many. Adapting to globally changing conditions is the challenge for all of us.

The following papers by Morley and Savic [8], and Rusforth et al. [9], deal with water scarcity. Morley and Savic offer an optimization approach to the "Lower Thames Control Diagram", a set of control curves subject to a large number of constraints. The diagram is used to regulate abstraction of water for the public drinking water supply for London, UK, and to maintain downstream environmental and navigational flows. The optimized configuration of the Lower Thames Control Diagram was adopted by the water utility and the environmental regulators and is currently in use. Rusforth et al. present a rigorous quantitative, systems-based model to measure a municipality's water portfolio security using multiple objectives. This simple model can be operationalized using readily available data to capture water security dimensions that go far beyond typical reliability and cost analysis. They used the Phoenix Metropolitan Area as a case study.

Horriche and Benabdallah [10], Lee and Kang [11], and Hattab et al. [12] further the discussion to the applications of groundwater management, multipurpose reservoir operations, and urban drainage, respectively. The first paper examines the impact of an artificial recharge site on groundwater level and salinity using treated domestic wastewater for the Korba aquifer (north eastern Tunisia). Groundwater flow and solute transport models are utilized in the identification of suitable areas for aquifer recharge. Lee and Kang, in their study, clarify relationships within the social and hydrological systems and quantitatively analyze the effects of a multi-purpose dam on the target society using a system dynamics simulation approach. Hattab et al. implement the soft system engineering and Analytic Network Process (ANP) approaches in a methodological framework to improve the understanding of the stakeholders within the sustainable urban drainage system and their key priorities, which leads to selecting the appropriate modeling technique according to the end-use application.

The three contributions by Rehana et al. [13], Agrawal et al. [14], and Sabbaghian and Nejadhashemi [15] bring uncertainty into the discussion of complex water resources systems management. Rehana et al. appraise the quantification of uncertainties in systems modeling in India and discuss various water resource management and operation models. The basic formulation of models for probabilistic, fuzzy, and grey/inexact simulation, optimization, and multi-objective analyses to water resource design, planning, and operations are very well presented in this work. Agrawal et al. present a study that includes identifying and quantifying the gap between people's perception of exposure and susceptibility to the risk, a lack of coping capacity and objective assessment of risk and resilience, as well as estimating an integrated measure of disaster resilience in a community. The proposed method has been applied to floods in the hope that the study will encourage a broader debate if a unified strategy for disaster resilience would be feasible and beneficial in Canada. Sabbaghian and Nejadhashemi present a risk-based consensus-based group decision-support system model for choosing the desirable urban water strategy. This model is successfully implemented for the Kashafroud urban watershed in Iran, for selecting the more desirable urban water strategy in 2040.

Stojkovic and Simonovic [16], Hooshyar et al. [17], and Aivazidou and Tsolakis [18] address various issues in managing complex water problems. Stojkovic and Simonovic study the impact of climate change on the management of a complex multipurpose water system and present a set of steps of the climate change impact analysis process. They used the Lim water system in Serbia (southeast Europe) as a case study. Furthermore, their study analyzed the uncertainty in the system outputs introduced by different steps of the modeling process. Hooshyar et al. deal with reservoir operations optimization under uncertainty. They introduce reinforcement learning, a simulation-based stochastic optimization approach that can effectively eliminate the curse of modeling that arises from the need to calculate a very large transition probability matrix. This paper presents a multi-agent approach combined with an aggregation/decomposition method. The method has been applied to a real-world five-reservoir problem, the Parambikulam–Aliyar Project in India. Aivazidou and Tsolakis present an interesting and unusual problem of wine–water footprint assessment to investigate the water dynamics of wine production in Italy and the wine sector's water efficiency. This research provides insights for practitioners in the Italian wine sector to develop water-friendly corporate schemes for enhancing the added value of their products.

The next two papers by Madani and Shafiee-Jood [19] and Ponnambalam and Mousavi [20] target a controversial development related to socio-hydrology as a "new science" of interaction between human and natural systems. Madani and Shafiee-Jood correctly point that the socio-hydrology studies show strong overlap with what has already been in the literature, especially in the water resources systems and coupled human and natural systems (CHANS) areas. Nevertheless, the work in these areas has been generally dismissed by the socio-hydrology literature. Their paper overviews some of the general concerns about originality, practicality, and contributions of socio-hydrology. It is argued that, while in theory, a common-sense approach about the need for considering humans as an integral component of water resources systems models can strengthen our coupled human-water systems research, the current approaches, and trends in socio-hydrology can make this interest area less inclusive and interdisciplinary. Ponnambalam and Mousavi state that coupled human–natural system models provide the practical approach needed for applications both in the descriptive science of socio-hydrology and in the prescriptive method of integrated water resources management. Since the introduction of socio-hydrology as a "new science" various responses and criticisms clearly indicating no novelty in the concept and presence of interaction between human activities and water systems in the literature over a number of decades. However, in this paper there are some issues like (i) treatment of integrated water resources management (IWRM) as a tool, not a process; (ii) a view of socio-hydrology as science and IWRM as an engineering approach (which is clearly wrong); (iii) stating that socio-hydrology promotes CHANS (the literature of socio-hydrology, unfortunately, had not originally admitted CHANS); and (iv) proposing CHANS as a modeling tool (which is problematic as CHANS is not a tool but an analysis approach/framework which takes advantage of many tools including system dynamics, economics, and others).

The Special Issue ends with my paper [6] that states that systems approaches based on simulation, optimization, and multi-objective analyses, in deterministic, stochastic, and fuzzy forms, have demonstrated great success in supporting effective water resources management in the last half of last century. In this paper, I explore the future opportunities that will utilize advancements in systems theory that might transform the management of water resources on a broader scale. The paper presents performance-based water resources engineering as a methodological framework to extend the systems approach's role in improved sustainable water resources management under changing conditions (with special consideration given to rapid climate destabilization).

3. Conclusions

The key messages we can extract from the submissions included in this Special Issue are quite broad and definitively not limited to what has been addressed with these contributions. It can be concluded that the water resources systems approach: (i) offers a very reachable portfolio of applications and a scientific interdisciplinary context for dealing with the complex practical issues of water management and prediction of the water resources future; (ii) is helping all those who are responsible for water resources management to organize water related information in order to distinguish between the noise and important information and improve the decision-making; (iii) provides the information necessary to understand resource flows and the larger water resources management context in close collaboration with the general public to understand the relationships between human behavior and environmental and economic impacts of water resources management decisions; (iv) is helping the improvement of planning and forecasting by articulation of assumptions, use of models, identification of feedback relationships, and monitoring system behavior; (v) offers the tools (simulation, optimization and multi-objective analysis) that are helping to improve the quality of decision-making; (vi) is contributing to the improvement in human behavior by using systems thinking; and (vii) leads to greater practical and safer risk management policies.

There are still remaining challenges necessary to respond to global changes that affect and alter the hydrologic cycle, and that define human relationships with natural systems. It is our hope that some of the ideas addressed in this collection of papers will help all of us in become more innovative, and increase our collaboration in securing solutions for a sustainable future.

Funding: This research received no external funding.

Acknowledgments: My time for working on this Special Issue was supported by the Discovery Grant from the Natural Sciences and Engineering Research Council of Canada. I would like to express my gratitude to all reviewers of the submissions included in the Special Issue, WATER Journal editors, and staff for their effort, time, and significant contributions to this publication.

Conflicts of Interest: The author declares no conflict of interest.

References

1. Maass, A.; Hufschmidt, M.; Dorfman, R.; Thomas, H.; Marglin, S.; Fair, G. *Design of Water-Resource Systems: New Techniques for Relating Economic Objectives, Engineering Analysis, and Governmental Planning*; Harvard Univ. Press: Cambridge, MA, USA, 1962.
2. Loucks, D.P.; Stedinger, J.R.; Haith, D.A. *Water Resources Systems Planning and Analysis*; Prentice Hall: Englewood Cliffs, NJ, USA, 1981.
3. Loucks, D.P.; van Beek, E. *Water Resources Systems Planning and Management: An Introduction to Methods, Models and Applications*; UNESCO: Paris, France, 2005; p. 680.
4. Simonovic, S.P. *Managing Water Resources: Methods and Tools for a Systems Approach*; UNESCO: Paris, France; Earthscan James & James: London, UK, 2009; p. 576. ISBN 978-1-84407-554-6.
5. Brown, C.M.; Lund, J.R.; Cai, X.; Reed, P.M.; Zagona, E.A.; Ostfeld, A.; Hall, J.; Characklis, G.W.; Yu, W.; Brekke, L. The future of water resources systems analysis: Toward a scientific framework for sustainable water management. *Water Resour. Res.* **2015**, *51*, 6110–6124. [CrossRef]
6. Simonovic, S.P. Systems Approach to Management of Water Resources—Toward Performance Based Water Resources Engineering. *Water* **2020**, *12*, 1208. [CrossRef]
7. Loucks, D.P. From Analyses to Implementation and Innovation. *Water* **2020**, *12*, 974. [CrossRef]
8. Morley, M.; Savić, D. Water Resource Systems Analysis for Water Scarcity Management: The Thames Water Case Study. *Water* **2020**, *12*, 1761. [CrossRef]
9. Rushforth, R.R.; Messerschmidt, M.; Ruddell, B.L. A Systems Approach to Municipal Water Portfolio Security: A Case Study of the Phoenix Metropolitan Area. *Water* **2020**, *12*, 1663. [CrossRef]
10. Jarraya Horriche, F.; Benabdallah, S. Assessing Aquifer Water Level and Salinity for a Managed Artificial Recharge Site Using Reclaimed Water. *Water* **2020**, *12*, 341. [CrossRef]
11. Lee, S.; Kang, D. Analyzing the Effectiveness of a Multi-Purpose Dam Using a System Dynamics Model. *Water* **2020**, *12*, 1062. [CrossRef]
12. El Hattab, M.H.; Theodoropoulos, G.; Rong, X.; Mijic, A. Applying the Systems Approach to Decompose the SuDS Decision-Making Process for Appropriate Hydrologic Model Selection. *Water* **2020**, *12*, 632. [CrossRef]
13. Rehana, S.; Rajulapati, C.R.; Ghosh, S.; Karmakar, S.; Mujumdar, P. Uncertainty Quantification in Water Resource Systems Modeling: Case Studies from India. *Water* **2020**, *12*, 1793. [CrossRef]
14. Agrawal, N.; Elliott, M.; Simonovic, S.P. Risk and Resilience: A Case of Perception versus Reality in Flood Management. *Water* **2020**, *12*, 1254. [CrossRef]
15. Javidi Sabbaghian, R.; Nejadhashemi, A.P. Developing a Risk-Based Consensus-Based Decision-Support System Model for Selection of the Desirable Urban Water Strategy: Kashafroud Watershed Study. *Water* **2020**, *12*, 1305. [CrossRef]
16. Stojkovic, M.; Simonovic, S.P. System Dynamics Approach for Assessing the Behaviour of the Lim Reservoir System (Serbia) under Changing Climate Conditions. *Water* **2019**, *11*, 1620. [CrossRef]
17. Hooshyar, M.; Mousavi, S.J.; Mahootchi, M.; Ponnambalam, K. Aggregation–Decomposition-Based Multi-Agent Reinforcement Learning for Multi-Reservoir Operations Optimization. *Water* **2020**, *12*, 2688. [CrossRef]
18. Aivazidou, E.; Tsolakis, N. A Water Footprint Review of Italian Wine: Drivers, Barriers, and Practices for Sustainable Stewardship. *Water* **2020**, *12*, 369. [CrossRef]
19. Madani, K.; Shafiee-Jood, M. Socio-Hydrology: A New Understanding to Unite or a New Science to Divide? *Water* **2020**, *12*, 1941. [CrossRef]
20. Ponnambalam, K.; Mousavi, S.J. CHNS Modeling for Study and Management of Human–Water Interactions at Multiple Scales. *Water* **2020**, *12*, 1699. [CrossRef]

Publisher's Note: MDPI stays neutral with regard to jurisdictional claims in published maps and institutional affiliations.

Commentary

From Analyses to Implementation and Innovation

Daniel P. Loucks

School of Civil and Environmental Engineering and Institute of Public Affairs Cornell University, Ithaca, NY 14850, USA; loucks@cornell.edu or dpl3@cornell.edu

Received: 10 March 2020; Accepted: 26 March 2020; Published: 30 March 2020

Abstract: Reviews of the water resource systems planning and management literature show considerable interest in methodological issues and less so in implementation experiences. This paper offers some thoughts on the use of our analysis tools in the political environment where water management decisions are typically made. This paper also addresses the challenge of going beyond analysis and synthesis to innovation. How can we extend our modeling methods so as to help ourselves become more creative in the identification of potentially improved infrastructure design and/or operating policies, and even of institutional changes, that we have otherwise not considered or thought of?

Keywords: systems analyses; water resources; planning; management; implementation; political processes; innovation; impact

---

1. Introduction

This commentary is addressed to all of us who develop and apply various quantitative modeling approaches designed to assist those responsible for managing water and related environmental resources. This includes those of us trained in various disciplines that offer different perspectives and contribute to identifying and analyzing alternative solutions in different ways, all aimed at forming a more comprehensive estimate of the impacts that could result from decisions that might be made.

Those of us who have been involved in the use of systems analysis methods are aware of the contribution these methods have made and are making in a wide range of applications, including agriculture, defense, ecosystem management, education, environmental protection, industry, law enforcement, medical care, resources management, transportation, and urban planning among others. Systems analyses have been most helpful in addressing issues dominated by natural and physical sciences and engineering. Yet such issues are usually addressed and resolved in a political environment. This is certainly the case when planning, designing, and operating infrastructure for managing water. This paper focuses on the implementation of systems analysis for informing the largely political processes of deciding how best to manage our water resources. If the purpose of our analyzing specific water resource systems is to implement change, then we, analysts, must get involved in and cater to the political processes in which water management decisions are typically made.

Most of us will agree that while systems analysis methods, and each of the disciplines they come from, have their limitations, they can introduce a certain objectivity into the political process of decision making. Progress in managing water more effectively requires knowledge from the natural, social and political sciences, economics and other disciplines. Of course, achieving change requires institutions and political alignments in addition to the insights derived from scientific knowledge. Yet such scientific objectivity can help achieve stakeholder acceptance of the identified options available and the inevitable tradeoffs among the goals they may wish to obtain [1]. Such analyses can address uncertainties, even uncertain uncertainties; they can estimate various impacts and tradeoffs among multiple system performance measures; and they can help reveal unexpected consequences of particular policies and actions. We can use systems analysis methods to help identify plans and policies that achieve a balance among multiple goals of multiple stakeholders. Simply stated, systems analysis methods have proven themselves to be useful for addressing large, complex water management challenges and opportunities. Results from such analyses can inform but, with rare exceptions, they have not proven very effective in substituting for those responsible for decision making.

The art of applying systems analysis tools, especially to water management issues, is itself inherently multi- and interdisciplinary. One can argue that it was borne in a multidisciplinary environment [2]. Systems analysis approaches are designed to focus on the performance of entire systems rather than each of their components, but just what components are or are not included in a particular water resource system depends on the management issues being addressed and the authorities given to the institutions involved. If the art of defining the system components and their interactions is done well, and in collaboration with those involved in decision making so as to enhance communication, gain trust, and ensure relevance, there is an excellent chance that the structured and objective nature of the systems approach will provide information considered useful by those involved in the decision-making process [3–5].

2. Mismatches

Water resource management issues arise when there are mismatches between what people want or desire and what they are getting or observing. There seems to be a continuing stream of such mismatches reported in the news media each day. They give proof that many of our water management problems have become very large and very complex, technically and politically, and that these mismatches can have substantial adverse consequences on our wellbeing as individuals and as communities, and also on our environment. Addressing and reducing these mismatches is a challenge given the uncertainties in supplies and demands. Without the aid of our analysis tools, it is considerably more difficult to deal with such problems simply by, but not excluding, intuition or hunches [6].

Consider some headlines that have made the news in recent months as reported in Circle of Blue <info@circleofblue.org>:

- Utilities in Colorado, US, prepare for water shortages amid the lowest mountain snowfall in 30 years.
- Volatile weather patterns cause rivers across Germany to overflow their banks.
- Southeast England may be at risk of water shortages following a year of dry weather.
- More than 200 flood alerts were in place across the UK, including several severe or "danger to life" warnings. Fifteen rivers across England's Midlands, Yorkshire, and Lancashire have reached their highest levels ever recorded, and an estimated 3300 English homes have been flooded. Several hundred homes in Wales were inundated as well.
- Somalia experiences its fourth consecutive failed rainy season, exacerbating the country's instability.
- Disputes between Texas and Colorado and New Mexico over the Rio Grande and between Florida and Georgia over the allocation of the water flowing from the Blue Ridge Mountains are being addressed by the U.S. Supreme Court.
- Water shortages play a role in ongoing unrest across Iran.
- Drought, flooding, and other natural disasters threaten half of U.S. military bases worldwide.
- Taps have been on the verge of running dry in several major global cities, including Cape Town, South Africa; Mexico City, Mexico; Melbourne, Australia; and Kabul, Afghanistan. the United Nations claims this will happen to 2/3rds of the globe by the year 2025.
- Almost one-fifth of the world's population, live in areas of physical scarcity, and 500 million people are approaching this situation.
- Almost one quarter of the world's population face economic water shortage due to inadequate infrastructure.

- Last year, the Mekong river's waters dropped to the lowest in a century. The water has changed to an ominous color and begun filling with globs of algae. Fish in the Mekong, the world's largest inland fishery, are emaciated.
- Glacier melt in western China increases, threatening the water supply of 1.8 billion people
- Tests results following a massive fish die-off in Iraq's Euphrates River show high levels of bacteria and heavy metals in the waterway.
- U.S. food trade increasingly depends on groundwater use that is not sustainable.
- Flooding and landslides in Belo Horizonte, Brazil, have killed over 50 people.
- Chemicals, including pesticide DDT, are found in the tissues of dolphins swimming in waters flowing to the Great Barrier Reef.
- Heavy flooding in Madagascar displaces at least 16,000 people.
- A vessel runs aground on the Danube river in northern Bulgaria due to low water levels, blocking a key shipping route.
- Ongoing research reveals the pervasiveness of polyfluoroalkyl substances (PFAS). These "forever chemicals" are estimated to be in the bloodstream of 99 percent of Americans, and some scientists believe that nearly all of the country's surface water is likely contaminated.

The list could go on. What is clear is that there are many places and times where widespread mismatches between the desired flows, levels, and qualities of water and what exists. The question is what to do about issues such as these. It is the responsibility of water managers to address these issues, and one way of identifying, analyzing, and evaluating alternative options is through the use of systems analysis methods. Yet such analyses by themselves will not change anything. To effect change, one has to perform such analyses in collaboration with those institutions having the responsibility and authority to make water management changes in specific situations. Analysts need to address the goals (as stated and as understood) of these institutions, recognizing that these goals can change during the time analyses are being performed. Lawyers are useful participants in such efforts. They can translate the results of our systems analyses into the legislation needed to enable changes. Skillful analysts are those who can work in a multidisciplinary environment that may include engineers, economists, ecologists, lawyers, planners, and politicians among others.

3. Water Resources Systems Analysis

No doubt everyone reading the papers in this series knows what systems analysis is, but I have to admit that when I began studying this subject, no one knew much about what that term meant, except for the fact that our military had a bunch of so-called "whiz kids" using systems analysis methods to 'win' the Vietnam war. (Clearly, systems analysis has its limitations!) I began studying this subject just as the Harvard Water Program published their first book [2] showing how optimization and simulation models running on computers could be used to address water resources management issues in ways that integrated economics, hydrology, engineering and political science perspectives. Pretty neat and pretty exciting!

Since then, we have been busy developing and applying many different types of modeling methods, each having its strengths and weaknesses. So far, we have not found one best modeling approach, and I am convinced that we will not. What we have been able to do because of improvements in both model solution algorithms and computer technology is to address increasingly more complex and comprehensive water resources management issues using a variety of methods. From the perspective of a scientist and researcher, a primary role of systems analysis approaches is to contribute to a better understanding of real-world water system performances, humans included, and how they can be improved. From the perspective of a water manager, the primary role of systems analysis methods is to provide quantitative information to help them do their job, i.e., support their decision-making processes [7].

Much of our water resources systems literature today focuses on new modeling approaches (the hammers), often selecting data from particular rivers or basins or urban areas (the nails) to illustrate how their hammers perform. This literature rarely addresses actual model implementations in an institutional environment. However, that does not mean those implementations are not taking place. Firms such as Danish Hydraulics Institute (DHI) and Deltares, and government agencies such as US Environmental Protection Agency (USEPA) and US Army Corps of Engineers-Hydrologic Engineering Center (USACE-HEC) here in the United States are heavily engaged in the implementation of their models and software. The same applies for developers and users of Aquatool [8], CalSim [9], IRAS [10], Riverware [11], WEAP [12,13], and other models used for planning and even real-time operation. Experiences using these models are rarely written up and published in professional journals so most of us cannot learn from those experiences. While I know all of us in this business of modeling water resource systems enjoy inventing new hammers and applying them, going the extra step of actually using them in a political decision-making environment is, in my view, even more fun, more challenging, and certainly very educational. We model developers would all benefit if more of these experiences were included in our literature, including the periodic reviews or assessments of the state of the art of water resources systems modeling such as in [14–37].

4. Implementing Systems Analysis

So, what about the use of systems analysis in support of institutions involved in addressing water management mismatches and making changes in the way water is managed? If I had to summarize my experiences using systems analysis within decision-making processes, I would have to admit that while the results of modeling almost always helped focus the debates on what decisions to make, the decisions themselves were not exactly as I would have predicted. The relative importance of various objectives or even the objectives themselves almost always changed during the planning and decision-making processes, and sometimes even immediately after those process ended. (For example, one month after the completion of a national water resources infrastructure development plan, the country's president died, and the new president and his new department ministers decided to discard that three-year effort carried out by a previous administration, giving the study to others do over again.) My conclusion based on my limited experiences over some five decades is that one should expect such surprises and be ready to adapt to them. All this is in part why, at least for me, these experiences have never been boring and indeed have taught me more than I could have imagined when I began studying for this profession [38].

The motivation to use systems analysis is to identify how to make something better, i.e., how to reduce mismatches. To the extent that the results of the analyses are implemented and improvement actually happens as predicted by the analyses is one measure of success. However, it is not the only one in my opinion. A more achievable measure of success is the extent that the results of systems analyses influences the debate on what to do to achieve improvement. An analysis can be superb technically but if no one pays any attention to its results when debates are taking place about what to do, I will judge it as being unsuccessful with respect to implementing change. Admittedly, it may make a great journal paper. If it has an impact on our research in systems analysis methods, it can certainly be considered successful with respect to that goal.

To affect change in how water is managed, however, we need to do our modeling and analyses so as to have buy in. For this, we need to produce results that are not only deemed useful, and timely, but also simultaneously enhance the salience, credibility, and legitimacy of the insights that they produce. To accomplish this requires, at a minimum, staying in close contact with those involved in decision making throughout the decision-making process. Further, that is not always easy. As just mentioned, goals, constraints, system boundaries, and even stakeholders can change during the period of analysis. An awareness of institutional goals and constraints is critical, and again these will likely change. Stated objectives may differ from what is really desired. Stakeholders and decision makers can change over time, and thus their goals may change. They may really not know what they will want until they know better what they can have and do, perhaps informed by information coming from various analyses taking place over time. Two-way communication between us, analysts, and our clients needs to be maintained throughout the period of developing, using and solving models if we hope to be useful in influencing the debate about what decisions to make.

5. Infrastructure

Water management is accomplished through the design and operation of water infrastructure that permits us to alter the temporal and spatial distribution of water and its quality and the benefits derived from the various uses of it. Infrastructure can include water and wastewater distribution and collection systems, treatment plants, surface and subsurface storage, pumps, canals, aqueducts, cisterns, rain gardens, flood protection measures, and facilities for generating hydropower, cooling, navigation, and rainwater harvesting. Most analyses of water resource systems are focused on addressing what, if any, infrastructure to develop and/or operate, where (siting), when (staging), to what extent (capacity), and why.

Water infrastructure can provide important benefits to society, but it can also generate adverse impacts as well. Today, the flows of water and sediment in over two-thirds of the world's major rivers are altered by dams, diversions, and levees. Close to 1000 new dams are planned or under construction just in South America, in Southeast Asia, and in China. This expansion of dams and associated infrastructure is driven by the need to better satisfy agricultural, domestic and industrial demands for reliable water supplies, for more energy, for recreational opportunities, for reduced risks of damages from droughts and floods, among other purposes. However, dams and levees, for example, can alter the geomorphology of rivers and the functioning of wetland ecosystems including fish habitats, and downstream deltas. Due to population increases and accompanying increases in demands for water, many more dams are being built and for sure, along with their benefits will be their adverse impacts [39].

Our water resources systems analysis literature is full of papers exploring the use of particular methods for identifying, analyzing and evaluating infrastructure and policies for, for example, responding to floods, droughts, and other catastrophic events, restoring ecological habitats, preventing pollution, meeting domestic and industrial water supply and water quality demands, generating power, providing recreational opportunities, meeting energy and agricultural demands, and informing and educating the public on water issues. Many of these water management issues are driven by a changing climate that is bringing us more frequent and more intense storms, floods, droughts, and corresponding land erosion and pollution [40]. Adding to these stresses on our water resource systems are the increasing demands for adequate, reliable, clean and inexpensive supplies of water and for reducing the discharges of a wide range of pollutants that are threating human and ecosystem health. Our early literature mostly focused on models with economic objectives and constraints. However, even if estimated net benefits may have to be positive, it seems to me that many agencies managing water today are less interested in improving economic efficiency or effectiveness as in minimizing the chances of being criticized for doing the wrong thing and in making sure they are spending the money allocated and available to them. This suggests we need to acknowledge not only economic and hydrologic uncertainty but also institutional and political behavior. We often tend to beat to death the former, because we can, and ignore the latter, because it is much harder. It is hard to know what political/institutional uncertainties to consider when we do not even know what they could be, let along their probabilities. It is hard to guess how objectives and constraints, and indeed human behavior, may change over time, and what future generations, many of whom are not yet alive, will want us to do for them as we develop infrastructure today that will exist for a long time. Further, when we do think we know what future generations would want from us, such as reduced greenhouse gas emissions, some alive today will object. There are no 'optimal' solutions no matter what goals are used to rank alternatives. We often have to settle for any change in policies that seems feasible and compatible with how implementing institutions work and that stakeholders will support. I know all of this is not new to anyone involved in practicing the art of water resources systems analysis or engineering. However, it reinforces the argument on how essential it is to perform these analyses in collaboration with our clients if we expect to inform and even influence their decisions. Decisions related to water resources planning and management are made by people in their institutional environments, not by models or algorithms, especially those developed and solved without client involvement.

6. Challenges

It is always fashionable to use the word 'complex' when describing the systems we are analyzing. Nevertheless, is fair to say that many of the water resource systems being analyzed in recent years really are complex. It is one thing to use optimization for identifying the dimensions of a least-cost n-sided ($n \geq 3$) water storage tank as part of an urban water storage and distribution network. This is not a complex problem. However, interestingly, the solution always shows that the minimum total cost results when two-thirds of the tank's total cost is associated with its sides. Knowing this means that we do not need to use optimization models for identifying the dimensions of least-cost tanks. Having such rules of thumb is rare for the systems we are typically asked to study. Without the use of systems analysis methods, it is hard to imagine how else we would estimate the values of all the design and operating decision variables and resulting performance measures we need to be able to improve system design and operation.

Consider, for example, part of the water distribution system for the US city of Houston, in Texas. Houston is a small city compared to some of the world's largest cities, but it is still complex. Its water and sewer system covers a service area of some 640 square miles and consists of over 7500 miles of drinking water and over 6500 miles of wastewater networks. Issues facing the managers of this system include handling the risk of being overwhelmed by floods and having to flush untreated wastewater into surface waters, thereby putting the public at risk of disease and infection. Prolonged power outages could further stress the system's ability to operate. Power outages could shut down some of the city's 380 pump stations. If there is no power, there is no water. In addition, over time, distribution networks break and when they do there is a risk of drinking water supplies becoming contaminated and people becoming sick. Every city in the world faces such challenges. Learning how to address such problems efficiently and effectively over time is clearly a complex challenge [41].

For another example, consider the US state of California. Hundreds of dams, many aquifers and pumping stations, and tens of thousands of kilometers of aqueducts, service a variety of agricultural, commercial, industrial, residential, environmental and ecological water demands throughout the state. These demands include water for instream flows, wetlands, and for maintaining cool or warm water temperatures depending on the local aquatic species of concern. This complex system is managed by numerous federal and state agencies and local water districts and suppliers. Some of these institutions have project management authority and others have regulatory authority. Their water management decisions impact millions of water users each day. They also impact the state's food, energy, industrial and public health sectors. One can be justified by calling even a small part of California's water resource system complex. Many within California have been involved in the development and implementation of models for analyzing and evaluating various alternative designs and operating policies of major parts of California's water resource infrastructure. They have done this in close collaboration with appropriate federal and state and local agencies and to their credit have had a beneficial impact on the management of the state's water [42,43].

These are just two examples of what I would call complex systems. They are made up of multiple interacting components, both physical and institutional, and have hundreds of decision variables that managers can assign values to when attempting to satisfy the multitude of objectives people want from the system. They must make their decisions under all the uncertainties of future water supplies and demands [44,45].

Adding to the complexity of many water resource systems are their links to the energy sector. The energy required to transport, clean, and heat water is estimated to consume approximately

13 percent of US's total national energy supply. As more communities turn to desalination and water recycling technologies the energy embedded in each unit of treated water will rise. At the same time, continued use of coal and nuclear material will further strain water supplies. The energy sector consumes more water for thermoelectric cooling than any other sector, at least in the United States. Rising energy use and prices will drive up the use and price of water and vice versa. Surely this will occur as the future demands for both water and energy increase.

Because of multiple challenges such as these, there has been a push toward using systems analysis methods to address how to develop and manage increasingly integrated water-energy-food-environmental resource systems. However, efforts to achieve greater integration often run up against the lack of institutions having overall responsibilities for managing water supply, wastewater treatment, stormwater management, flood control, energy and agriculture linkages, and habitat restoration especially if these issues cross political boundaries. The lack of such institutions does not negate the value of learning how better to manage water and related resources, including working with nature and thereby benefiting from sustainable ecosystem services that a healthy environment can provide. Who knows, perhaps such studies will motivate institutional cooperation if not change.

7. From Analysis to Innovation?

Systems analysis tools give us some useful tools for helping us plan, design, manage and operate more effective systems. However, they cannot generate suggestions on what we should also be thinking about in addition to what we have been thinking about and including in our models. Our analysis methods do not have the creative capacities that our human brains have. When computer technology became available for developing and using menu-driven graphics interfaces allowing direct interaction between the models and the model users, some of us got excited about the potential of directly involving stakeholders and decision makers in the analyses of various water resource systems. Many of us believed that generating pictures that could show the impact of various design and management decisions or assumptions they might want to explore, and providing the interactivity that would allow them to explore and get estimates of the impacts of different decisions they might make, would give them a better understanding of the system being modeled and how it might work. Such interactive visual displays would also let them explore how they might improve system performance, and indeed help them think about options that could be 'outside the box.' In other words, help them innovate.

We even got fancy with respect to performing sensitivity analyses and displaying uncertainty. Our displays were clear, understandable, and colorful. Sometimes, we witnessed users even believing what they were seeing, ignoring the fact that the models creating the graphical and pictorial displays were approximations of reality. It was fun developing and using such tools in a participatory environment [46–48]. Of course, today almost every model used to analyze water resource systems incorporate interactive, graphics-based, interfaces. They remain approximations of reality. (Sound and smell and touch are coming!)

But what we, modelers and analysts, have not done yet is to figure out how to make our models suggest planning and management options other than what they are programmed to consider. Right now, our models can only inform us about a system we have defined in some general way. To return to our simple tank example, any model for finding the dimensions of a tank is not going to suggest alternatives that might negate the need for that tank. Similarly, if we are modeling a proposed reservoir, say on the Mekong or Nile Rivers, in addition to learning how fast it may fill with sediment under different hydropower production and sediment management policies, would it not be nice if our model could suggest other sites, other designs and other options, such as the use of solar panels for generating power, that might be preferable to what has been modeled, and show the appropriate impacts of those panels regarding costs, power production and sediment and fish passage [49].

In short, models can analyze and synthesize but they cannot yet innovate. They cannot suggest different systems boundaries or components. They cannot suggest different components or policies or options that we have not already included in them. Yet humans can think and create new ideas that are outside the scope of any particular model. How can we develop models that help humans do this? Some are exploring the use of games. Some are suggesting interactive evolutionary computation, maybe coupled to artificial intelligence (and even to the use of LSD). Also waiting to be used in more creative and innovative ways are massive data, Google search engines and their ability to access all the information available on the Internet, Google Earth, voice recognition (that kids take for granted when asking their cell phones questions), parallel cloud computing, and even three-dimensional virtual reality environments that stakeholders can step into and interact with [50,51]. (We can dream!).

I am not at all optimistic about our ability to model and predict human behavior, but with such enhanced interfaces, stakeholders—including decision makers—involved in the use of our computer-based models can predict how they might react to and behave given some possible future scenario. We, model builders and analysts, can then observe what questions are asked and what decisions are made by real humans before and during simulations of a water system. If nothing else, this information should improve our water resources planning and management in the future.

8. The Future

A crisis in water is no longer an abstraction for many. Climate change, underfunded and aging infrastructure, outdated or limited management approaches and institutional authorities, and the pressures of urbanization and the growing demands for food and energy are stressing water infrastructure worldwide. Aging infrastructure, growing populations and shifting patterns of settlements, and increasing costs are all making water management one of leading infrastructure challenges in many regions of our planet. A changing climate is skewing precipitation patterns that guided earlier engineering and making water scarcity and water-related natural disasters topics of concern almost everywhere.

Populations, especially in urban centers, are often impacted by inadequate water quality. The most serious clean water issues occur in water bodies where older combined sewers, diffuse non-point urban stormwater, and growing runoff of agricultural pollutants have remained largely uncontrolled. Improvements to many of our receiving waters in North America, such as Chesapeake Bay, the Everglades, the Great Lakes, the Gulf of Mexico, Puget Sound, and the San Francisco Bay-Delta, will require multistakeholder strategies using insights derived from the application of our best and most appropriate systems analysis tools.

Losses in water-related biodiversity continues, from pressures to divert critical water supplies to agriculture and urban uses, and from the loss of wetlands and other critical water habitats. Learning to use green infrastructure—and the services that such functioning ecosystems offer, from fisheries and recreation to water purification and flood protection—is increasingly gaining acceptance in water resource management agencies. Disputes over regulating base flow water levels essential to commercial and non-commercial fish species are common throughout much of the world.

Changes in our climate are now altering hydrologic cycles and impacting how water is being managed. The variability of water supplies has increased and hence the reliability of those supplies has decreased. These trends will no doubt continue into the future. Warmer temperatures and changing precipitation patterns are reducing annual snowpacks and increasing evaporation, reducing the performance of reservoirs and the services provided by watersheds. Increasing weather extremes will stress not only humans but also wildlife and natural systems. Sea level rise and higher intensity storms will steadily increase risks of coastal and inland flooding. Adapting to these impacts will challenge all of us. We face these challenges at a time when the limited financial resources available to pay for the needed water infrastructure calls for management decisions that are more cost and socially effective, perhaps in addition to being just politically feasible. We, analysts, should be able to contribute to this effort. Our most compelling contribution is our systematically gathered, analyzed and objectively interpreted information. Our job, as I see it, is to identify alternative plans, designs or policies, that are politically feasible and identify the tradeoffs, if any, with respect to what is expected or desired by all stakeholders. This includes effectively communicating this information to those engaged in the political decision-making process of policy implementation. The political environment in which we use our analyses to inform water managers can dictate the choice of the systems analysis methods we use and how our technical support influences the decision-making process and its outcomes.

Finally, one can analyze the scientific aspects of water resource systems in increasing detail all day, every day, but if these quantitative analyses do not take into account the qualitative stakeholder, decision maker and institutional biases, emotions and opinions, the result will likely fail to influence how water will be managed. We, modelers, need to figure out how the information we produce can have a greater beneficial impact in the real, as opposed to just the published, world. The answer is not in paying attention to either the current research on modeling methods or on aspects that enhance real-world implementation and impact, it is in considering both of these activities together [52].

Funding: This research received no external funding.

Conflicts of Interest: The author declares no conflict of interest.

References

1. Polasky, S.; Kling, C.L.; Levin, S.A.; Carpenter, S.R.; Daily, G.C.; Ehrlich, P.R.; Heal, G.M.; Lubchenco, J. Role of economics in analyzing the environment and sustainable development. *Proc. Natl. Acad. Sci. USA* **2019**, *116*, 5233–5238. [CrossRef]
2. Maass, A.; Hufschmidt, M.M.; Dorfman, R.; Thomas, H.A., Jr.; Marglin, S.A.; Fair, G.M. Design of Water-Resource Systems: New Techniques for Relating Economic Objectives. In *Engineering Analysis, and Governmental Planning*; Harvard University Press: Cambridge, MA, USA, 1962. [CrossRef]
3. Cash, D.W.; Adger, W.N.; Berkes, F.; Garden, P.; Lebel, L.; Olsson, P.; Pritchard, L.; Young, O. Scale and cross-scale dynamics: Governance and information in a multilevel world. *Ecol. Soc.* **2006**, *11*, 8. [CrossRef]
4. Cash, D.W.; Clark, W.C.; Alcock, F.; Dickson, N.M.; Eckley, N.; Guston, D.H.; Jäger, J.; Mitchell, R.B. Knowledge systems for sustainable development. *Proc. Natl. Acad. Sci. USA* **2003**, *100*, 8086–8091. [CrossRef] [PubMed]
5. Daw, T.M.; Coulthard, S.; Cheung, W.; Brown, K.; Abunge, C.; Galafassi, D.; Peterson, G.D.; McClanahan, T.R.; Omukoto, J.O.; Munyi, L. Evaluating taboo trade-offs in ecosystems services and human well-being. *Proc. Natl. Acad. Sci. USA* **2015**, *112*, 6949–6954. [CrossRef] [PubMed]
6. Kahneman, D.; Klein, G. Conditions for intuitive expertise: A failure to disagree. *Am. Psychol.* **2009**, *64*, 515. [CrossRef] [PubMed]
7. Moallemi, E.A.; Zare, F.; Reed, P.M.; Elsawah, S.; Ryan, M.J.; Bryan, B.A. Structuring and evaluating decision support processes to enhance the robustness of complex human–natural systems. *Environ. Model. Softw.* **2020**, *123*, 104551. [CrossRef]
8. Andreu, J.; Capilla, J.; Sanchis, E. AQUATOOL, a generalized decision-support system for water-resources planning and operational management. *J. Hydrol.* **1996**, *177*, 269–291. [CrossRef]
9. Draper, A.J.; Munévar, A.; Arora, S.K.; Reyes, E.; Parker, N.L.; Chung, F.I.; Peterson, L.E. CalSim: Generalized model for reservoir system analysis. *J. Water Resour. Plan. Manag. ASCE* **2004**, *130*, 480–489. [CrossRef]
10. Matrosov, E.S.; Harou, J.J.; Loucks, D.P. A computationally efficient open-source water resource system simulator-Application to London and the Thames Basin. *Environ. Model. Softw.* **2011**, *26*, 1599–1610. [CrossRef]
11. Zagona, E.; Fulp, T.J.; Shane, R.; Magee, T.; Goranflo, M. RIVERWARE: A generalized tool for complex reservoir system modeling. *J. Am. Water Resour. Assoc.* **2001**, *37*, 913–929. [CrossRef]
12. Stockholm Environment Institute (SEI); Water Evaluation and Planning System (WEAP). *User Guide*; Stockholm Environment Institute: Somerville, MA, USA, 2005; ISSN 1913–1844. E-ISSN 1913-1852.
13. Yates, D.; Sieber, J.; Purkey, D.; Huber-Lee, A. WEAP21, A demand-priority-and preference-driven water planning model, Part 1: Model characteristics. *Water Int.* **2005**, *30*, 487–500. [CrossRef]
14. Afshar, A.; Massoumi, F.; Afshar, A.; Mariño, M.A. State of the Art Review of Ant Colony Optimization. *Appl. Water Resour. Manag.* **2015**, *29*, 3891–3904. [CrossRef]
15. Brown, C.M.; Lund, J.R.; Cai, X.; Reed, P.M.; Zagona, E.A.; Ostfeld, A.; Hall, J.; Characklis, G.W.; Yu, W.; Brekke, L. The future of water resources systems analysis: Toward a scientific framework for sustainable water management. *Water Resour. Res.* **2015**, *51*, 6110–6124. [CrossRef]

16. Burges, S.J. Water resource systems planning in USA: 1776–1976. *J. Water Resour. Plann. Manage. Div.* **1979**, *105*, 91–111.

17. Cohon, J.L.; Marks, D.H. A review and evaluation of multiobjective programming techniques. *Water Resour. Res.* **1975**, *11*, 208–220. [CrossRef]

18. Harou, J.J.; Pulido-Velazquez, M.; Rosenberg, D.E.; Medellin-Azuara, J.; Lund, J.R.; Howitt, R.E. Hydro-economic models: Concepts, design, applications and future prospects. *J. Hydrol.* **2009**, *375*, 627–643. [CrossRef]

19. Labadie, J.W. Optimal Operation of Multireservoir Systems: State-of-the-Art Review. *J. Water Resour. Plan. Manag.* **2004**, *130*, 93–111. [CrossRef]

20. Lund, J.R. Prospects for systems analysis in an age of changing systems: Thoughts from a Saturday afternoon. In *World Environmental and Water Resources Congress Proceedings*; Environment and Water Resource Institute: Washington, DC, USA, 2010.

21. Maier, H.R.; Kapelan, Z.; Kasprzyk, J.; Kollat, J.; Matott, L.S.; Cunha, M.C.; Dandy, G.C.; Gibbs, M.S.; Keedwell, E.; Marchi, A. Evolutionary algorithms and other metaheuristics in water resources: Current status, research challenges and future directions. *Envion. Model. Softw.* **2014**, *62*, 271–299. [CrossRef]

22. Nandalal, K.D.W.; Simonovic, S.P. *State-of-the-Art Report on Systems Analysis Methods for Resolution of Conflicts in Water Resources Management*; Division of Water Sciences, UNESCO: Paris, France, 2002.

23. Nicklow, J.; Reed, P.; Savic, D.; Dessalegne, T.; Harrell, L.; Chan-Hilton, A.; Karamouz, M.; Minsker, B.; Ostfeld, A.; Singh, A.; et al. State of the Art for Genetic Algorithms and Beyond in Water Resources Planning and Management. *J. Water Resour. Plan. Manag. ASCE* **2010**, *136*, 412–432. [CrossRef]

24. Razavi, S.; BTolson, A.; Burn, D.H. Review of surrogate modeling in water resources. *Water Resour. Res.* **2012**, *48*, W07401. [CrossRef]

25. Reed, P.M.; Hadka, D.; Herman, J.D.; Kasprzyk, J.R.; Kollat, J.B. Evolutionary multiobjective optimization in water resources: The past, present, and future. *Adv. Water Resour.* **2013**, *51*, 438–456. [CrossRef]

26. Rogers, P.P.; Fiering, M.B. Use of systems analysis in water management. *Water Resour. Res.* **1986**, *22*, 146S–158S. [CrossRef]

27. Rosenberg, D.E.; Madani, K. Water Resources Systems Analysis: A Bright Past and a Challenging but Promising Future. *J. Water Resour. Plann. Manag.* **2014**, *140*, 407–409. [CrossRef]

28. Simonovic, S.P. Tools for Water Management: One View of the Future. *Water Int.* **2000**, *25*, 76–88. [CrossRef]

29. Simonovic, S.P. Reservoir systems analysis: Closing gap between theory and practice. *J. Water Resour. Plann. Manag.* **1992**, *118*, 262–280. [CrossRef]

30. Simpson, A.R.; Dandy, G.C.; Murphy, L.J. Genetic algorithms compared to other techniques for pipe optimization. *J. Water Resour. Plann. Manag.* **1994**, *120*, 423–443. [CrossRef]

31. Solomatine, D.P. Genetic and other global optimization algorithms comparison and use in calibration problems. In *Proceedings of the 3rd International Conference on Hydroinformatics*; Babovic, V., Larsen, L.C., Eds.; Balkema: Rotterdam, The Netherlands, 1998; pp. 1021–1028.

32. Sulis, A.; Sechi, G.M. Comparison of generic simulation models for water resource systems. *Environ. Model. Softw.* **2003**, *40*, 214–225. [CrossRef]

33. Waage, M.D.; Kaatz, L. Nonstationary Water Planning: An Overview of Several Promising Planning Methods. *J. Am. Water Resour. Assoc.* **2011**, *47*, 535–540. [CrossRef]

34. Wurbs, R.A. *Comparative Evaluation of Generalized River/Reservoir System Models*. TR-282; Texas Water Resources Institute: College Station, TX, USA, 2005.

35. Wurbs, R.A. *Computer Models For Water Resources Planning And Management, USA*; IWR Report 94-NDS-7; Army Corps of Engineers, Institute for Water Resources, Water Resources Support Center: Alexandria, VA, USA, 1994.

36. Wurbs, R.A. *Optimization of Multiple-Purpose Reservoir System Operations: A Review of Modeling and Analysis Approaches*; U.S. Army Corps of Engineers, Hydrologic Engineering Center: Davis, CA, USA, 1991; p. 34.

37. Yeh, W.W.-G. Reservoir Management and Operations Models: A State-of-the-Art Review. *Water Resour. Res.* **1985**, *21*, 1797–1818. [CrossRef]

38. Loucks., D.P.; Saito, L. *Adventures in Managing Water: Real World Engineering Experiences*; ASCE Press: Reston, VA, USA, 2019; ISBN 978-0-7844-1533-7 (print), ISBN 978-0-7844-8227-8 (PDF), ISBN-10: 0784415331.

39. Leslie, J. After a Long Boom, an Uncertain Future for Big Dam Projects, Yale Environment 260, Yale School of Forestry & Environmental Studies. Available online: https://e360.yale.edu/features/after-a-long-boom-an-uncertain-future-for-big-dam-projects (accessed on 27 November 2018).

40. USGCRP. *Climate Science Special Report: Fourth National Climate Assessment, Volume I.*; Wuebbles, D.J., Fahey, D.W., Hibbard, K.A., Dokken, D.J., Stewart, B.C., Maycock, T.K., Eds.; U.S. Global Change Research Program: Washington, DC, USA, 2017; p. 470. [CrossRef]

41. Menard, K.E. Houston-Texas A Big City with a Growing Thirst for Drinking Water, Texas AWWA, All Things H2O, Issue. Available online: https://www.tawwa.org/blogpost/1203603/258815/Houston-Texas-A-Big-City-with-a-Growing-Thirst-for-Drinking-Water (accessed on 5 March 2016).

42. Draper, A.J.; Jenkins, M.W.; Kirby, K.W.; Lund, J.R.; Howitt, R.E. Economic-engineering optimization for California water management. *J. Water Resour. Plann. Manag.* **2003**, *129*, 155–164. [CrossRef]

43. Jenkins, M.W.; Lund, J.R.; Howitt, R.E.; Draper, A.J.; Msangi, S.M.; Tanaka, S.K.; Ritzema, R.S.; Marques, G.F. Optimization of California's Water System: Results and Insights. *J. Water Resour. Plan. Manag. Asce* **2004**, *130*, 271–280. [CrossRef]

44. Haimes, Y.Y. *Modeling and Managing Interdependent Complex Systems of Systems*; John Wiley & Sons: Hoboken, NJ, USA, 2018; ISBN 978-1-119-17367-0.

45. Marchau, V.A.; Walker, W.E.; Bloemen, P.J.; Popper, S.W. *Decision Making under Deep Uncertainty From Theory to Practice*; Springer: Berlin/Heidelberg, Germany, 2019; ISBN 978-3-030-05251-5. ISBN 978-3-030-05252-2 (eBook). [CrossRef]

46. Assaf, H.; van Beek, E.; Borden, C.; Gijsbers, P.; Jolma, A.; Kaden, S.; Kaltofen, M.; Labadie, J.W.; Loucks, D.P.; Quinn, N.W.; et al. Generic simulation models for facilitating stakeholder involvement in water resources planning and management: A comparison, evaluation, and identification of future needs in environmental modelling. In *Software and Decision Support: The State of the Art and New Perspective*; Jakeman, A., Voinov, A., Rizzoli, A.E., Chen, S., Eds.; IDEA Book Series; Elsevier: Amsterdam, The Netherlands, 2008; pp. 229–246.

47. Langsdale, S.; Beall, A.; Bourget, E.; Hagen, E.; Kudlas, S.; Palmer, R.; Werick, W. Collaborative modeling for decision support in water resources: Principles and best practices. *J. Am. Water Resour. Assoc.* **2013**, *49*, 629–638. [CrossRef]

48. Palmer, R.N.; Cardwell, H.E.; Lorie, M.A.; Werick, W. Disciplined planning, structured participation, and collaborative modeling: Applying shared vision planning to water resources. *J. Am. Water Resour. Assoc.* **2013**, *49*, 614–628. [CrossRef]

49. Wild, T.B.; Reed, P.M.; Loucks, D.P.; Mallen-Cooper, M.; Jensen, E.D. Balancing Hydropower Development and Ecological Impacts in the Mekong: Tradeoffs for Sambor Mega Dam. *J. Water Resour. Plann. Manag.* **2019**, *145*, 05018019. [CrossRef]

50. Redpath, S.M.; Keane, A.; Andrén, H.; Baynham-Herd, Z.; Bunnefeld, N.; Duthie, A.B.; Frank, J.; Garcia, C.A.; Månsson, J.; Nilsson, L.; et al. Games as Tools to Address Conservation Conflicts. *Trends Ecol. Evol.* **2018**, *33*, 415–426. [CrossRef]

51. Speelman, E.N.; van Noordwijk, M.; Garcia, C. Gaming to better manage complex natural resource landscapes. In *Coinvestment in Ecosystem Services: Global Lessons from Payment and Incentive Schemes*; Namirembe, S., Leimona, B., van Noordwijk, M., Minang, P., Eds.; World Agroforestry Centre (ICRAF): Nairobi, Kenya, 2017.

52. Reuss, M. Coping with uncertainty: Social scientists, engineers, and federal water resources planning. *Nat. Resour. J.* **1992**, *32*, 101.

Article

Water Resource Systems Analysis for Water Scarcity Management: The Thames Water Case Study

Mark Morley [1] and Dragan Savić [1,2,*]

[1] KWR Water Research Institute, 3433 PE Nieuwegein, The Netherlands; mark.morley@kwrwater.nl
[2] Centre for Water Systems, College of Engineering, Mathematics and Physical Sciences, University of Exeter, Exeter EX4 4QF, UK
* Correspondence: dragan.savic@kwrwater.nl or d.savic@exeter.ac.uk

Received: 17 May 2020; Accepted: 18 June 2020; Published: 20 June 2020

Abstract: Optimisation tools are a practical solution to problems involving the complex and interdependent constituents of water resource systems and offer the opportunity to engage with practitioners as an integral part of the optimisation process. A multiobjective genetic algorithm is employed in conjunction with a detailed water resource model to optimise the "Lower Thames Control Diagram", a set of control curves subject to a large number of constraints. The Diagram is used to regulate abstraction of water for the public drinking water supply for London, UK, and to maintain downstream environmental and navigational flows. The optimisation is undertaken with the aim of increasing the amount of water that can be supplied (deployable output) through solely operational changes. A significant improvement of 33 Ml/day (1% or £59.4 million of equivalent investment in alternative resources) of deployable output was achieved through the optimisation, improving the performance of the system whilst maintaining the level of service constraints without negatively impacting on the amount of water released downstream. A further 0.2% (£11.9 million equivalent) was found to be realisable through an additional low-cost intervention. A more realistic comparison of solutions indicated even larger savings for the utility, as the baseline solution did not satisfy the basic problem constraints. The optimised configuration of the Lower Thames Control Diagram was adopted by the water utility and the environmental regulators and is currently in use.

Keywords: water resource modelling; multiobjective optimisation; river abstraction

---

1. Introduction

The field of systems analysis has often been associated with the advent of operations research during and after the Second World War, while its application to water resource systems advanced more significantly from around the 1970s, when computers became widely available [1]. The parallel development in computational hydraulics and hydrology, which was also stimulated by the advent of modern information and communication technology, led to the emergence of the aligned discipline of hydroinformatics in the 1990s [2]. Both systems analysis and hydroinformatics embrace not only technological issues, such as scientific methods and the application of data, models and decision support tools, but also much wider questions of the role of the discipline in addressing societal challenges [2]. Water security, resilience, governance and ethical issues are just a few of those societal challenges that are also affected by growing climate, population and uncertainty concerns. The complexity of water issues, often involving incomplete, contradictory and changing requirements, together with the involvement of stakeholders holding multiple and opposing views, give water challenges a "wicked" (ill-defined) character [3,4]. The wicked nature of water resource challenges also meant that a multitude of methods for optimising the planning and management of water resources developed over the years [5,6] were not fully adopted in practice [7].

One of the challenges most often encountered in water resource systems planning and management is how to define operating rules for multiple sources requiring an integrated vision, thus accounting for interrelations and interdependencies among complex system components [5–8]. The widespread reporting of the use of simulation and optimisation methods shows that systems analysis tools are being used in practice [8,9]. However, despite this vast wealth of literature, publications reporting on practical applications of such tools, their impact and the experiences of analysts and clients are rare.

Each water service provider in England and Wales must produce a water resources management plan (WRMP), which is updated every five years. Such plans aim to ensure "sufficient supply of water to meet the anticipated demands of its customers over a minimum 25-year planning period, even under conditions where water supplies are stressed" [10]. This paper presents a case study in which systems analysis tools were used to develop a constituent of a WRMP for a water service provider, Thames Water, considering the complexities and requirements of such a plan. An optimisation tool was developed to redesign a key component of the water management strategy for the River Thames in such a way as to maximise the capacity of the system to supply drinking water whilst ensuring the maintenance of strict environmental criteria regarding the quantity of water left in the river as it flows into its tidal reach.

2. Materials and Methods

Thames Water abstracts water from the lower reaches of the River Thames for the purpose of public water supply via a number of large reservoirs to the west of London. Transfers are also made to reservoirs in the Lea Valley found to the northeast of London. Left unconstrained, these abstractions could have a deleterious effect on the downstream environment. Accordingly, these are undertaken in agreement with the English Environment Agency (EA) environmental regulator, under Section 20 of the Water Resources Act 1991 [11]. This agreement describes the Lower Thames Control Diagram (LTCD), which is used to control the level of abstraction permitted as a function of current reservoir storage. Thames Water seeks to optimise the LTCD, with a view to maximising the deployable output of the system as a whole. Deployable output is considered to be the maximum output capacity (i.e., demand that could be supplied) of one or more commissioned water sources that can achieve a prescribed level of service as constrained by factors such as, inter alia, hydrological yield, licence constraints and treatment and transport and pumping capacity.

In addition, a second optimisation scenario was evisaged in which aggregate could be extracted from an existing reservoir to facilitate additional storage capacity for the system. This was to be run as a separate analysis to determine what impact such a change would have on the deployable output of the system as a whole.

2.1. Lower Thames Control Diagram

The LTCD controls abstraction principally by defining a target environmental and navigational flow that must reach the tidal reaches of the Thames at Teddington Lock: the Teddington target flow (TTF). The TTF matrix is illustrated in Figure 1 where each month/operating band has a minimum flow target.

As can be seen, when reservoir storage is full, Thames Water are obligated to ensure that a minimum of 800 Ml/day is discharged into the tidal reach. This figure diminishes as reservoir storage becomes lower, with the constraints becoming more relaxed in the late spring and early summer months.

The solid lines on the LTCD represent the points at which the various demand-saving measures, agreed with the environmental (EA) and economic (Ofwat) regulators and outlined in the appropriate act and statutory instruments [11,12], are implemented:

- Level 1: intensive media campaign.
- Level 2: sprinkler/unattended hosepipe ban and enhanced media campaign.
- Level 3: temporary use ban, ordinary drought order (non-essential use ban).
- Level 4: emergency drought order (e.g., standpipes and rota cuts).

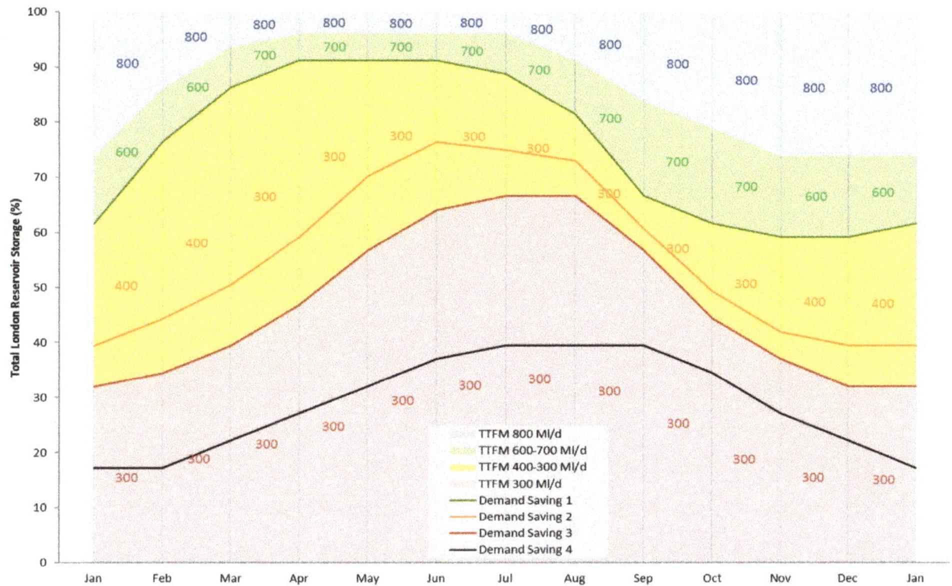

Figure 1. The Lower Thames Control Diagram (LTCD).

In addition, the "crossing" of these demand-saving lines also triggers the implementation of further schemes, such as transfers of water from neighbouring water resource zones and the use of the Thames Gateway desalination plant at Beckton [13].

For the purposes of this analysis, the existing LTCD [14], which dates back to 1980 and was last updated in 1997, is considered to give a deployable output of 2285 Ml/day. The shape of the curves was derived by iteratively applying a water resource model over the historical draw-down record and adjusting the profiles to account for violations of the level of service constraints.

2.2. Constraints

The deployable output (DO) is defined as being the maximum demand that the system can supply whilst meeting the terms of the level of service. The level of service criteria, measured over a time horizon of 100 years, agreed with the regulator for the system are:

- Level 1 events should occur at a frequency of no more than 1 in 5 years.
- Level 2 events should occur at a frequency of no more than 1 in 10 years.
- Level 3 events should occur at a frequency of no more than 1 in 20 years.
- Level 4 events are considered unacceptable and thus any solution must not allow such an event.

The permitted occurrence of Level 2 and Level 3 events is complicated by the impact of the Flood and Water Management Act 2010 [15], which stipulates that there should be periods of 14 and 56 days of public consultation, respectively, in advance of these measures being implemented. Accordingly, it is required that 14 days elapse between a Level 1 and a Level 2 event starting, and 56 days between the start of Level 2 and Level 3 events. The existing LTCD DO of 2285 Ml/day did not consider these additional constraints. As at present, the lines defining the implementation of the demand-saving levels were to be considered coincident with the boundaries between the respective TTF bands.

Further constraints were agreed with the environmental regulator for the production of the new LTCD, which included ensuring that the boundary between the TTF800 and TTF600-700 bands (coloured blue and green, respectively, in Figure 1) should be no higher than its current implementation. In addition, the definition of the Level 4 curve is changed to represent 30 days of storage at the prevailing DO and thus this line will represent a greater storage capacity for higher demand scenarios; the revised form of the Level 4 curve for the baseline scenario is shown as the horizontal line in Figure 2.

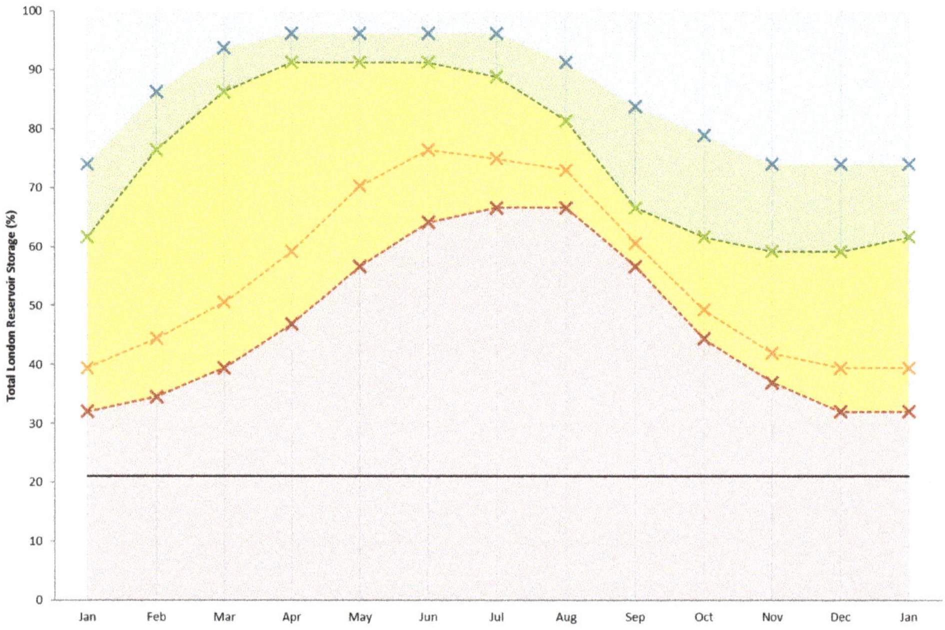

Figure 2. Annotated LTCD showing the 48 decision variables needed to define its shape.

The addition of each of these constraints increases the complexity of the problem to such an extent that it is difficult to imagine an efficient mechanism for deriving a workable solution, let alone a good one, without the use of optimisation tools.

2.3. Aquator Model

As part of this research project, an Aquator [16] water resources model was made of the whole Thames Water resource and supply area. Aquator is widely used in the UK water industry and has been used as a platform for a software application: AquatorGA [17]. This optimisation tool has been used in a number of projects in which it acts as a controller for the Aquator modelling package. The Aquator model simulates the daily operation of the system, applying the rules and constraints of the LTCD. Uncertainty in future inflows is accommodated by running the model for a given present-day DO against historic inflow data from 1920 to 2010. These inflows could be substituted by stochastically generated ensembles, if required. The model is executed by the optimisation algorithm repeatedly for a given DO and curve profile combination and is used to determine whether the combination (a) is feasible, and (b) meets the constraints of the maximum number of level of service events that occur in each category over the 90-year time horizon. The model is able to operate in two modes: a simplified cut-down model, for the purposes of optimisation, and a full mode, for validating the results as a post-process, which takes approximately three times longer to run. Tests showed that the differences in the accuracy of the two modes of the model were of the order of 1 or 2 Ml/day. Even so, the cut-down model required around 1 h to run for the historic inflow data.

2.4. Genetic Algorithm Optimisation

Genetic algorithms (GAs) are a powerful optimisation technique which can be applied to a wide variety of problems without any prerequisite knowledge of the problem domain. They perform a directed search of the decision space, which also contains a stochastic component, based on the "survival of the fittest" principle. The methodology takes advantage of the simulation model, i.e., Aquator in this case, ensuring that each potential solution is tested using a realistic representation of the water resource system being analysed. The most important advantage of GA over any other optimisation techniques is its flexibility in simulating different decision variables, objectives and constraints, due to the fact that any potential solution can be assessed directly in the model without the need for the derivation of specific mathematical properties (e.g., linearity) or expressions (e.g., derivatives), which present the main drawbacks to classic optimisation methods. A multiobjective GA [18] that can easily handle multiple constraints was used as part of the AquatorGA software.

Two objectives were specified for the production of the new LTCD: to maximise the deployable output of the system and to minimise the complexity of the produced curves in order to make them acceptable to practitioners by reducing their jaggedness. Although this latter objective is, strictly speaking, not a genuine operational requirement, this objective was included as a result of the discussions with the client and consideration of the practicability of the implementation of the solution. Past applications of the AquatorGA software demonstrated that practitioners find it easier to relate to and explain control rules and curves when they are presented as smooth curves rather than more jagged ones, even though these may be perfectly valid solutions and represent mathematically "better" results.

The shape of the LTCD is represented by 48 decision variables representing the monthly values for each of the four profile curves, as shown in Figure 2. Each variable was defined with a nominal precision of 1 decimal place and was permitted to vary between the level of the Level 4 line and the current boundary between the TTF800 and TTF600-700 bands. In order to accommodate the curve complexity objective, each of the 48 curve shape decision variables was coupled with a Boolean decision variable, which determined whether the point was considered as part of the curve or not. In this way, by "switching off" the curve points, the optimisation can easily simplify the shape of the curves.

One further decision variable was used to define the requested DO for the solution, hence the unusual situation where the DO was both an objective a decision variable. This approach was adopted because the total demand on the system was, along with the curve profile shapes, an input value submitted to the Aquator simulation model. The long run-times of the Aquator model meant that it was important that the number of infeasible solutions evaluated was minimised. To this end, once a feasible set of profiles had been identified, the DO decision variable was gradually increased in subsequent generations in order to determine the maximum valid DO for the combination of curve profiles specified. This was achieved by dynamically constraining the allowed range of the DO decision variable. If a feasible solution was subsequently found to have a valid higher DO, then the minimum value of the DO decision variable for this solution would be set to the new high DO. Similarly, if the evaluation of a higher DO proved not to be feasible, then the maximum value of the DO would be pegged to that higher value, so that higher values could no longer considered for that solution. Over time, the population of solutions gradually migrated to their true DO values. "Immature" solutions whose maximum DO has yet to be determined were protected from being removed from the population.

The combination of decision variables and constraints gives rise to a solution space consisting of 48 curve shape decisions, each of which can take on 1000 different values (0–100% at 1 decimal place = 1000^{48} options) plus 48 boolean decisions (2^{48} options) plus a single integral decision variable representing DO which is allowed to vary between 1800 and 2350 Ml/day (550 options), which gives $1000^{48} \times 2^{48} \times 550 = 1.5 \times 10^{161}$ possible solutions to the problem. The use of a genetic algorithm allowed this huge space to be efficiently sampled and evaluated, using of the order of 120,000 solutions. Nevertheless, with each solution taking around 1 h to simulate on a high-specification PC (2015), it was necessary to employ some form of parallelisation in order to reduce the optimisation run-times to a manageable length. To this end, the AquatorGA software used in this optimisation included a distributed-processing system in order to militate against the extended run-times that are a common issue when optimising evolution algorithms applied to hydroinformatics problems. The software employs the industry standard message passing interface (MPI) protocol to execute many Aquator simulation models in parallel. This system permits the concurrent evaluation of a large number of potential solutions either on local processors or to other computers on a local area network.

For the purposes of this optimisation, the software was deployed across a cluster of five workstations, each equipped with two Intel Xeon E5645 CPU packages, which comprise six cores running at 2.4 GHz for a total of 60 processor cores. In addition, this hardware architecture can take advantage of hyper-threading technology, which improves the performance of identical threads running on multiple cores by around 10–20%. Accordingly, the run-time of the optimisation model was reduced, in total, from around 13 years to around 3 weeks when deployed to 120 virtual processor cores.

3. Results

A multiple objective optimisation produces a gamut of results distributed between the competing objectives. This allows the end user to select a solution which meets their requirements, rather than being presented with a single solution. This optimisation resulted in a trade-off between the maximum DO of the system versus the complexity of the profile curves obtained. Figure 3 illustrates the least complex profile curve set from the optimisation results, in which the curves are collapsed to two straight lines and a greatly simplified upper band shape. This solution demonstrates a DO of 2144 Ml/day.

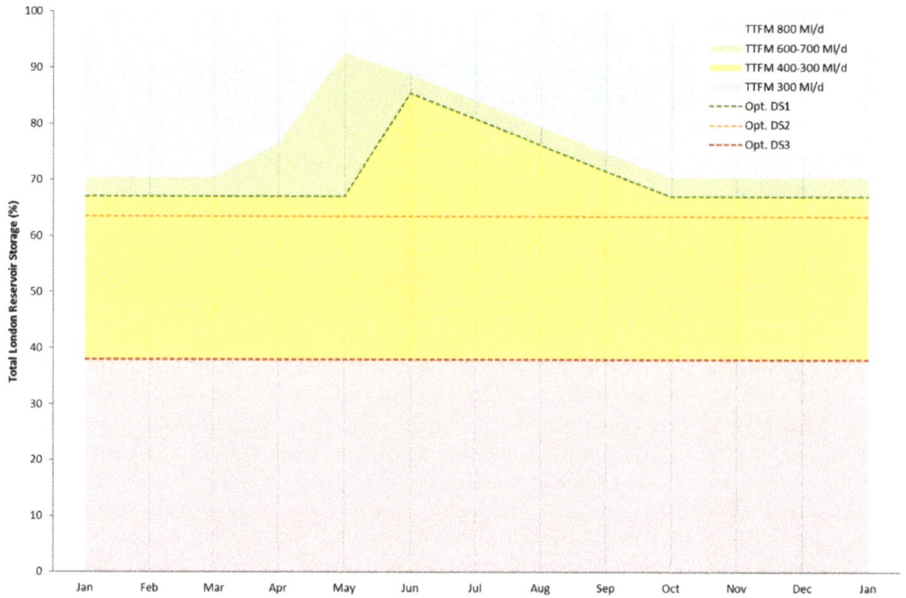

Figure 3. The simplest solution obtained for the LTCD which results in DO of 2144 Ml/day.

The highest DO/most complex curve result can be seen in Figure 4. This solution represents a DO for the system of 2308 Ml/day. It is interesting to note that the overall shape of the profile curves obtained is very similar to that of the original LTCD.

A second scenario was considered in which the total storage capacity of the London system was expanded by approximately 3% (6000 Ml) through the dredging (removal) from the reservoir of aggregate which had accumulated over time. The optimisation was rerun to take account of the increased storage and the flexibility this might add to the operation of the system. This result is seen in Figure 5.

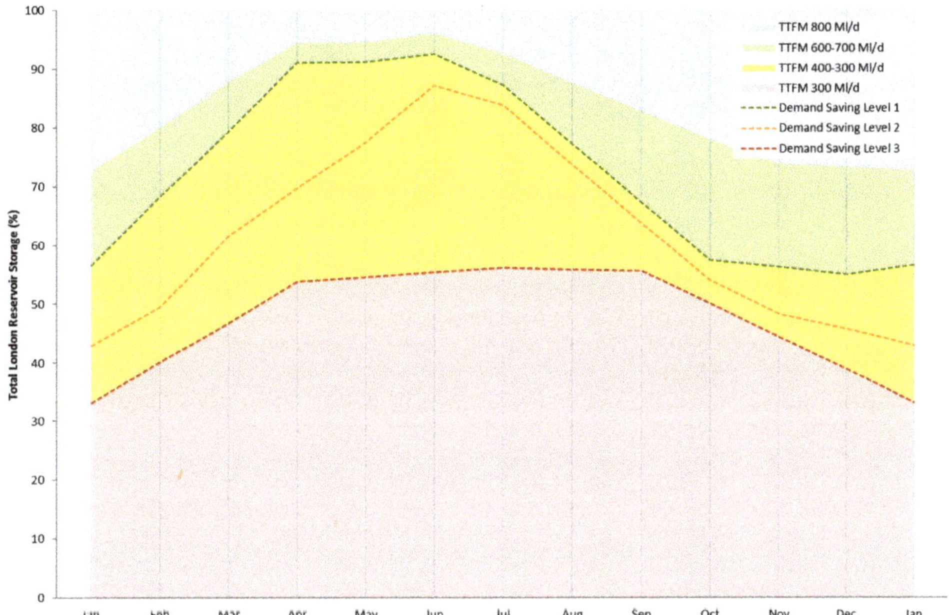

Figure 4. The most complex solution obtained for the LTCD which results in DO of 2308 Ml/day.

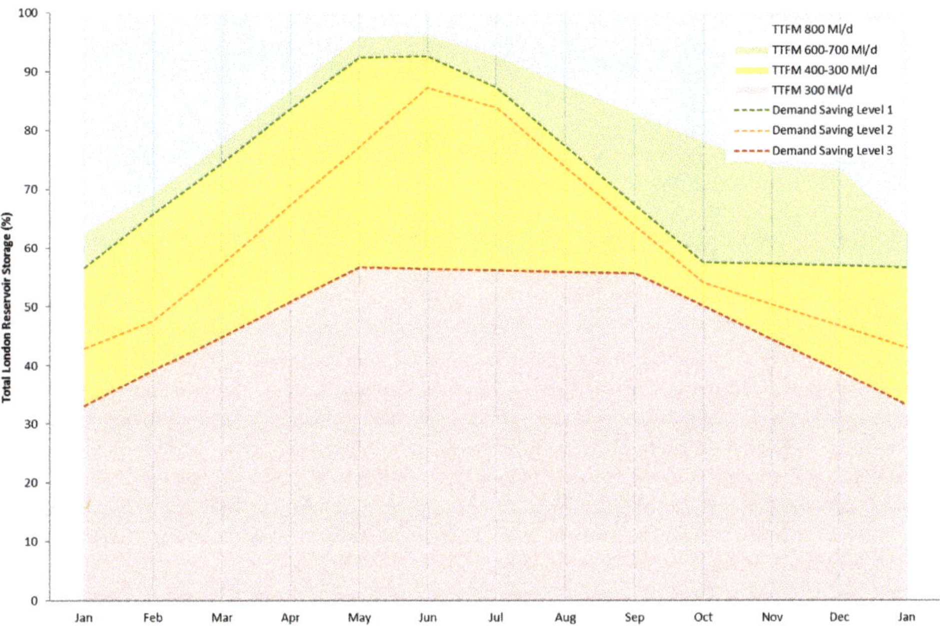

Figure 5. The most complex solution obtained for the LTCD with expanded storage of 6 Ml, which results in DO of 2335 Ml/day.

As shown, a small increase in total storage results in a slightly different shape of LTCD, but one which results in DO of 2335 Ml/day.

4. Discussion

The application of a GA optimisation tool to the Lower Thames Control Diagram resulted in the realisation of a significant increase of around 1% (33 Ml/day) in the deployable output of the London Water Resource Zone. The optimised LTCD for the scenario in which storage had not been increased (2308 Ml/day) was adopted by Thames Water, approved for use by the environmental regulator in late 2016 and continues to be in use to date [19]. However, it is interesting to note that the baseline solution (Table 1) did not satisfy the key constraints, which made it not valid for implementation, but also not easily comparable with the optimised solutions. If, for example, the simple optimised solution is used as a baseline, the realistic improvement in DO afforded by the selected optimised (complex) solution or the optimised solution with the increased storage amounts to an increase of 164 Ml/day (7.6%) and 191 Ml/day (8.9%), respectively. Considering also that London is in a drought-prone area and that Thames Water invested £270 million [20] in a desalination plant (Beckton) with a nominal capacity of 150 Ml/day that began operating in 2011, the selected solution would save the utility almost £60 million in equivalent capital expenditure. The two more realistic increases in DO would result in savings of £295 and £344 million, respectively.

Table 1. Summary of LTCD optimisation results.

LTCD Version	Deployable Output (Ml/day)	% Change
Baseline	2285 [1]	n/a
Optimised: simple)	2144	−6.2%
Optimised: complex	2308	1.0%
Optimised: increased storage	2335	1.2%

[1] Baseline LTCD does not respect the constraints for Level 2/Level 3 events relating to public consultation lead-times. If these constraints are considered, the DO is some 200 Ml/day lower.

Table 1 summarises the results obtained for the LTCD optimisation:

The use of an optimisation tool, particularly one with such long run-times, afforded a good opportunity for incorporating feedback from the client into the optimisation whilst it was still ongoing. One requirement that emerged during the optimisation was that each of the bands representing the different Teddington target flows should be present in the final solution. Early results saw the GA collapsing the 600–700 Ml/day band out of existence, something that was thought unlikely to satisfy the regulator. Accordingly, a minimum storage percentage for each band was incorporated into the optimisation while it was running.

The environmental objective for this study is embodied in the Teddington target flow matrix, detailing how much water must remain in the river at Teddington Lock. The key influence on this matrix was the maintenance of levels for the purposes of fisheries, particularly for Atlantic salmon [21] returning to the river to spawn. There is also a lesser component for the maintainance of navigation. In the absence of a numerical model or criteria to allow the comparison of potential solutions for this objective, it is not possible to undertake Pareto or qualitative multi-criteria analysis to compare solutions [22] which would allow further investigation of conflicts and synergies in policy choices. In lieu of such a possibility, the target flow matrix is applied as a hard constraint in the optimisation model, such that the status quo must be met or exceeded, with no way of quantifying what benefit any excess water might accrue for the environmental objective.

5. Conclusions

Optimal operational policies should, as in this case, be formulated as a multi-objective problem, i.e., one with more than one objective. In this way, instead of a single optimal solution, this approach leads to multiple solutions: a set of efficient or non-dominated solutions, also known as Pareto-optimal solutions, that represent the optimal trade-off curve between the objectives. Each solution is optimal in that it can only be improved for one objective, at the expense of another. The Pareto set gives a decision maker more flexibility in the selection of a suitable alternative. Each solution along the front is considered to be equally optimal. In this instance, the second objective (curve complexity) is largely a cosmetic consideration; however, it was one that was strongly valued by those in the water company. For each of the two optimisations undertaken, eight candidate solutions were presented to the client for their final selection.

The existence of a group of solutions, rather than a single one, offers additional advantages from the engineering point of view, such as increased sensitivity analysis possibilities and selection according to priorities, such as the attitude of the practitioner toward risk. It is possible to extend the application of this approach to water resources management, such that further objectives might be considered, including differential costs for supply sources, assessing infrastructure options to achieve a given DO, maximising different level of service requirements (or conversely, minimising water shortages), etc.

Author Contributions: Conceptualization, M.M. and D.S.; methodology, M.M.; software, M.M.; validation, M.M. and D.S.; formal analysis, D.S.; investigation, M.M.; resources, D.S.; data curation, M.M.; writing—original draft preparation, M.M. and D.S.; writing—review and editing, M.M. and D.S.; visualization, M.M.; supervision, D.S.; project administration, M.M.; funding acquisition, D.S. All authors have read and agreed to the published version of the manuscript.

Funding: This research received no external funding.

Acknowledgments: The authors would like to express their gratitude to Kevin Mountain of Thames Water Utilities Ltd. and Chris Green, Will Clark and Peter Edgely for their longstanding collaboration in optimising water resource models with Aquator.

Conflicts of Interest: The authors declare no conflict of interest.

References

1. Brown, C.M.; Lund, J.R.; Cai, X.; Reed, P.M.; Zagona, E.A.; Ostfeld, A.; Hall, J.; Characklis, G.W.; Yu, W.; Brekke, L. The future of water resources systems analysis: Toward a scientific framework for sustainable water management. *Water Resour. Res.* **2015**, *51*, 6110–6124. [CrossRef]
2. Vojinović, Z.; Abbott, M.B. Twenty-five years of hydroinformatics. *Water* **2017**, *9*, 59. [CrossRef]
3. Freeman, D.M. Wicked water problems: Sociology and local water organizations in addressing water resources policy. *J. Am. Water Resour. Assoc.* **2000**, *36*, 483–491. [CrossRef]
4. Reed, P.M.; Kasprzyk, J. Water resources management: The myth, the wicked, and the future. *J. Water Resour. Plan. Manag.* **2009**, *135*, 411–413. [CrossRef]
5. Simonović, S.P. Reservoir systems analysis: Closing gap between theory and practice. *J. Water Resour. Plan. Manag.* **1992**, *118*, 262–280. [CrossRef]
6. Labadie, J.W. Optimal operation of multireservoir systems: State-of-the-art review. *J. Water Resour. Plan. Manag.* **2004**, *130*, 93–111. [CrossRef]
7. Quinn, J.D.; Reed, P.M.; Giuliani, M.; Castelletti, A. What is controlling our control rules? Opening the black box of multireservoir operating policies using time-varying sensitivity analysis. *Water Resour. Res.* **2019**, *55*, 5962–5984. [CrossRef]
8. Macian-Sorribes, H.; Pulido-Velazquez, M. Inferring efficient operating rules in multireservoir water resource systems: A review. *Wiley Interdiscip. Rev. Water* **2020**, *7*, e1400. [CrossRef]
9. Loucks, D.P. Managing water as a critical component of a changing world. *Water Resour. Manag.* **2017**, *31*, 2905–2916. [CrossRef]
10. Water UK. *Water Resources Long Term Planning Framework (2015–2065)*; Water: London, UK, 2016.
11. Water Resources Act. 1991. Available online: http://www.legislation.gov.uk/ukpga/1991/57/contents (accessed on 15 May 2020).
12. The Water Use (Temporary Bans) Order. 2010. Available online: http://www.legislation.gov.uk/uksi/2010/2231/contents/made (accessed on 15 May 2020).
13. Thames Gateway Water Treatment Works. Available online: https://www.thameswater.co.uk/help-and-advice/water-quality/where-our-water-comes-from/thames-gateway-water-treatment-works (accessed on 15 May 2020).
14. Thames Water. *Water Resources Management Plan. 2015–2040*; Thames Water Utilities Ltd.: Reading, UK, 2014.

15. Flood and Water Management Act. 2010. Available online: http://www.legislation.gov.uk/ukpga/2010/29/contents (accessed on 15 May 2020).

16. Hydro-Logic Aquator. 2020. Available online: https://hydro-int.com/en-gb/products/hydro-logic-aquator (accessed on 15 May 2020).

17. Vamvakeridou-Lyroudia, L.S.; Morley, M.S.; Bicik., J.; Green, C.; Smith, M.; Savić, D.A. AquatorGA: Integrated Optimisation for Reservoir Operation using Multiobjective Genetic Algorithms. In *Integrated Water Systems: Proceedings 10th International Conference Computing and Control for the Water Industry*; Boxall, J., Maksimović, Č., Eds.; Taylor & Francis: Sheffield, UK, 2010; pp. 493–500.

18. Deb, K.; Tiwari, S. Omni-Optimizer: A Generic Evolutionary Algorithm for Single and Multi-Objective Optimization. *Eur. J. Oper. Res.* **2008**, *185*, 1062–1087. [CrossRef]

19. Thames Water. *Water Resources Management Plan. 2019. Section 4: Current and Future Water Supply—April 2020*; Thames Water Utilities Ltd.: Reading, UK, 2019; p. 47.

20. Loftus, A.; March, H. Financializing desalination: Rethinking the returns of big infrastructure. *Int. J. Urban Reg. Res.* **2016**, *40*, 46–61. [CrossRef]

21. Hughes, S.N.; Willis, D.J. The fish communities of the River Thames: Status, pressures and management. In *Management and Ecology of River Fisheries*; Cowx, I.G., Ed.; Wiley-Blackwell: Oxford, UK, 2008; pp. 55–70.

22. van der Voorn, T.; Svenfelt, A.; Bjornberg, K.E.; Faure, E.; Milestad, R. Envisioning carbon-free land use futures for Sweden: A scenario study on conflicts and synergies between environmental policy goals. *Reg. Environ. Chang.* **2020**, *20*, 1–10. [CrossRef]

Article

A Systems Approach to Municipal Water Portfolio Security: A Case Study of the Phoenix Metropolitan Area

Richard R. Rushforth [1,*], Maggie Messerschmidt [2] and Benjamin L. Ruddell [1]

[1] School of Informatics, Computing, and Cyber Systems, Northern Arizona University, Flagstaff, AZ 86011, USA; benjamin.ruddell@nau.edu
[2] ICF, 9300 Lee Highway, Fairfax, VA 22031, USA; messerschmidt.maggie@gmail.com
* Correspondence: richard.rushforth@nau.edu

Received: 23 April 2020; Accepted: 29 May 2020; Published: 10 June 2020

Abstract: We present a rigorous quantitative, systems-based model to measure a municipality's water portfolio security using four objectives: Sustainability, Resilience, Vulnerability, and Cost (SRVC). Water engineers and planners can operationalize this simple model using readily available data to capture dimensions of water security that go far beyond typical reliability and cost analysis. We implement this model for the Phoenix Metropolitan Area under several scenarios to assess multi-objective water security outcomes at the municipal-level and metropolitan area-level to water shocks and drought. We find the benefits of adaptive water security policies are dependent on a municipality's predominant water source, calling for a variegated approach to water security planning across a tightly interrelated metropolitan area. Additionally, we find little correlation between sustainability, resilience, and vulnerability versus cost. Therefore, municipalities can enhance water security along cost-neutral, adaptive policy pathways. Residential water conservation and upstream flow augmentation are cost-effective policies to improve water security that also improve sustainability, resilience, and vulnerability and are adequate adaptations to a short-term Colorado River shortage. The Phoenix Metropolitan Area's resilience to drought is higher than that of any of its constituent municipalities, underscoring the benefits of coordinated water planning at the metropolitan area-level.

Keywords: water policy; water portfolio planning; water resources management; systems assessment; adaptive capacity

1. Introduction

Facing growing urban water demand and nonstationary water availability due to climate change, a key challenge for municipal water planning is the development of theoretically and empirically robust frameworks that are actionable for decision-making [1]. Metropolitan areas (MAs) complicate water security planning for multiple reasons. Municipalities within a MA may differ in history, political and economic power or structure, or demographics or have distinct locational advantages within the conurbation. Consequently, these municipal characteristics may influence water rights seniority, the ability to finance or build new infrastructure, acquire new water rights, or receptivity toward water resource cooperation. These conditions create the potential for zero-sum water decisions amongst municipalities within a MA. Further, water provision may occur through a mixture of private, public, and quasi-public water utilities that may not align with municipal boundaries [2], adding complexity to water portfolio planning. Therefore, water portfolio security is a systems-level characteristic manifest at multiple adaptive decision scales from the municipal scale to the MA scale.

Traditionally, cost has been the dominant criterion for assessing water portfolios—subject to minimum water quality and reliability criteria. This standard approach, while broadly successful, struggles in the presence of confounding factors. First, decision pathway externalities—i.e., one agent's choice changes the landscape of available choices for other agents or create irreversible, non-resilient, locked-in infrastructure pathways [3,4]. Second, discounting future economic and social costs using time-value-of-money accounting fails to account for sustainability considerations fully. Third, non-market benefits are difficult to monetize or undervalued—e.g., natural climate and nature-based solutions [5,6], ecosystem services, green infrastructure solutions with multiple benefits, and the risk to human life from critical water infrastructure failure. Fourth, Knightian risks arising from a nonstationary shifting climate, inaccurate actuarial data, or the occurrence of unforeseeable Black Swan disasters belie traditional cost and reliability metrics [7]. Finally, standard methods break down when there are multiple independent planning timescales and decision boundaries. When we neglect these confounding factors, we develop water portfolios biased toward cost efficiency and presume rational technocratic decision-making at the expense of increased vulnerability, along with decreased sustainability and resilience [8–10].

Generality, transferability, simplicity, and communicability have limited the impact of water portfolio options modeling [11]. Previous work evaluating water portfolios options have developed precise plans for efficient long-term capital infrastructure investments but rely on elaborate modeling and require highly-detailed past, current, and forecasted financial and engineering data [3,11–14]. Others have evaluated the reliability, resilience, and vulnerability of water systems [15]. By contrast, other water portfolio options models de-emphasize long-term prediction and instead emphasize flexible, short-term adaptation [16], which has the practical benefit of (1) having lightweight requirements for data and inputs that are already available in most municipalities and (2) no requirement for accurate prediction of the long-term future.

For robust water portfolio security planning, precise systems-level metrics of sustainability, resilience, and vulnerability must complement and contextualize traditional financial costs and engineering reliabilities. Toward this end, we propose augmenting the standard cost-based (C) approach with sustainability (S), resilience (R), and vulnerability (V) metrics: This is the SRVC model for quantitative, systems-based water portfolio options analysis. We quantify sustainability as the length of time a water portfolio can provide water before a stress event transitions to a water shock or triggers adaptation. We use an ecologically based approach of source diversity to measure water resources portfolio resilience [17] that we define as the ability of a water system to function while enduring internal or external change [18]. Our short-term, event driven approach is in contrast to water portfolio planning studies that have focused on long-term optimization to meet reliability goals [19] and planning under uncertainty [20]. We measure vulnerability as a municipality-specific demand-to-availability metric [21] that measures demand to total available municipal water allocation—i.e., pressure on a municipality's legal water allocation. We calculate costs as the net of benefits and costs to acquire and convey new water sources and ecosystem services. Finally, as our framework only measures the outcome of water portfolio configurations, we assume municipalities comply with standards and regulations in all scenarios.

We have developed the SRVC model to provide a short-term, event-driven, policy-based complement to long-term predictive water portfolio planning. Using the SRVC model, we assess water security in the Phoenix metropolitan area (PMA) across SRVC metrics in response to multiple short-term water shock scenarios. Since the SRVC model is inherently a multiscale, systems-based approach, we assess water security at two scales: the municipal scale for 12 PMA municipalities and at the MA scale. We developed ten scenarios to evaluate different Phoenix SRVC configurations and answer the following research questions: (1) How long can a municipality or the MA sustain day-to-day water supply in response to a supply shock? (2) What is the diversity and resilience of available water options to municipalities and the MA? (3) How does vulnerability to a shock differ among individual municipalities and the MA as a whole? (4) How might municipalities develop cost-neutral/affordable policy options that are more sustainable, resilient, and less vulnerable?

2. Materials and Methods

2.1. Water SRVC Model Data Requirements

Information for decision-making has a cost. Collecting data and building decision support models are expensive, and fully dynamic, detailed systems modeling is potentially cost prohibitive for typical municipal water decision-makers. Moreover, there are disadvantages to dynamic complex systems modeling: in the presence of feedback and imprecise input data, errors can propagate and prevent robust predictions [22]. As a result, lower-dimensional models have significant advantages in terms of cost and robustness for complex systems. The SRVC model is not a dynamic systems model, but it is a systems model that captures the basic system geometry and decision/adaptive pathway constraints. Table 1 presents decision agent (municipality) data requirements for the SRVC model.

Table 1. Water Sustainability, Resilience, Vulnerability, and Cost (SRVC) model data requirements.

Requirement	Description
1.	The identity of each online and potential water source and its capacity limitations.
2.	The operating and capital costs of each online or potential water supply option, where cost includes monetized estimates of risk and insurance against risk.
3.	The time delay required to develop each potential water source.
4.	Sustainability of each water source measured as storage depletion at a given consumption rate.
5.	Current and future Demand-to-Availability on online and potential water sources
6.	Curtailment and conservation options available.
7.	Quantities of water supply and consumption defining each planning scenario, e.g., climate change and growth in population.
8.	The legal, hydrological, and infrastructural capacity and utilization of that capacity to use each online and potential water source.
9.	Water supply deficit contingency plans and pathways.

2.2. Measuring Shock, Stress, Sustainability, Vulnerability, and Adaptation

Water SRVC model results rely explicitly on the system decision and an adaptation period, Δt. A system's Δt is the time constraint for adaptive policy decision making; it is a discrete interval related to the duration of extreme events and disturbances, the velocity of adaptive policy and infrastructure decisions to cope with the stress, system storage capacity at customary usage rates, and the sustainability of alternative system states or adaptive options. Available adaptive options within the decision timeframe, such as switching to groundwater storage credits to deal with short term drought, are endogenized in the calculation of system resilience and vulnerability. Adaptive options unavailable within the decision timeframe, such as building a new desalination plant to respond to a one-year drought, are exogenized and considered alongside other exogenous drivers in long-term planning scenario development. Stress becomes shock if demand exceeds total online capacity, after adaptive options are exercised.

The Δt therefore determines the set of viable adaptive options we consider. For the Water SRVC model, Δt is 1 year, which excludes adaptive options that are unsustainable beyond that timeframe (Sustainability $S < \Delta t$ is unsustainable), such as reliance on a three-day reservoir to adapt to a one-year drought. A one-year Δt focuses water stress events to unplanned failure of a major water supply infrastructure—e.g., canal or dam failure, or policy decision, or forest fire, or an extreme drought lasting between one and several years that triggers a mandated water curtailment. The 1-year timeframe also matches the reporting period for municipal water data and storage capacity of surface water reservoirs, and annual groundwater usage is insignificant compared to aquifer storage at this Δt. A decadal or centurial Δt would change these considerations significantly as groundwater storage exhaustion is relevant water shock, and major infrastructural reworking is an adaptive option. Additionally, at that Δt surface, water drought may be relevant, depending on drought duration. In this SRVC model case study, adaptive options that have a $\Delta t \gg 1$ year are viable such as purchasing new permanent water rights or developing new physical water supply infrastructure. These adaptive options take much longer than Δt to implement because acquiring and utilizing long-term water rights has capital costs and/or legal and construction time delays, which make responding to a one-year water emergency infeasible. Similarly, a seasonal or daily Δt yields a different adaptive option portfolio and changes water security considerations by emphasizing operational measures, engineering reliability standards, and emergency management. Naturally, municipalities must develop adaptive options for all Δt, which is why the SRVC model explicitly considers timescale.

We now present a simple visual derivation to illustrate how to determine Δt and related quantities (Figure 1). This time plot begins with an initial time t_0 when the system departs from a prior equilibrium state and t_e is the time (t) elapsed since t_0: $t_e = t - t_0$ Both capacity and demand "ramp" at average rates (or slopes) $\overline{m}_C$ and $\overline{m}_D$. When Demand (D) approaches Capacity (C) from below, the system becomes increasingly stressed and increasingly vulnerable to disruptions in capacity. When the Magnitude (M) of D exceeds C, a deficit (F) exists ($F = D - C$, for positive F), a functionally damaging shock (k) ensues that reduces function by some percentage X (i.e., $F_x : F_{50}$ corresponds to 50% damage to function). The shock point in time (t_k) and magnitude (M_k) exists where the capacity and demand curves are equal. A shock point can occur for multiple reasons—canal or dam failure, or policy decision, or forest fire, as mentioned previously—and could cause a supply shock, demand shock, or both. After the shock begins, real demand drops to meet capacity, at the cost of degraded function. Degradation persists until time t_x, when function returns to normal after a shock creating a deficit F_x; the duration of this shock is t_r: $t_r = t_x - t_k$, for positive t_r. The Vulnerability (V) of the system to shock is indexed as the ratio of demand to capacity ($DTC = D/C$), and the ratio for a shock of severity $X\%$ is DTC_x.

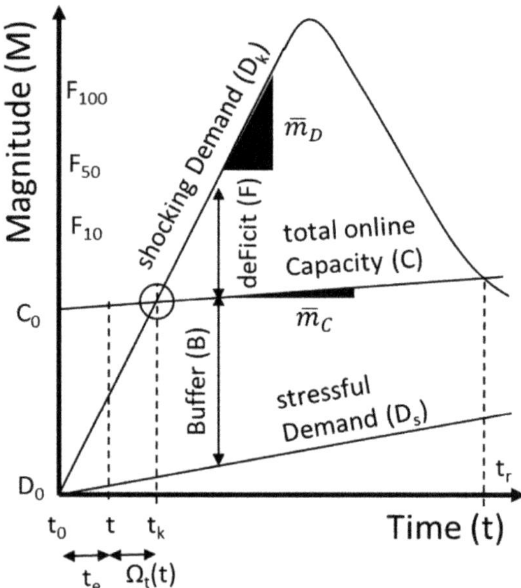

Figure 1. Illustration of the geometry of a shock, including the role of buffers and deficits, the transition from stress to shock (and recovery) at some future point in time, and ramp rates of demand and of capacity. Note that excess demand is curtailed to equal capacity during a shock, creating damage and that deficits can therefore only exist as a hypothetical or planning concept. Linear ramping is illustrated, but this is usually a valid assumption only in the short term. This figure has decreasing capacity and constant demand, but it is conceptually similar to the figure as drawn.

The sustainability (S) of this system state may be estimated as the length of time (t) before a shock ensues ($S(t) = t_k - t$, where $S_t = \Delta t$ at t_0). By convention we choose the initial $S(0) = t_0$ to define the system's sustainability. Over the long term as t becomes large, capacity must adapt (or ramp) at least as fast as demand ($\overline{m}_C \geq \overline{m}_D$) or demand must decrease to adapt to new capacity constraints ($\overline{m}_D \geq \overline{m}_C$) to support sustainable system function without shocks (Equation (1)). Over the short term, a Buffer (B), ($B(t) = C(t) - D(t)$, for positive B), allows demand to increase faster than capacity until the buffer is exhausted at t_k. Buffers and capacity include excess capacity and reservoirs (see Section 2.3 below).

$$S(t) = \frac{B_0}{\overline{m}_D - \overline{m}_C} - t_e \tag{1}$$

This geometry yields notable findings that we now discuss. Buffers are the key to absorbing unexpected increases in DTC, but they can also breed complacency in adaptive decision making because stress can become shock when $B > 0$ and $dC/dD < 1$. If we do not sense that we are consuming a buffer, shocks appear by surprise after demand overshoots the carrying capacity; this is the essence of classic sustainability models [23]. If we fail at t0 to take adaptive action to ramp up capacity and/or ramp down demand, this delay increases DTC, decreases buffer, and eliminates some of our adaptive options because we have less time. In this way, the SRVC model definition of sustainability comports with the classic Brundtland Report sustainability definition [24]. Sustainability of a water resource or water portfolio decreases when $D > C$, which can occur if $\overline{m}_D$ outstrips $\overline{m}_C$ to exhaust B_0, if $\overline{m}_C$ becomes negative relative to a constant $\overline{m}_D$ to exhaust B_0, or adaptation occurs too late, and t_e becomes greater than t_k. By definition, if any these conditions occur, system function would degrade for current and future generations.

2.3. The Components of Municipal Water Portfolio Adaptive Capacity

A municipality's capacity is the lesser of demand or the capacity of the system's hydrologic, legal, infrastructural, financial, and economic components. There are five qualitatively different measures of water capacity available to a municipality from each independent water source (*i*) during a Δt. First, Total Online Capacity (C) is available without significant capital cost or delay, including both the utilized and unutilized (or available) capacities, with capital cost judged relative to marginal operating cost, and delay relative to Δt. Second, Online Capacity (U) is the utilized portion of capacity from the source and is available without capital cost or delay. Third, Available Capacity (A) is the unutilized capacity from a source and is available without significant capital cost or time delay relative to the time constant; reservoirs are usually of this type. Fourth, Potential Capacity (P) is capacity from an existing or new source available with a delay shorter than the adaptation period but with significant, non-negligible cost. Fifth, Adaptive Capacity (O) for a water source is $O_i = A_i - P_i$ and is all water supply available during the adaptation period, which may or may not have a time delay and/or capital costs. These five water capacity metrics are interrelated for each municipality and for each scenario. We only consider a single adaptation period in this paper, but for serial adaptations $t_2, t_3, \ldots, t_n$, each adaptation depends on previous adaptations. At this point, the SRVC model begins to resemble the pathway mathematics [4,14]. Equation (2) and Figure 2 show the relationship between municipal water capacity metrics.

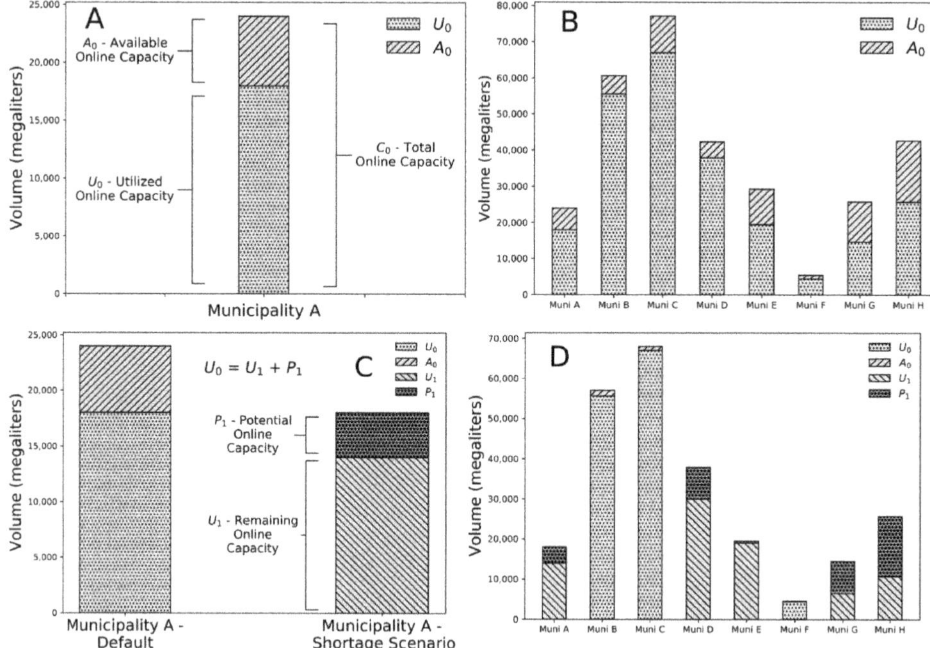

Figure 2. Graphical explanation of water source capacity and utilization before (0) and after (1) adaptation. (**A**) A municipality uses some portion of its legal, hydrological, and infrastructural capacity from a water source. (**B**) Multiple municipalities share a water source, and each uses and reserves some portion of the water source. (**C**) If there is a mandatory water curtailment, post-curtailment total capacity may be less than pre-curtailment utilization, creating a structural water deficit made whole by securing a "potential" source of water to fill the deficit (the potential source could be a conservation offset). (**D**) In a metropolitan area, a water curtailment scenario may affect each municipality differently, such that some municipalities still have available water from baseline sources, but other municipalities face structural water deficits.

For scenario (Sa), the subscript zero (0) represents the baseline or equilibrium water supply portfolio at the start of the decision timeframe (BASE scenario), and subscript one (1) represents the post-adaptation water supply portfolio at the end of Δt. The Supplemental Information contains C, U, and A scenario data for the Phoenix Metropolitan Area (Tables S1–S7 in supplementary materials).

$$C_{Sa,c} = \sum_i U_{Sa,i,c} + \sum_i A_{Sa,i,c} \tag{2}$$

For a metropolitan area, the total online capacity across all cities for the metropolitan area is,

$$C_{Sa} = \sum_c \sum_i C_{Sa,i,c} \tag{3}$$

If Total Online Capacity exceeds Demand at current water prices ($C \geq D$) after a scenario's curtailments or supply shortages, we assume no development of potential capacity, P, because a municipality, in order to meet a short-term demand shock, would not expend capital to bring online potential sources meant for meeting long-term demand growth. However, if $C < D$, then a structural water deficit (positive $F = D - C$) exists, such that a municipality must bring P online, beginning with the lowest cost P. Within the SRVC model, municipalities share P proportionally to structural water deficit magnitude. If F is positive, the lowest cost adaptation may be the conversion of pre-adaptation available sources A_0 into post-adaptation sources UA_1, which shrinks post-adaptation available sources A_1 such that $A_1 = A_0 - UA_1$. If $A_1 > F_0$, the utilized portion of the potential supply after adaptation, UP_1, is set equal to the deficit, such that $UP_1 = F_0$, and $P_1 = P_0 - UP_1$. If $A_1 < F_0$, the post-adaptation Shock Deficit $SF_1 = F_0 - UP_1$. After adaptation, the municipality's utilization is $U_1 = U_0 + UP_1 + UA_1$, its capacity is $C_1 = U_1 + A_1$, and its options are $O_1 = A_1 + P_1$. Bringing P online strands capital cost with exception for leasing water rights.

Measuring Municipal Water Portfolio Adaptive Resilience

We need to adopt a diversification-based approach to resilience based on keeping as many independent options as possible open [25]. Premature commitment or investment in a backup water source may backfire if that source becomes compromised, stranding valuable capital, and damaging our financial capacity to adapt further [4,11]. Reversibility that avoids long-term commitment, and retains a diversity of independent options, creates resilience. This definition of resilience mirrors existing definitions of ecological resilience that implement a normalized Shannon diversity index (R) to measure structural diversity [17]. Using this definition, municipal water portfolio resilience to a water stress event depends on the number of independent water sources that are online and accessible during an event and on the relative abundance of those sources (Figure 3) and thus maintain function while enduring external shocks [18]. A municipality with only one water source and no potential options for new sources during an unexpected water stress event has no resilience to that event, regardless of how sustainable, abundant, and cost effective that source may be. In this model, resilience is therefore a measure of the ability to maintain flexible decision options during current and future events.

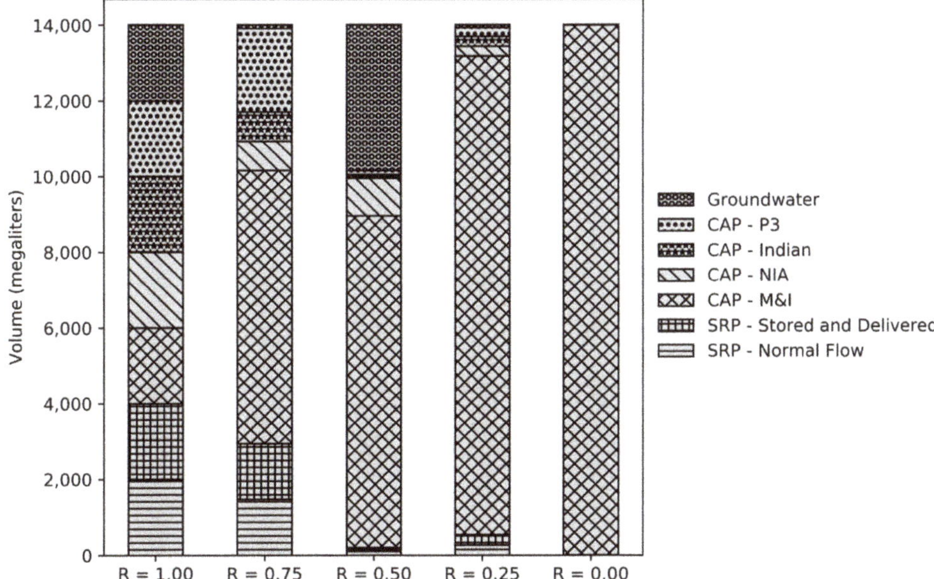

Figure 3. A hypothetical example showing change in resilience of a water resources portfolio, as measured by a normalized Shannon Diversity Index. As one water source (CAP M&I in this case) begins to dominate the portfolio of supply options, (left to right), R approaches zero, and a municipality loses the ability to adapt to an unexpected shortage if that shortage severely impacts the primary supply.

In Phoenix, the SRP and CAP water wholesalers deliver water from the same physical source at a continental scale: Western U.S. surface water flows originating from mountain snowmelt within the Colorado River Basin. However, SRP and CAP independently operate separate storage and delivery infrastructures that have multiple, distinct types of water rights. Within both systems, the distinct types of water rights have their own discrete bundle of use restrictions, seniority, and vulnerabilities. Therefore, owing to the governance the SRP and CAP systems, there is a great deal of independence and decorrelation between the two water systems.

Resilience is computed over a municipality's adaptive water supply options, O, using a normalized Shannon Index (R) for each municipality, c, and normalized between 0 and 1 using the number (n) of water sources (i) in its portfolio (Figure 3). p is the fractional proportion of the municipality's adaptive water supply options that lie with each independent source.

$$R^O = \frac{-\sum_{i \in I} p(O_i / \sum_i O_i) \times \ln(O_i / (\sum_i O_i))}{\ln n} \tag{4}$$

Since we normalize R^O using the maximum theoretical number of water sources available to a municipality, the method enumerating those sources represents an important assumption. There are two choices for the definition of the set of potential sources, i: "local" and "global" [26,27]. The local set includes only the sources that are accessible to a specific municipal decision agent c. The global set includes all sources that are accessible to any municipality within the metropolitan area. The correct choice for an "apples to apples" comparison between the resilience of municipalities within a population is the global set, because this choice reduces the resilience of municipalities with access to a small number of independent sources relative to municipalities with access to more independent sources. If the local set is used, the normalization for each municipality's R^O differs, rendering each municipality's R^O incomparable.

We could calculate resilience based on only available options A. Calculating R using A provides an indicator of resilience that considers only online backup water sources that do not incur significant capital costs or time delays. This is a conservative resilience calculation that neglects many of our adaptive options and is analogous to engineering or emergency response resilience calculation rather than an adaptive resilience. Calculating R using O is less conservative because it includes expensive adaptive options that involve capital investment, but this resilience will usually be higher because it includes all viable short-term options. It is clear that, based on this logic, resilience not only implies a "to" but also implies a "cost"—i.e., resilience *to* water supply deficits *at a cost*.

2.4. DTC, an Index of Municipal Water Portfolio Vulnerability

We calculate municipal water portfolio Vulnerability (V) as the ratio of demand-to-capacity (DTC), which is mathematically identical to the consumption-to-availability ratio (CTA). However, these two metrics are subtly different: CTA is an index of stress while DTC is a measure of vulnerability. In theory CTA should not exceed a value of unity; however, in practice, $CTA > 1$ is common. This is because of the way "availability" is defined in the typical CTA metric; availability is normally calculated at an arbitrary timescale that is unrelated to the adaptation time constant of the system and may not include reservoirs, potential capacity, P, or available capacity, A. This version of the Water SRVC model employs DTC. With respect to a water shock, pre-adaptation demand can exceed capacity but not post-adaptation; post-adaptation demand would decrease to meet capacity. In other words, the adaptation of last resort is a reduction in demand through rationing or pricing. Because DTC cannot exceed a value of unity without incurring functional damage, DTC is a vulnerability index—that is, vulnerability to damaging shock—rather than a stress index like CTA. CTA indexes water stress, not vulnerability, because vulnerability implies the risk of real social, environmental, or economic damage or impact following an event [28].

Local Reserve Margin (LRM) in power systems engineering provides a sound analog to the Water SRVC model's DTC-based vulnerability index. For example, LRM less than 15% (i.e., $DTC > 0.85$) significantly increases the likelihood of system demand failure during power stress events, leading to brownouts or blackouts (i.e., shocks). Notwithstanding the vastly different adaptation timeframes, the analogy between vulnerability, DTC, and LRM holds. Therefore, we believe that this DTC-based index is robust across a broad class of supply portfolio analysis contexts, not only in municipal water supply. Curtailing demand to meet supply during a shock creates functional damage, both in the power grid and in water supply.

For each municipality (c) and scenario (Sa), we can specify DTC for each water source, i, yielding multiple water resource vulnerabilities for each municipality. Therefore, $CTA_{Sa,c}$, which is also a municipality's Vulnerability to a water shock (V_c), is the volumetric weighted average of $DTC_{Sa,c,i}$ (Equation (5)). Further, this calculation is scalable to the metropolitan area, which yields the metropolitan area's vulnerability to the same.

$$V_{Sa,c} = DTC_{Sa,c} = \sum_i \left[(U_{Sa,i,c}/C_{Sa,i,c}) \times (U_{Sa,i,c}/\sum_i U_{Sa,i,c}) \right]_i \tag{5}$$

2.5. Water Supply Portfolio Cost

Cost in the Water SRVC model is the cost of bringing water supplies online for adaptation and to cover F. These adaptive water supply costs include capital costs (or water prices), legal costs, operational costs, and infrastructure construction. Therefore, adaptive cost is the delta between existing costs and the levelized costs of adaptive options at an annual Δt timeframe (Equation (6)). To calculate the cost of adaptation, we compute volumetrically weighted average costs of all water sources used by the municipality under the scenario. We assumed that each municipality would purchase the least expensive available water source per unit volume until exhaustion and then purchase the next least expensive water source, and so forth. The baseline cost of each municipality's water portfolio is based on SRP surface water, CAP surface water, and groundwater cost data. Baseline per unit water costs varied drastically due to unique contractual conditions for each municipality's water resources portfolio. Adaptive cost data to cover structural water deficits is based on Phoenix Active Management Area (AMA) water market data [29].

$$\begin{aligned} Cost_{Sa} = \quad & \sum_i (U_{Sa,i} \times \$AFY_{Sa,i,c,operational}) \\ & + \sum_i (UP_{Sa,i} \times \$AFY_{Sa,i,levelized}) \end{aligned} \tag{6}$$

In the SRVC model, the cost metric is used as information to support and weigh decisions S, R, V outcomes and is not intended to be used a sole or overriding criterion for water portfolio decision-making due to the potential constraints and instability to water utility financing [30]

3. Case Study: The Phoenix Metropolitan Area Water Portfolio

Water decision-making and governance in the Phoenix metropolitan area (PMA) occurs at multiple scales ranging from the municipality to regional water authorities and to federal agencies. This study focuses on the PMA municipalities of Avondale, Chandler, Gilbert, Glendale, Goodyear, Mesa, Peoria, Phoenix, Scottsdale, Tempe, and Tolleson, which comprise 85% of PMA population in Maricopa County, Arizona [31]. PMA municipalities access shared, rivalrous water sources that have varying degrees of excludability depending on law and infrastructure. Water sources include the Central Arizona Project (CAP) (40%), which conveys Colorado River water; the Salt River Project (SRP) (36%), which stores and delivers water from the Salt-Verde watershed; groundwater (7%), which is a significant gross water supply with smaller net consumption because of aquifer recharge with treated effluent and groundwater banking through long-term storage credits (LTSCs); and treated effluent (17%), which is used primarily for parks and golf courses, or utilized by non-municipal power generators. As described in the subsequent sections, all water sources have legal and infrastructural constraints on their use (Table 2). Only Goodyear and Tolleson do not have access to all water sources; Goodyear does not have access to SRP water and Tolleson does not currently have access to CAP water. An advantage of studying a metropolitan area with municipalities that have uneven access to regional water sources is that it allows for better understanding of how and to what extent SRVC metrics are interrelated.

Table 2. Online and potential water sources for Phoenix metropolitan area (PMA) municipalities at the annual decision timescale.

Current Online Water Sources	Contractual Water Categories	Description
Salt River Project	Normal Flow	The "Normal Flow" a city would receive from the Salt River had there been no surface water storage system built along the Salt River.
	Stored and Developed Water	Water delivered to On-Project Lands as defined by the 1910 Kent Decree.
Central Arizona Project	Municipal and Industrial (M&I) Subcontracts	Water leased directly to municipalities by the Central Arizona Water Conservation District
	Indian Contracts	Water originally decreed for agricultural purposes on Native American reservations in Central and Southern Arizona but later obtained by municipalities by lease or exchange
	Non-Indian Agricultural Subcontracts	Water originally decreed for agricultural purposes Central and Southern Arizona but later obtained by municipalities by lease or exchange
	Central Arizona Project (CAP) Priority Water	Water obtained by municipalities via an exchange with the Wellton-Mohawk Irrigation District located in Yuma County for main stem Colorado River water
Groundwater	—	
Potential Online Water Sources	**Contractual Water Categories**	**Description**
Type 2 Groundwater Rights	—	Grandfathered groundwater right established before Arizona's 1980 Groundwater Management Act
Arizona Water Banking Authority (AWBA)	—	Purchased Long-Term Storage Credit of surface water banked by AWBA
Flow Augmentation	—	Thinning overgrown forests in the Salt-Verde Watershed.
		Investing in irrigation efficiency in the Verde Valley and pledging water savings to in-stream flow.

CAP and SRP delivery curtailment decisions occur each year after measurements of spring snowmelt and projections of next year's spring snowmelt. Each August, the U.S. Bureau of Reclamation uses Lake Mead elevation projections to decide for or against a shortage declaration, which would curtail CAP deliveries the following calendar year [32]. During the previous SRP curtailment, the SRP governing board decided to curtail deliveries in the fall for reductions the following calendar year [33]. Given these decision-making and response timeframes, municipalities may have less than a year to successfully adapt to water curtailments before they become a water shock. The Water SRVC model utilizes these decision-making criteria as triggers for water curtailment scenarios. To successfully adapt within CAP and SRP decision-making timeframes, PMA municipalities must exercise options for available water rights or free up water from "soft" elastic demand through conservation measures. New, non-traditional water options may result from increasing upstream flows in the Salt-Verde watershed through forest thinning, agricultural efficiency upgrades, and farm fallowing. Given these factors, the PMA water supply portfolio has multiple decision boundaries and planning timescales [34,35].

PMA municipalities have several annual-timescale adaptive water policy and supply options. Supply-side options include redeeming LTSCs; short-term or long-term CAP water leases with Indian and non-Indian agricultural; long-term contracts to purchase Colorado River water from mainstem agricultural users; lease or purchase existing non-irrigation groundwater rights (called Type 2 groundwater rights) within the PMA; and theoretically, interbasin groundwater transfers by retiring agricultural groundwater surrounding the PMA. Additionally, flow augmentation projects that leave agricultural flows in rivers and aquifers upstream of the PMA—fallowing, crop switching, and irrigation efficiency—can increase Verde River flows [36]. These additional flows have potential to act as backup water supplies during drought. Demand-side policy options to offset lost supply include outdoor water use restrictions, incentivizing mesic-to-xeric landscape conversions, and infrastructure upgrades to reduce lost and unaccounted-for water. In addition to traditional supply-side and demand-side options, nature-based options include expanding forest thinning in the Salt-Verde River watershed, specifically the Four Forest Regional Initiative (4FRI) region. Forest thinning could increase upstream runoff by up to 20% and increase reservoir water storage levels along the Salt-Verde system in a cost-competitive fashion [37]. Mechanical forest thinning offers a limited benefit compared to direct flow augmentation because runoff increases may last only for 6-years without continuous forest thinning [37,38]. However, forest thinning has additional social and economic benefits such as supporting rural economic development [39]. Notably, one drawback to upstream nature-based options is that they do not benefit municipalities unless additional flows are guaranteed, which requires legal assurances and certainty regarding flow increases from nature-based projects.

Regional climate models for the Colorado River basin project decreased flows due to increased evaporation (over 75%) at the headwaters [40,41]. Within Arizona, downscaled global climate models (GCMs) indicate the potential for increased temperatures and decreased runoff in the Salt-Verde watershed. Specifically, these models indicate increased mean annual temperature by 2.4 to 5.6 °C and decreased winter precipitation, potentially reducing mean annual runoff in the Salt-Verde watershed to 77.4% ± 24.0% of normal [42]. Longer, hotter, dry periods will increase fire likelihood in overgrown ponderosa pine forests in the Salt-Verde watershed [43]. Forest fires in the Salt-Verde watershed have the potential to acutely affect SRP water supply volume and quality [44]. Given these projections, future configurations of the PMA water resources portfolio include severe and unexpected water stress.

3.1. Surface Water Supply Options

Surface water delivered to PMA municipalities—either SRP or CAP—represent a bundle of "paper water" types that are physically comingled but legally distinct.

SRP conjunctively manages a surface water reservoir system and groundwater to deliver water to Avondale, Chandler, Gilbert, Glendale, Mesa, Peoria, Phoenix, Scottsdale, Tempe, and Tolleson. SRP provides surface water to municipalities as either: (1) Normal Flow water or (2) Stored and Developed water. Normal Flow water is the most senior water right among SRP water types and the river flow to each SRP municipality under natural river conditions (prior to the SRP reservoir system). Stored and Developed water is a junior SRP water right and a combination of surface water, pumped groundwater, and LTSCs. SRP allocates Stored and Developed water to municipalities based on the area "on-project" lands, which is the area of municipality land within the original SRP service area, using a multiplier of 3 acre-feet per acre of on-project land (9.144 ML ha^{-1}). SRP can reduce the Stored and Developed water supply multiplier in response to severe water shortage. In 2003 and 2004, the SRP governing board voted to reduce deliveries by 33% in response to ongoing regional drought [33,45]. Previously, the SRP governing board decided to curtail deliveries when reservoirs, which are along the Salt and Verde rivers, were at 27% capacity. At the beginning of 2016, SRP reservoir capacity was at 57%. For SRP, annual deliveries average 945,195 ML year^{-1} while the 30-year median runoff into SRP reservoirs is 659,093 ML year^{-1} [46]. At the current drawdown rate of 254,709 ML year^{-1}, the SRP reservoir capacity will dip below the 27% capacity benchmark for water delivery curtailment by 2020.

In the PMA, CAP water has most exposure to water stress due to its junior status on Colorado River [47,48]. CAP water right types are Municipal and Industrial (M&I) Subcontracts, water leased directly to municipalities by the Central Arizona Water Conservation District; M&I Priority, which is CAP water obtained by municipalities for the purposes of M&I use as a result of a lease from the Gila River Indian Community; CAP Agricultural Priority, which is water originally decreed for agricultural purposes on Native American reservations in Central and Southern Arizona but later obtained by municipalities by lease or exchange; CAP Non-Indian Agriculture, which is water originally decreed for agricultural purposes Central and Southern Arizona but later obtained by municipalities by lease or exchange; and CAP Priority water (P3), which is water obtained by municipalities via an exchange with the Wellton-Mohawk Irrigation District located in Yuma County for main stem Colorado River water, thus giving these water rights priority over other types of CAP water [49]. Among CAP water rights types, P3 & M&I water rights have the highest priority, followed by Agricultural and NIA water rights [50,51]. In the SRVC model, surface water delivered to PMA municipalities is considered in C_0 and A_0 (see Tables S1–S8 in supplementary materials).

3.2. Groundwater Supply and Storage Options

The Groundwater Management Act of 1980 (GMA) established the Phoenix Active Management Area (AMA), among other AMAs, and capped groundwater pumping to eventually achieve a safe-yield condition, where natural groundwater recharge exceeds groundwater withdrawals. In essence, the GMA codified a groundwater sustainability goal for the Phoenix AMA. ADWR regulates municipal groundwater pumping within the Phoenix AMA [52]. PMA municipalities can bank excess water as LTSCs via managed aquifer recharge. LTSCs are convertible from storage credits to real water to lessen the impact of Colorado River shortages [53] as the 100-year average of the total groundwater banked [54–63]. Therefore, if a municipality has 123,348 kL banked as LTSCs, that municipality can convert up to 1233.48 kL year^{-1} of LTSCs to real water. In the SRVC model, groundwater delivered to PMA municipalities is considered in C_0 and A_0 (see Tables S1–S8 in supplementary materials).

3.3. Effluent Water Supply Options

PMA municipalities utilize treated effluent for turf irrigation and aquifer recharge in the PMA [52]. A notable exception is the City of Phoenix that has contracted its effluent for cooling the Palo Verde Nuclear Generating Station [64]. ADWR calculates municipal effluent generation as 40% of potable demand, although ADWR may revise this number to accommodate instream recharge and direct potable reuse. The Arizona Department of Environmental Quality (ADEQ) enforces greywater and wastewater reuse rules to facilitate fit-for-purpose reuse, which assigns varying stipulations for reuse of different water classes. In the SRVC model, effluent water supply is considered in C_0 and A_0 (see Tables S1–S8 in supplementary materials).

3.4. Conservation and Nature-Based Water Supply Options

Water conservation, ecosystem management, improving upstream irrigation efficiency, and upstream fallowing farms are potential new sources of water. For example, since landscape irrigation comprises 40% of water use in the City of Phoenix [52], residential landscape water conservation represents a large volume of convertible soft, elastic water demand. Nature-based water sources developed by non-governmental organizations, such as The Nature Conservancy, through irrigation efficiency projects, farm fallowing, crop switching, or mechanical forest thinning leave water in the Verde River for riparian ecosystems and downstream users. Mechanical thinning of ponderosa pine forests in the Salt-Verde watershed reduces the probability of forest fires and increases watershed runoff. TNC provided a rough estimate of localized flow increases (but not systemic downstream increases) based on preliminary modeling and field work conducted by The Nature Conservancy in Arizona's staff and subcontractors farm-based projects; these preliminary findings show an additional 54,110 ML year^{-1} of in-stream flow in the Verde River [65]. Mechanical forest thinning has the potential to increase Salt-Verde watershed runoff by 26% [37]. In the SRVC model, water added to the Salt-Verde river system through nature-based solutions is considered an adaptive option.

3.5. Water SRVC Model Assumptions

The Water SRVC model utilizes legally assured, secured water rights to structure model logic, water rights access, water rights allocation, water rights utilization, and water curtailment adaptation. Fundamental to the Water SRVC model is that municipalities access shared water sources that are rivalrous, with varying degrees of excludability [66]. The Water SRVC models consider only the PMA municipal and metropolitan boundaries because water sharing occurs at this scale. Owing to the excludability of the region's water sources and the multiple decision boundaries, each Water SRVC model outcome yields a different policy pathway.

With these assumptions and constraints, we developed the Water SRVC model with 2016 serving as the base year. We obtained CAP delivery data to municipalities by water type from CAP delivery reports [49]. Municipal and Industrial CAP deliveries are senior and protected among CAP water types and minimally affected by the Water SRVC scenarios [51]. We obtained SRP delivery data by water type from annual reports on water withdrawal and use data within the service district [46]. Additionally, we obtained reservoir capacity and runoff data from SRP [67]. We obtained groundwater, groundwater storage, and effluent data from ADWR [54–63]. Water supply costs were obtained from a confidential market analysis produced for The Nature Conservancy by WestWater Research [29]. The market analysis collected and summarized observed water rights prices by asset class in the Phoenix AMA and other regional water markets in Arizona, in order to estimate the value of agricultural and urban water in the Verde River Basin. Due to the confidentiality of this data, we cannot directly publish the findings from this market analysis; however, we can publish the ranges of observed water-rights prices and the conclusions derived from the confidential data. That said, historically, Colorado River water tends to be more expensive than local sources, such as groundwater, stored groundwater, and Salt River Project water (Table 3).

Table 3. Cost and potential capacity of PMA adaptive water options.

Adaptive Water Option	Phoenix Active Management Area (AMA) AWBA Balance	CAP Lease (Short-Term)	CAP Lease (Long-Term)	Type 2 Lease	Colorado River
Water Source	AWBA	CAP	CAP	Groundwater	Colorado River
Total Annualized Cost [a] ($ ML^{-1} year^{-1})	100–500	100–500	100–500	100–500	100–500
Potential Capacity [b], P_0 (ML)	1,373,503 [b]	~729,643 [c]	~729,643 [c]	38,376 [d]	972,857 [e]

[a] Cost data adapted from [29]; a range of 100–500 is displayed instead of the precise numbers used in this study as these market cost data are private, sensitive, proprietary, and confidential; Colorado River and CAP Short-Term Lease options have the highest costs among these five options. [b] Potential Capacity data calculated from [49]. [c] [68]. [d] [69]. [e] [70].

When a modeled curtailment created a structural water deficit, municipalities exercise water portfolio options according to cost—starting with underutilized existing online water sources and then the lowest cost, single-use potential source. While exact cost estimates are not shown as they are confidential, Colorado River water tends to be more expensive than local sources. Further, due to a lack of available data, we assume that operations, maintenance and repair costs (OM&R) remain unchanged pre- and post-adaptation.

3.6. Water SRVC Model Scenarios

We evaluated one baseline scenario, two Colorado River shortage declaration scenarios, two SRP shortage scenarios, two nature-based flow augmentation scenarios, one municipal conservation, one scenario with simultaneous Colorado River and SRP shortage declarations, and one scenario with municipal conservation in response to simultaneous Colorado River and SRP shortage declarations. For simplicity, the municipal decision scenarios assume that municipalities take identical actions with respect to conservation policies, namely municipalities reduce consumption by reducing soft, elastic demand such as landscape irrigation. This initial SRVC model limits scenarios to a single stage of adaptive decision-making over one year (Table 4). Considering long-term cascades of multiple decisions over multiple years and under changing conditions remains future work.

Table 4. Water security planning scenarios.

Scenario	Description
Baseline Scenario (Base)	Current normal operations of the SRP and CAP water systems for PMA municipalities [46,54–63].
Colorado River Current Drought Contingency Plan (CRB-D)	The U.S. Bureau of Reclamation declares a shortage on the Colorado River due to the summer elevation of Lake Mead dropping below 1075 ft. above mean sea level. CAP curtails water deliveries to municipalities according to the current drought contingency plan where Arizona takes deepest water cuts [48,50,51].
Colorado River Draft Drought Contingency Plan (CRB-C)	The U.S. Bureau of Reclamation declares a shortage on the Colorado River due to the summer elevation of Lake Mead dropping below 1075 ft. above mean sea level. CAP curtails water deliveries to municipalities according to the draft contingency plan currently discussed by Arizona, California, and Nevada. Arizona takes deeper water cuts, but California and Nevada also take water cuts when the elevation of Lake Mead drops below 1075 ft. above mean sea level [48,50,51].
Drought in the Salt River Watershed (SRD)	The SRP governing board reduces the Stored and Delivered Water multiplier to from 3 acre-feet per acre (9.144 ML ha^{-1}) of on-project lands to 2 acre-feet per acre (6.096 ML ha^{-1}) due ongoing drought in the Salt River Watershed. The scenario is based on the 2003 decision by the Salt River Project to reduce surface water deliveries to customers [33,45].
Increased Verde River Watershed flows from Nature-Based Conservation (VR-NBF)	Augmentation of Verde River flow from various water conservation programs in the Verde River watershed, such as farm retirement and voluntary fallowing [65].
Increased Verde River flows due to Upstream Forest Thinning (VR-FT)	Mechanically thinning ponderosa pine forests increases runoff by 20%, up to 18.1 to 42.9 million m3 per year on average over a 15-year thinning period [37,38]. However, increased runoff declines after a six-year period [37].
Residential Conservation (PMA-RC)	Municipalities reduce water consumption by 10%, meeting or exceeding Phoenix Active Management Area 4th Management Plan Gallons Per Capita Day (GPCD) Targets. We chose a 10% threshold because Tolleson, due to its small size, is not part of the Phoenix AMA GPCD targets [71].
Prolonged Drought in Salt-Verde Watershed (SRD-P)	The local impacts of climate change may permanently reduce Salt-Verde River watershed runoff. SRP estimates the 30-year median runoff is 659,093 ML year^{-1}. Downscaled GCM models estimate a 23% runoff reduction [42]; streamflow data between 1996–2015 show a 64% reduction versus the long-term average [42,44,46].
Combination of CRB-D and SRD (SRD-CRB-D)	A combination of scenarios CRB-C and SRD.
Combination of CRB-C, SRD, and PMA-RC (SRD-CRB-RC)	A combination of scenarios CRB-C, SRD, and PMA-RC.

4. Results

4.1. Structural Water Deficits

The PMA has a diverse water supply portfolio with variegated water sources and rights. Therefore, there ought to be an uneven response to structural water deficits across municipalities under the various curtailment scenarios. Indeed, structural water deficits vary dramatically depending on the scenario and the municipality, ranging from a surplus or no deficit to 27% water supply deficits for a single municipality (Tolleson) and across several orders of magnitude across the PMA for a single scenario. Notably, Avondale, Goodyear, and Peoria experience no deficits under any scenario due to their large excess capacity relative to current demand.

Of the ten scenarios developed, seven focused on municipal-level impacts while the remaining three focused on the impacts of flow augmentation programs in the Salt-Verde watershed. The municipal-focused scenarios were the Baseline Scenario (BASE), CAP Shortage Default (CRB-D), CAP Shortage Draft Plan (CRB-C), SRP Shortage (SRD), Municipal Conservation (PMA-RC), CAP Default + SRP Shortage (SRD-CRB-D) and CAP Default + SRP Shortage + Conservation (SRD-CRB-RC). Five of the seven municipal-level scenarios implemented water curtailments on either the SRP or CAP (CRB-D, CRB-C, SRD, SRD-CRB-D, and SRD-CRB-RC) and four of the seven scenarios created structural water budget deficits in municipality water budgets (CRB-C, SRD, SRD-CRB-D, and SRD-CRB-RC).

The CRB-D and CRB-C scenarios compare the outcomes of Colorado River shortage declaration under the current shortage plan and the draft contingency shortage plan. Due to how the different shortage plans impact the different CAP water types, initial cuts to CAP under current shortage guidelines did not induce structural water deficits in municipal water portfolios. However, the draft contingency plan creates a 31,018 ML year^{-1} deficit at the PMA scale with Phoenix (22,453 ML year^{-1}) most affected by an order of magnitude, followed by Mesa (3344 ML year^{-1}), Chandler (2363 ML year^{-1}), and Scottsdale (2292 ML year^{-1}). LTSCs and Type 2 groundwater leases are adaptive water options to cover a structural water deficit created by the CRB-D and CRB-C scenarios.

Despite being the less visible, regional water shortage scenario, a curtailment of Salt River supplies would have more impact on the PMA. The SRD scenario creates a 77,257 ML year^{-1} metropolitan area scale structural water deficit. In terms of water volume, the SRD scenario most affects Phoenix (62,193 ML year^{-1}) followed by Gilbert (11,194 ML year^{-1}), Tempe (2245 ML year^{-1}), and Tolleson (1624 ML year^{-1}); however, Tolleson is most affected in relative terms (27%). Due to the larger structural water deficits, the Arizona Water Banking Authority becomes a strategic short-term consideration for PMA municipalities.

Achieving a 10% water conservation goal across the PMA—whether through voluntary or mandatory programs—created a surplus of 68,561 ML year^{-1}. Specifically, for Phoenix, Gilbert, Tempe, and Tolleson, municipalities with the largest absolute and relative structural water deficits in CRB-D, CRB-C, and SRD, municipal water conservation can prevent or significantly reduce modeled structural water deficits.

For PMA municipalities, groundwater provides a strategic buffer against structural water deficits and induced water curtailments. This buffer is primarily due to existing LTSCs and the short-term leasing of Type 2 groundwater rights. Therefore, purchasing banked groundwater from, for example, the Arizona Water Banking Authority, is a strategic short-term consideration for PMA municipalities to guard against hydrologically- or legally-induced shortfalls in their water portfolios.

4.2. Sustainability

Given the high inter-annual variability in precipitation in the Salt-Verde watershed [38], our adaptive water policy decisions today could provide a vital and necessary buffer between wet winters by doubling the SRP system's sustainability. Without a wet winter to recharge SRP reservoirs, the best-case sustainability scenario along the SRP system is a six-year delay before delivery curtailments. At the current runoff levels in the Salt-Verde, which is approximately 246,696 ML year^{-1} less than 30-year median runoff, the SRP reservoir system will hit 27% capacity, which triggered the previous curtailment, within a 4-year period. However, a combination of municipal conservation in the PMA and flow augmentation in the Salt-Verde watershed (VR-NBF, VR-FT, PMA-RC) can postpone when SRP reservoir capacity reaches historical curtailment levels. The consumption avoided through municipal conservation programs is 1.4× to 3.4× larger than the expected structural water deficits modeled in Scenarios CRB-C, SRD, SRD-CRB-D, and SRD-CRB-C. Additionally, the potential increase

in annual runoff resulting from flow augmentation in the Salt-Verde watershed is approximately 0.8×
to 2.1× larger than the expected structural water deficits modeled. The sustainability of the SRP system
has six years of sustainability, corroborating [72]. A SRP shortage would affect the water portfolios of
Phoenix, Gilbert, Tolleson, and Tempe the most. Consequently, the municipalities would benefit most
from SRP system sustainability.

The sustainability scenarios evaluated the impact of drought on SRP reservoir capacity and how
far into the future SRP could postpone SRP water delivery curtailments before a shock ensues that
requires damaging water rationing (Table 5).

Table 5. Sustainability impact (S, ΔS) of water conservation and flow augmentation options based on a
2016 baseline year.

Scenario	Δ Salt River Flow (ML year^{-1})	Curtailment Year (S, ΔS)
BASE	0	2020 (4 year, -)
VR-NBF	3556	2021 (5 year, +1)
VR-FT	6413	2021 (5 year, +1)
PMA-RC	69,435	2022 (6 year, +2)
VR-NBF, VR-FT, PMA-RC	111,409	2022 (6 year, +3)
SRD-P	−162,143	2020 (4 year, 0)
VR-NBF, PMA-RC, SRD-P	−69,689	2020 (4 year, 0)

4.3. Resilience

There is a stark difference in resilience outcomes between scenarios, policy choices, and municipalities.
The resilience outcomes of CAP-dominated water portfolios were most impacted by scenarios CRB-D,
CRB-C, SRD-CRB-D, and SRD-CRB-RC. Conversely, the resilience outcomes of SRP-dominated water
portfolios were most impacted by SRP drought scenarios. (Figure 4). Municipal water conservation
and flow augmentation are potential options to increase water portfolio resilience. This increase in
resilience results from the creation of new water types (augmented flow) and increased surface water and
groundwater water availability as a strategic buffer. Reductions to water portfolio options created large
decreases in municipal water portfolio resilience; especially, if a shock reduced the number of available
water options from many options to one or two options.

Notably, a Colorado River shortage reduces resilience at both the individual municipality
level and the overall metropolitan area. Without the creation of new water sources, conservation,
or policy adaptations, groundwater storage credits are the last choice to cover structural water deficits.
For example, the Lower Colorado River draft drought contingency plan can increase the municipality's
resilience by forcing the creation of new options that were not available or not considered viable options,
before a shortage declaration. The Supplemental Information contains the full table of resilience results
(Table S8).

4.4. Vulnerability

Phoenix SRVC model results show modest vulnerability differences between cities and between
scenarios (Figure 5). Some scenarios yield minor changes in vulnerability (CRB-D, CRB-C),
some scenarios yield increased vulnerability for all municipalities (SRD, SRD-CRB-D), one scenario
decreases vulnerability for all municipalities (PMA-RC), and one scenario produces strong increases
and strong decreases in vulnerability for different municipalities (SRD-CRB-RC). On a scale from
zero to one, a change in vulnerability of +/− 0.10 is large in absolute terms and represents a water
buffer of approximately 123,348 ML year^{-1}. These differences are smaller in relative terms than the
differences in sustainability and resilience shown in earlier sections. Municipalities with CAP-dominant
water portfolios were most affected by scenarios CRB-D, CRB-C, SRD-CRB-D, and SRD-CRB-RC.
Conversely, SRP-dominant water portfolios were most affected by scenarios SRD, and SRD-CRB-D,

and SRD-CRB-RC. Importantly, municipal conservation as a response to mandated water curtailments, for most municipalities, soften the impact of water curtailments and reduced vulnerability to water stress. The Supplemental Information contains a full table of vulnerability results (Table S8).

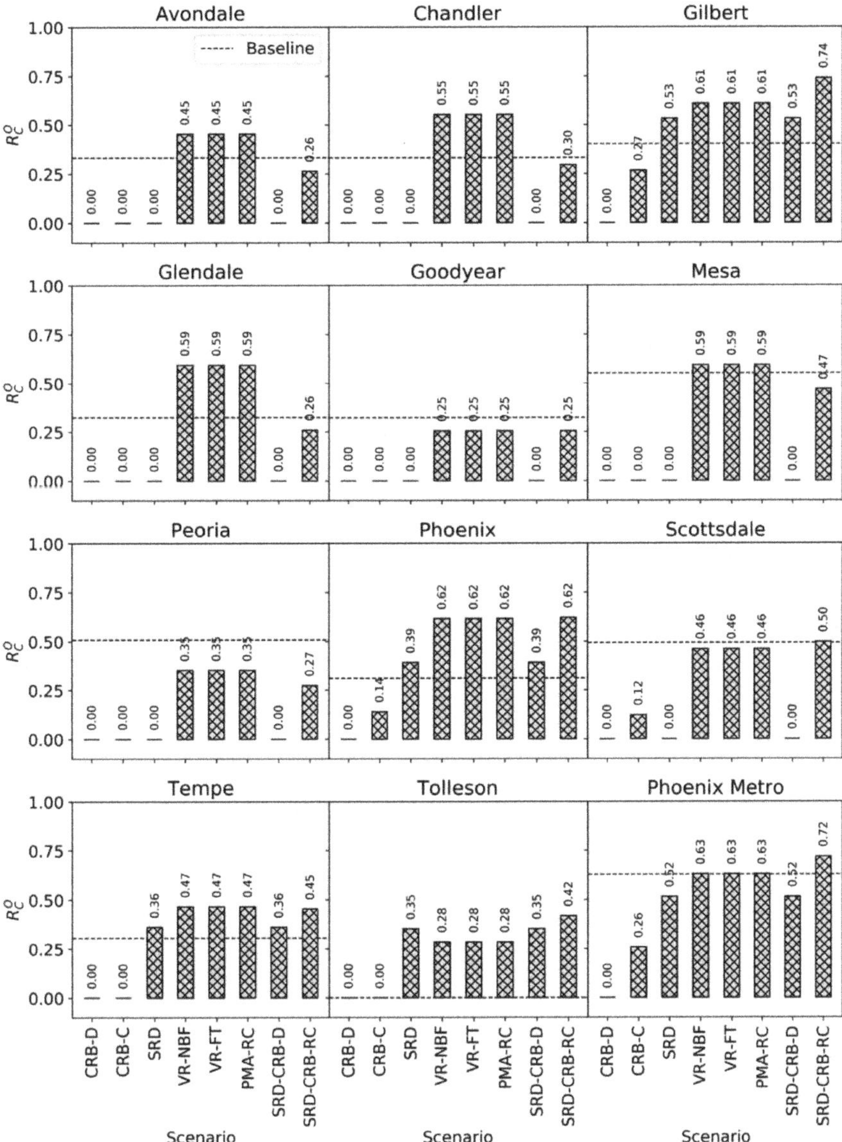

Figure 4. The resilience (R_c^O) and change from baseline resilience (ΔR_c^O) of a municipality's water portfolio. Overall, ΔR_c^O is large based on each municipality's portfolio choices.

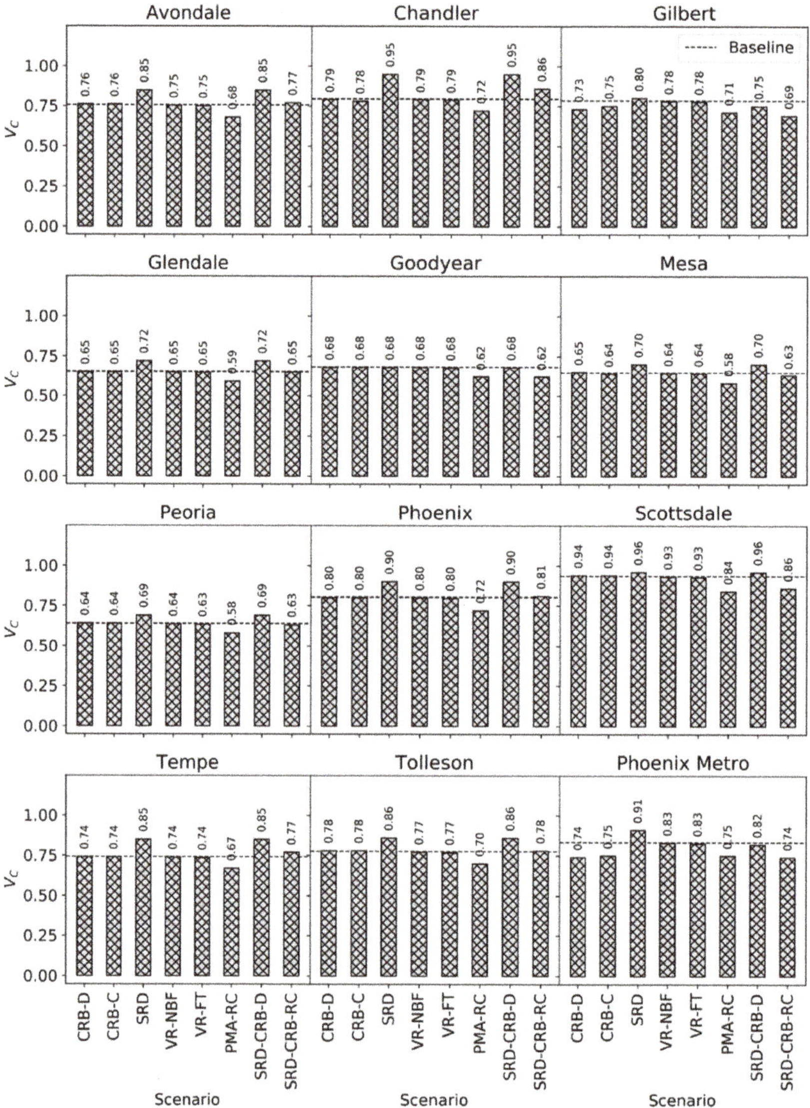

Figure 5. The vulnerability of a municipality's water supply (V_C) and change from the baseline scenario (ΔV_c). Overall, ΔV_c is small based on each municipality's portfolio choices.

4.5. Water Supply Costs

Cost changes to a municipal water portfolio pre- and post-adaptation varied based on the difference in cost between existing water contracts and new water contracts (Figure 6). For some municipalities, conveying existing LTSCs were cheaper than CAP supplies in CRB-D, CRB-C, SRD-CRB-D, and SRD-CRB-RC. Municipalities with a heavy reliance on SRP supplies—such as Phoenix, Gilbert, Tempe, and Tolleson—faced the sharpest per unit water cost increases due to switching from a low-cost source (SRP) to sources with higher relative cost. Phoenix and Tempe show modest changes in cost depending on their policy decisions, and other municipalities have little to no change in cost as a function of their policy decisions. In general, cost did not vary much across scenarios and water supply options. We expected this because we excluded from the scope of analysis the adaptive water supply options with dramatically higher costs than the current supplies (e.g., desalination). Upstream flow augmentation and residential conservation solutions are preferentially chosen by the model based on cost because they are adequate to cover structural water deficits at lower per-gallon costs to other adaptive water supply options. As explained in Table 3, the market cost data is not released in the study due to its sensitive nature, so we are only able to present these aggregated results and not the raw numbers used to calculate the results. These aggregated costs are calculated using the methods explained in Section 2.5.

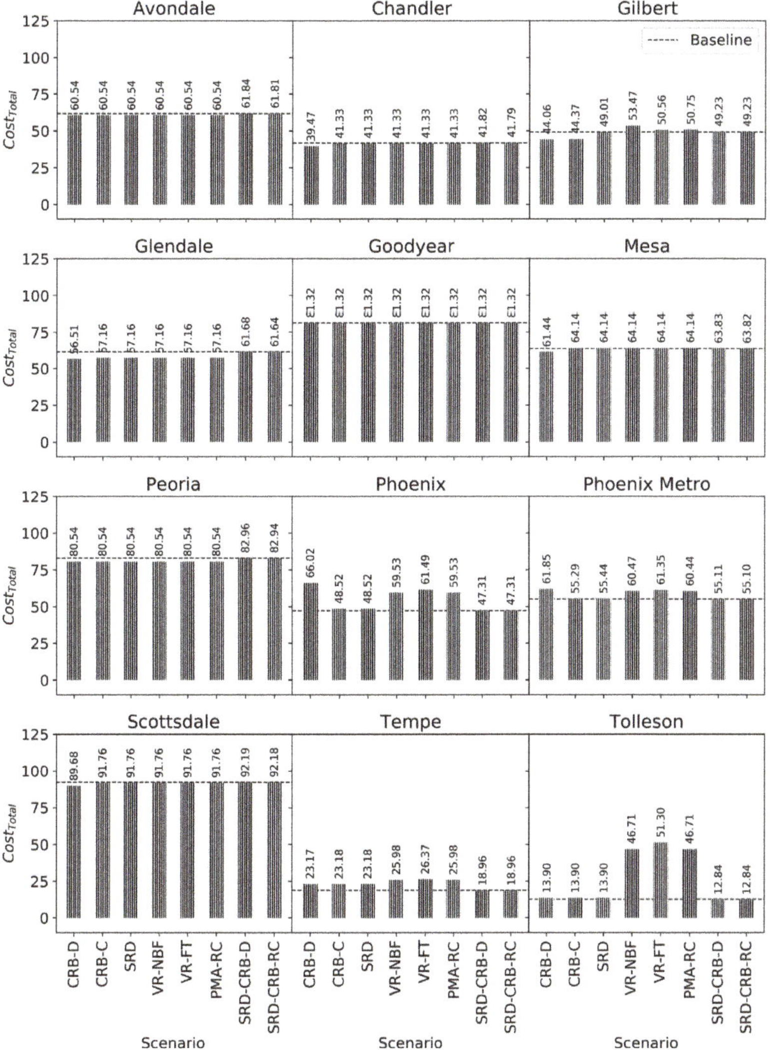

Figure 6. Aggregated cost data: the volume weighted cost ($ ML^{-1}) to obtain new water to fill the structural water deficits and change from baseline cost, for each modeled scenario, using methods explained in Section 2.5. Data shown in Table S2. Only cities that must switch their water sources to totally new sources (like Tolleson) at market prices during the event incur large cost increases.

A municipality's cost to bring potential sources online during drought is proportionate to the volume of water required. Some municipalities would face much larger relative deficits and much larger cost escalations than others, depending on the scenario considered. The escalation of cost is predictable at a significant R^2 of 0.94 using the equation $\text{Cost}_1/\text{Cost}_0 = 2.785\ F/C_0$ (Figure 7).

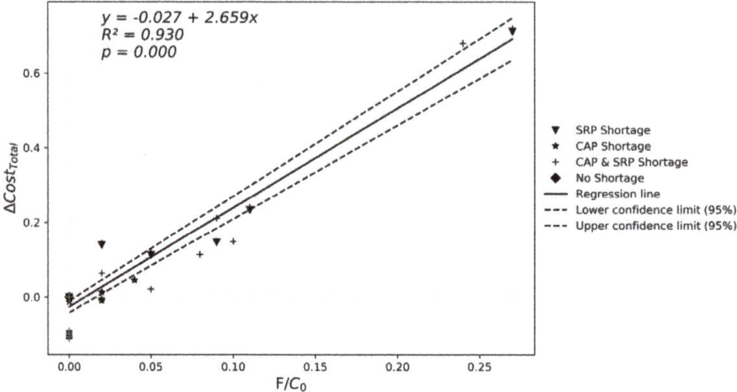

Figure 7. As water budget deficits grow, so does the cost to access available water sources and to bring online potential water sources to fill the water deficit (Data in Table S8).

4.6. Tradeoffs between S, R, V, and C

In the PMA, water portfolio vulnerability and resilience (V/R) have a weak significant relationship ($R^2 \sim 0.1$) where increasing resilience decreases vulnerability (Figure 8). Just as many municipalities and scenarios fall outside the confidence bounds of this relationship, as fit within the bounds. There is even less of a relationship between S/R or S/V.

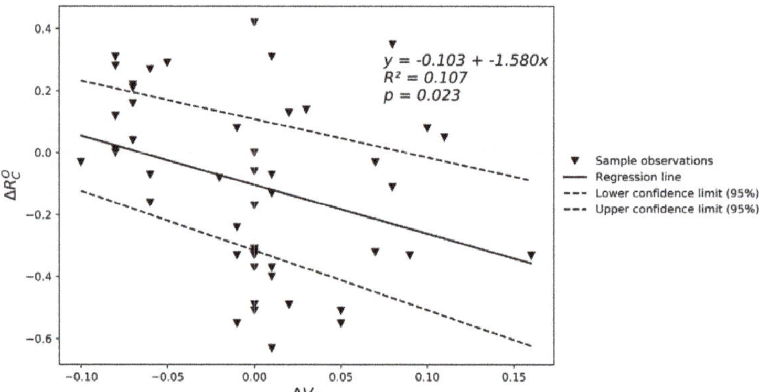

Figure 8. ΔR_C^O plotted against ΔV_C for all scenarios. As vulnerability to water stress increases, the resilience of available water sources decreases—but with important exceptions. Points in the upper left quadrant indicate win-win scenario outcomes where municipal water portfolio vulnerability decreases and municipal water portfolio resilience increases. Points in the lower right quadrant indicate lose-lose scenario outcomes where municipal water portfolio vulnerability increases and municipal water portfolio resilience decreases. Points plotted in the upper right and lower left quadrant indicate win-lose scenarios where municipal vulnerability increases while municipal resilience increases or vice versa. See Table S8 for data.

More importantly, there is little relationship between cost and the other three dimensions of sustainability, resilience, and vulnerability (Figure 9). These four water security dimensions are mostly independent. This means it is possible for a municipality to find win-win options (and lose-lose options). A municipality can gain sustainability or reduce vulnerability at little to no cost. An exception is that resilience does appear to come at an increased cost on average but only weakly and with many exceptions (Figure 9, center).

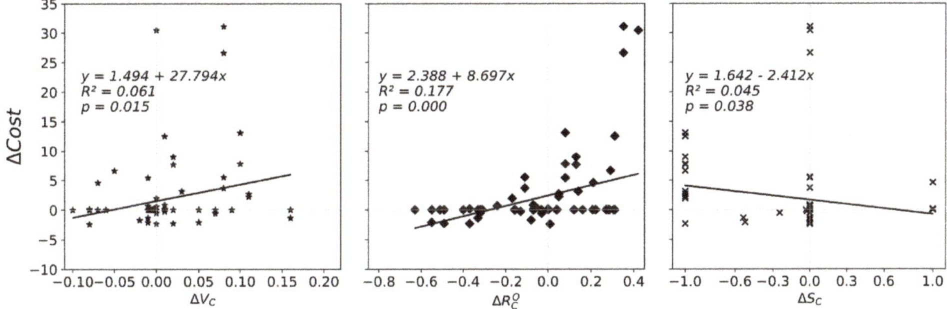

Figure 9. $\Delta Cost$ compared with ΔV_C, ΔR_C^O, and ΔS_C, with each municipality and scenario representing a mark on the scatter plot. Relationships between portfolio cost and the other dimensions of water security are weak or absent. See Table S8 for data.

The exceptions matter, because there is no "average" municipality in the Phoenix metro. If we isolate each of the twelve municipalities' adaptive scenarios and regress C against S, V, and R for each municipality, only eight out of thirty-six relationships have an R^2-value greater than 0.25 and for most relationships the slope is at or around 0 (Table 6). This means every municipality needs to independently contemplate its adaptation decisions and water portfolio positioning and reach its own conclusions; following other municipalities may lead to poor decisions.

Table 6. Municipal-level Vulnerability, Resilience, and Sustainability cost regressions.

Municipality	V vs. C		R_c^{A+P} vs. C		ΔS vs. $\Delta Cost_{total}$	
	Slope	R^2	**Slope**	R^2	**Slope**	R^2
Avondale	−0.02	0.06	−0.02	0.12	−0.15	0.02
Chandler	0.01	0.01	−0.01	0.02	0.28	0.25
Gilbert	0.00	0.08	0.01	0.12	−0.09	0.36
Glendale	0.00	0.03	−0.01	0.19	0.18	0.04
Goodyear	−4.25	0.07	1.46	0.02	0.00	0.00
Mesa	0.00	0.01	0.00	0.01	0.16	0.23
Peoria	0.00	0.02	−0.01	**0.33**	−0.18	0.02
Phoenix	0.00	0.23	0.01	**0.43**	−0.08	**0.53**
Phoenix Metro	0.00	0.00	0.01	0.60	−0.03	0.07
Scottsdale	0.00	0.01	−0.01	0.01	0.18	0.26
Tempe	0.01	**0.31**	0.01	**0.66**	−0.35	**0.74**
Tolleson	0.00	**0.46**	0.00	0.29	−0.01	0.09

5. Discussion

This paper develops and implements a systems-based Sustainability, Resilience, Vulnerability, and Cost (SRVC) model for water portfolio security assessment that adds systems-level S, R, and V dimensions to the traditional cost-and-reliability-based models. The SRVC model adds to a rich literature on systems-based analysis of multi-stakeholder systems including the management of municipal solid waste [73], streamflow analysis [74], wastewater management [75], and stormwater quality and management [76]. The results are promising and have implications for municipal water portfolio planning and preparation for unpredictable annual-timescale disruptions. In the PMA, we find that the S, R, and V dimensions are independent of each other and from cost. Therefore, municipalities should be able to develop cost-neutral policy options that are more sustainable and resilient and less vulnerable. For Phoenix metropolitan area municipalities, potential water portfolio reconfigurations in response to an annual-timescale shock are differentiated along resilience and sustainability metrics, rather than vulnerability or cost metrics. However, this finding does not hold evenly across all Phoenix metropolitan area municipalities. The details matter, and adaptive options must complement each municipality's unique water portfolio. Municipal water portfolios dominated by Salt River Project deliveries have different responses to unforeseeable infrastructure failures, regional drought, or forest fires than portfolios relying primarily on Central Arizona Project water. The SRVC model adds the information necessary for a municipality to identify cost-neutral adaptive water portfolio configurations that are more sustainable, more resilient, and/or less vulnerable.

After a Colorado River shortage declaration, which would curtail Central Arizona Project deliveries to the Phoenix metropolitan area, we find that resilience decreases as groundwater, in the form of long-term storage credits, is the only remaining unused supply option. Future updates to this study will also have to include modeled and/or observed increases to baseline OM&R costs as a result of a shortage declaration. After stored groundwater, a combination of residential conservation and nature-based flow augmentation programs can compensate for Colorado River drought and restore water portfolio resilience without building new infrastructure to supply greywater. These options only become more cost effective as OM&R costs along the CAP system increase after a water shortage declaration.

The most cost-effective adaptive options to improve S, R, and V—and mitigate the effects of climate change, drought, and water curtailment on water security—are water conservation and flow augmentation programs in the Salt-Verde watershed. Residential water conservation programs are impactful win-win adaptive policy options in cities with comparatively high per-capita consumption. Relying on conservation programs to enhance S, R, and V presents several challenges. For example, an upstream conservation program to enhance streamflow will depend on willing farmers, and willingness to participate may present challenges. Further, monitoring the contributions of conservation to river flow augmentation is difficult given a number of factors contributing to flow levels; however, headgates help in accounting for flow returned to the river. For municipalities, monitoring changes in household-level water consumption is potentially easier through metering but may harden demand over time, reducing the effectiveness of future conservation programs. Previous research in the Los Angeles metropolitan area found that mandatory water restrictions and price increases reduced water consumption more effectively than voluntary reductions [77]. Additionally, conservation provides a smaller marginal benefit for municipalities with hardened residential demand. Municipalities must retain the benefits of conservation rather than fall prey to efficiency paradoxes (i.e., growth offsetting conservation gains) to successfully realize S, R, and V gains [78].

Flow augmentation in the Salt-Verde watersheds is a promising adaptive option. However, the potential for flow increases remains uncertain, especially for forest thinning programs. Nonetheless, increased runoff and in-stream flows can create co-benefits that include creating socio-ecological public value from recreational activities and forest thinning jobs while reducing the probability of forest fire and leaving more water for riparian ecosystems.

The role that the Central Arizona Project has played in facilitating water transfers (by trading on the conveyance of Colorado River water) between Phoenix metropolitan area cities can serve as an example for how the Salt River Project might help the cities adapt to drought. However, instead of shifting demand around Arizona by allowing cities to consume water already slated for agricultural use, the Salt River Project could play a role in conveying to cities the water that is made available through conservation, flow restoration, and the provision of flow supporting instream ecosystem services. However, no market for flow augmentation projects currently exists. Phoenix, Gilbert, and Tempe are among the largest users of Salt River Project water in the metropolitan area, so they would be key participants in a market that would allow municipalities to purchase additional flows by funding upstream watershed restoration and nature-based flow augmentation. Municipalities that do not currently use the Salt River Project system have no initial incentive to participate in such a market but might choose to participate if the market resulted in access to new backup water supplies or water sharing agreements in partnership with the Salt River Project. Given that Phoenix municipalities are already willing to pay for mainstream Colorado River water, nature-based projects should be a cost-competitive option even with a small Salt River Project conveyance and management fee. Third parties, such as NGOs, have potential roles in this system as intermediaries between PMA municipalities and individual farmers, irrigation districts, and forest thinning operations by administering programs. Such mechanisms could provide municipalities with a market or program to invest in upstream environmental flows, improve rural economies, and mitigate ecosystem damage while cost-competitively boosting urban water security, leveraging the water supply system to create system-wide environmental and water security benefits [79].

Future research can consider a wider range of adaptive options and future scenarios such as green infrastructure growth policies, a wider range of population and economic growth scenarios, and longer-term shocks beyond the approximately annual event timescale. Future work could consider cascading SRVC impacts from chained adaptive decisions (t_2 and subsequent) and multiple decision timeframes. Future work should introduce multiple-objective optimization to identify the best adaptive pathways for each municipality. Future work may distinguish between recoverable and irrecoverable or stranded capital costs to bring potential sources online because this distinction is relevant for decision pathways involving multiple adaptive steps and for the financial feasibility of adaptation. This paper does not consider the decision maker's ability to pay, which is not a factor for PMA municipalities because all options on the table are affordable at current prices given healthy budgetary and bonding status. However, future versions of this method will address this by adding the ability to pay into the methodology. Future work could apply this model to critical infrastructure planning problems beyond the scope of water policy and address power systems, transportation systems, and food systems [80].

6. Conclusions

The resilience of the entire PMA is greater than its constituent municipalities. Given this, cooperation during long-term planning and also on emergency response to water shortages and droughts is an important win-win adaptive strategy for these municipalities, corroborating [14]. Diversification of supply is also an important means of enhancing water security through resilience, although diversification could come at a cost to some municipalities. Municipalities with water surpluses generated by conservation could potentially enter short-term water leases with other PMA municipalities with structural water deficits, providing a municipality a revenue stream and economic incentive to reach demand management goals. Historically, municipalities have paid higher prices than other government entities (Federal and state) in water rights markets [81]. This should provide motivation to overcome the financial and legal disincentives against municipal utility water conservation [82]. Intra-metropolitan and intra-watershed fragmentation of water delivery and water decision systems demonstrably damages resilience in a context of shared water sources; fragmentation hurts small municipalities with narrow portfolios (e.g., Goodyear) more than the largest municipalities that have diverse portfolios (e.g., Phoenix).

Supplementary Materials: The following are available online at http://www.mdpi.com/2073-4441/12/6/1663/s1. Table S1: Total Online Capacity, Utilized Online Capacity, Available Capacity, and Structural Water Deficit of Phoenix Metropolitan Area Municipalities–Baseline (ML). Table S2: Total Online Capacity, Utilized Online Capacity, Available Capacity, and Structural Water Deficit of Phoenix Metropolitan Area Municipalities–CRB-D (ML). Table S3: Total Online Capacity, Utilized Online Capacity, Available Capacity, and Structural Water Deficit of Phoenix Metropolitan Area Municipalities–CRB-C (ML). Table S4: Total Online Capacity, Utilized Online Capacity, Available Capacity, and Structural Water Deficit of Phoenix Metropolitan Area Municipalities–SRD (ML). Table S5: Total Online Capacity, Utilized Online Capacity, Available Capacity, and Structural Water Deficit of Phoenix Metropolitan Area Municipalities–PMA-RC (ML). Table S6: Total Online Capacity, Utilized Online Capacity, Available Capacity, and Structural Water Deficit of Phoenix Metropolitan Area Municipalities–SRD-CRB-D (ML).

Table S7: Total Online Capacity, Utilized Online Capacity, Available Capacity, and Structural Water Deficit of Phoenix Metropolitan Area Municipalities–SRD-CRB-RC (ML). Table S8: The system statistics for each scenario.

Author Contributions: Conceptualization, R.R.R. and B.L.R.; methodology, R.R.R. and B.L.R.; software, R.R.R.; validation, R.R.R. and B.L.R.; formal analysis, R.R.R.; investigation, R.R.R.; data curation, R.R.R.; writing—original draft preparation, R.R.R.; writing—review and editing, R.R.R., M.M., B.L.R.; visualization, R.R.R.; supervision, B.L.R.; project administration, B.L.R.; funding acquisition, B.L.R. All authors have read and agreed to the published version of the manuscript.

Funding: The authors acknowledge funding from The Nature Conservancy in Arizona and the Northern Arizona University TRIF. The Nature Conservancy in Arizona commissioned a study of the effects of flow restoration on downstream urban water users' water security and owns original data products upon which the study was partially based. This material is based upon work supported by the National Science Foundation under Grant No. ACI-1639529 and USDA Grant No. 2017-08812. The findings and opinions of this paper are those of the authors and not necessarily of the funding agencies.

Acknowledgments: The authors would like to thank the funding agencies the gracious input from the peer reviewers.

Conflicts of Interest: The authors declare no conflict of interest. The funders had no role in the design of the study; in the collection, analyses, or interpretation of data; or in the decision to publish the results; however, M.M. was employed by the funder (The Nature Conservancy) during the writing of the manuscript.

References

1. Hering, J.G.; Waite, T.D.; Luthy, R.G.; Drewes, J.E.; Sedlak, D.L. A Changing Framework for Urban Water Systems. *Environ. Sci. Technol.* **2013**, *47*, 10721–10726. [CrossRef]

2. Pincetl, S.; Porse, E.; Cheng, D. Fragmented Flows: Water Supply in Los Angeles County. *Environ. Manag.* **2016**, *58*, 208–222. [CrossRef] [PubMed]

3. Herman, J.D.; Reed, P.M.; Zeff, H.B.; Characklis, G.W. How Should Robustness Be Defined for Water Systems Planning under Change? *J. Water Resour. Plan. Manag.* **2015**, *141*, 04015012. [CrossRef]

4. Zeff, H.; Kasprzyk, J.; Herman, J.D.; Reed, P.M.; Characklis, G.W. Navigating financial and supply reliability tradeoffs in regional drought management portfolios. *Water Resour. Res.* **2014**, *50*, 4906–4923. [CrossRef]

5. Griscom, B.; Adams, J.; Ellis, P.W.; Houghton, R.A.; Lomax, G.; Miteva, D.A.; Schlesinger, W.H.; Shoch, D.; Siikamäki, J.V.; Smith, P.; et al. Natural climate solutions. *Proc. Natl. Acad. Sci. USA* **2017**, *114*, 11645–11650. [CrossRef]

6. Kabisch, N.; Frantzeskaki, N.; Pauleit, S.; Naumann, S.; Davis, M.; Artmann, M.; Haase, D.; Knapp, S.; Korn, H.; Stadler, J.; et al. Nature-based solutions to climate change mitigation and adaptation in urban areas: Perspectives on indicators, knowledge gaps, barriers, and opportunities for action. *Ecol. Soc.* **2016**, *21*. [CrossRef]

7. Taleb, N.N. *The Black Swan: The Impact of the Highly Improbable*; Random House Trade Paperbacks: New York, NY, USA, 2010; Available online: https://books.google.com/books?id=GSBcQVd3MqYC (accessed on 28 June 2017).

8. Brede, M.; De Vries, B.J. Networks that optimize a trade-off between efficiency and dynamical resilience. *Phys. Lett. A* **2009**, *373*, 3910–3914. [CrossRef]

9. Holling, C.S. Engineering Resilience versus Ecological Resilience. In *Engineering within Ecological Constraints*; Peter, S., Ed.; National Academies Press: Washington, DC, USA, 1996.

10. Ulanowicz, R.E.; Goerner, S.J.; Lietaer, B.; Gomez, M.D.R. Quantifying sustainability: Resilience, efficiency and the return of information theory. *Ecol. Complex.* **2009**, *6*, 27–36. [CrossRef]

11. Brown, C.M.; Lund, J.R.; Cai, X.; Reed, P.M.; Zagona, E.; Ostfeld, A.; Hall, J.; Characklis, G.W.; Yu, W.; Brekke, L. The future of water resources systems analysis: Toward a scientific framework for sustainable water management. *Water Resour. Res.* **2015**, *51*, 6110–6124. [CrossRef]

12. Matrosov, E.S.; Huskova, I.; Kasprzyk, J.; Harou, J.; Lambert, C.; Reed, P.M. Many-objective optimization and visual analytics reveal key trade-offs for London's water supply. *J. Hydrol.* **2015**, *531*, 1040–1053. [CrossRef]

13. Reed, P.M.; Brooks, R.P.; Davis, K.J.; Dewalle, D.R.; Dressler, K.A.; Duffy, C.J.; Lin, H.; Miller, D.A.; Najjar, R.G.; Salvage, K.M.; et al. Bridging river basin scales and processes to assess human-climate impacts and the terrestrial hydrologic system. *Water Resour. Res.* **2006**, *42*. [CrossRef]

14. Zeff, H.; Herman, J.D.; Reed, P.M.; Characklis, G.W. Cooperative drought adaptation: Integrating infrastructure development, conservation, and water transfers into adaptive policy pathways. *Water Resour. Res.* **2016**, *52*, 7327–7346. [CrossRef]

15. Hashimoto, T.; Stedinger, J.R.; Loucks, D.P. Reliability, resiliency, and vulnerability criteria for water resource system performance evaluation. *Water Resour. Res.* **1982**, *18*, 14–20. [CrossRef]

16. Rosenberg, D.E.; Tarawneh, T.; Lund, J.R.; Abdel-Khaleq, R. Modeling integrated water user decisions in intermittent supply systems. *Water Resour. Res.* **2007**, *43*. [CrossRef]

17. Walker, B.; Kinzig, A.; Langridge, J. Original Articles: Plant Attribute Diversity, Resilience, and Ecosystem Function: The Nature and Significance of Dominant and Minor Species. *Ecosystems* **1999**, *2*, 95–113. [CrossRef]

18. Allenby, B.; Fink, J. Social and ecological resilience: Toward inherently secure and resilient societies. *Science* **2000**, *24*, 347–364.

19. Kirsch, B.; Characklis, G.W.; Dillard, K.E.M.; Kelley, C. More efficient optimization of long-term water supply portfolios. *Water Resour. Res.* **2009**, *45*. [CrossRef]

20. Kasprzyk, J.; Reed, P.M.; Kirsch, B.R.; Characklis, G.W. Managing population and drought risks using many-objective water portfolio planning under uncertainty. *Water Resour. Res.* **2009**, *45*, 12401. [CrossRef]

21. Boulay, A.-M.; Bare, J.; De Camillis, C.; Döll, P.; Gassert, F.; Gerten, D.; Humbert, S.; Inaba, A.; Itsubo, N.; Lemoine, Y.; et al. Consensus building on the development of a stress-based indicator for LCA-based impact assessment of water consumption: Outcome of the expert workshops. *Int. J. Life Cycle Assess.* **2015**, *20*, 577–583. [CrossRef]

22. Nicklow, J.W.; Reed, P.M.; Savic, D.; Dessalegne, T.; Harrell, L.; Chan-Hilton, A.; Karamouz, M.; Minsker, B.; Ostfeld, A.; Singh, A.; et al. State of the Art for Genetic Algorithms and Beyond in Water Resources Planning and Management. *J. Water Resour. Plan. Manag.* **2010**, *136*, 412–432. [CrossRef]

23. Meadows, D.; Randers, J.; Meadows, D. *Limits to Growth: The 30-Year Update*; Chelsea Green Publishing: Hartford, VT, USA, 2004.

24. World Commission on Environment and Development. *Our Common Future*; Oxford University Press: Oxford, UK, 1987.

25. Paté-Cornell, E. On "Black Swans" and "Perfect Storms": Risk Analysis and Management When Statistics Are Not Enough. *Risk Anal.* **2012**, *32*, 1823–1833. [CrossRef] [PubMed]

26. Ruddell, B.L.; Kumar, P. Ecohydrologic process networks: 1. Identification. *Water Resour. Res.* **2009**, *45*. [CrossRef]

27. Ruddell, B.L.; Kumar, P. Ecohydrologic process networks: 2. Analysis and characterization. *Water Resour. Res.* **2009**, *45*. [CrossRef]

28. Boulay, A.-M.; Bare, J.; Benini, L.; Berger, M.; Lathuillière, M.J.; Manzardo, A.; Margni, M.; Masaharu, M.; Núñez, M.; Pastor, A.V.; et al. The WULCA consensus characterization model for water scarcity footprints: Assessing impacts of water consumption based on available water remaining (AWARE). *Int. J. Life Cycle Assess.* **2017**, *23*, 368–378. [CrossRef]

29. Messerschmidt, M.; The Nature Conservancy of Arizona, Tucson, AZ, USA. Personal Communication—Verde River Water Market Analysis. 2016.

30. Smith, R.; Kasprzyk, J.; Basdekas, L. Experimenting with Water Supply Planning Objectives Using the Eldorado Utility Planning Model Multireservoir Testbed. *J. Water Resour. Plan. Manag.* **2018**, *144*, 04018046. [CrossRef]

31. US Census Bureau. Annual Estimates of the Resident Population: April 1, 2010 to July 1, 2016. Available online: https://factfinder.census.gov/faces/tableservices/jsf/pages/productview.xhtml?pid=PEP_2016_PEPANNRES&src=pt (accessed on 3 May 2017).

32. Brean, H. Lake Mead Likely to Skirt Shortage Line for Another Year. Las Vegas Review-Journal, August 16. Available online: https://www.reviewjournal.com/local/local-nevada/lake-mead-likely-to-skirt-shortage-line-for-another-year/ (accessed on 3 May 2017).

33. McKinnon, S. SRP Will Extend Water Ration. *The Arizona Republic*, 9 September 2003.

34. Ruddell, B.L.; Adams, E.A.; Rushforth, R.; Tidwell, V.C. Embedded resource accounting for coupled natural-human systems: An application to water resource impacts of the western U.S. electrical energy trade. *Water Resour. Res.* **2014**, *50*, 7957–7972. [CrossRef]

35. Rushforth, R.; Adams, E.A.; Ruddell, B.L. Generalizing ecological, water and carbon footprint methods and their worldview assumptions using Embedded Resource Accounting. *Water Resour. Ind.* **2013**, *1*, 77–90. [CrossRef]

36. Brochure from the Nature Conservancy. Subject: Summary of River Projects Proposed Verde Valley 2017. Personal Communication with M. Messerschmidt, 2018. 2016.

37. Robles, M.D.; Marshall, R.M.; O'Donnell, F.C.; Smith, E.B.; Haney, J.A.; Gori, D.F. Effects of Climate Variability and Accelerated Forest Thinning on Watershed-Scale Runoff in Southwestern USA Ponderosa Pine Forests. *PLoS ONE* **2014**, *9*, e111092. [CrossRef]

38. Baker, M.B. Effects of Ponderosa Pine Treatments on Water Yield in Arizona. *Water Resour. Res.* **1986**, *22*, 67–73. [CrossRef]

39. Moreno, H.A.; Gupta, H.V.; White, D.; Sampson, D.A. Modeling the distributed effects of forest thinning on the long-term water balance and streamflow extremes for a semi-arid basin in the southwestern US. *Hydrol. Earth Syst. Sci.* **2016**, *20*, 1241–1267. [CrossRef]

40. Seager, R.; Ting, M.; Held, I.; Kushnir, Y.; Lu, J.; Vecchi, G.A.; Huang, H.-P.; Harnik, N.; Leetmaa, A.; Lau, N.-C.; et al. Model Projections of an Imminent Transition to a More Arid Climate in Southwestern North America. *Science* **2007**, *316*, 1181–1184. [CrossRef] [PubMed]

41. Seager, R.; Ting, M.; Li, C.; Naik, N.; Cook, B.; Nakamura, J.; Liu, H. Projections of declining surface-water availability for the southwestern United States. *Nat. Clim. Chang.* **2012**, *3*, 482–486. [CrossRef]

42. Ellis, A.; Hawkins, T.; Bailing, R.; Gober, P. Estimating future runoff levels for a semi-arid fluvial system in central Arizona, USA. *Clim. Res.* **2008**, *35*, 227–239. [CrossRef]

43. Westerling, A.L.; Hidalgo, H.G.; Cayan, D.R.; Swetnam, T.W. Warming and Earlier Spring Increase Western U.S. Forest Wildfire Activity. *Science* **2006**, *313*, 940–943. [CrossRef] [PubMed]

44. Gober, P.; Kirkwood, C.W. Vulnerability assessment of climate-induced water shortage in Phoenix. *Proc. Natl. Acad. Sci. USA* **2010**, *107*, 21295–21299. [CrossRef] [PubMed]

45. Janofsky, M. Arizona Awakens to Drought as Lakes Shrink and Harm Spreads. *The New York Times.* 27 January 2003. Available online: http://www.nytimes.com/2003/01/27/us/arizona-awakens-to-drought-as-lakes-shrink-and-harm-spreads.html (accessed on 28 June 2017).

46. Salt River Project. *2016 Annual Report: Annual Water Withdrawal and Use Irrigation District*; Arizona Department of Water Resources: Phoenix, AZ, USA, 2017.

47. Gleick, P.H. The Effects of Future Climatic Changes on International Water Resources: The Colorado River, the United States, and Mexico. *Policy Sci.* **1988**, *21*, 23–39. [CrossRef]

48. MacDonnell, L.J.; Getches, D.H.; Hugenberg, W.C. The Law of the Colorado River: Coping with Severe Sustained Drought. *J. Am. Water Resour. Assoc.* **1995**, *31*, 825–836. [CrossRef]

49. Central Arizona Project. Water Deliveries—Calendar Year 2017. 2017. Available online: https://www.cap-az.com/departments/water-operations/deliveries (accessed on 3 May 2017).

50. Webinar by Danial Bunk. Management of Colorado River Reservoirs and Current State of the System. 18 May 2016. Available online: http://www.cap-az.com/documents/shortage/Introduction-Colorado-River-Briefing.pdf (accessed on 3 May 2017).

51. Webinar by Ted Cooke, General Manager, Central Arizona Project May 18 2016. Colorado River Shortage Briefing: Arizona's Plans in Action. Available online: http://www.cap-az.com/documents/shortage/Introduction-Colorado-River-Briefing.pdf (accessed on 3 May 2017).

52. Jacobs, K.L.; Holway, J.M. Managing for sustainability in an arid climate: Lessons learned from 20 years of groundwater management in Arizona, USA. *Hydrogeol. J.* **2004**, *12*, 52–65. [CrossRef]

53. Megdal, S.B.; Dillon, P.J.; Seasholes, K. Water Banks: Using Managed Aquifer Recharge to Meet Water Policy Objectives. *Water* **2014**, *6*, 1500–1514. [CrossRef]

54. Arizona Department of Water Resources. Decision and Order No. 86-002003.0001 (City of Avondale Designation as Having and Assured Water Supply): Attachment A. 2010. Available online: https://infoshare.azwater.gov/docushare/dsweb/HomePage (accessed on 3 May 2017).

55. Arizona Department of Water Resources. Decision and Order No. 86-002009.0001 (City of Chandler Designation as Having and Assured Water Supply): Attachment A. 2010. Available online: https://infoshare.azwater.gov/docushare/dsweb/HomePage (accessed on 3 May 2017).

56. Arizona Department of Water Resources. Decision and Order No. 86-002019.0001 (City of Goodyear Designation as Having and Assured Water Supply): Attachment A. 2010. Available online: https://infoshare.azwater.gov/docushare/dsweb/HomePage (accessed on 3 May 2017).

57. Arizona Department of Water Resources. Decision and Order No. 86-002023.0001 (City of Mesa Designation as Having and Assured Water Supply): Attachment A. 2010. Available online: https://infoshare.azwater.gov/docushare/dsweb/HomePage (accessed on 3 May 2017).

58. Arizona Department of Water Resources. Decision and Order No. 86-002030.0001 (City of Phoenix Designation as Having and Assured Water Supply): Attachment A. 2010. Available online: https://infoshare.azwater.gov/docushare/dsweb/HomePage (accessed on 3 May 2017).

59. Arizona Department of Water Resources. Decision and Order No. 86-002043.0001 (City of Tempe Designation as Having and Assured Water Supply): Attachment A. 2010. Available online: https://infoshare.azwater.gov/docushare/dsweb/HomePage (accessed on 3 May 2017).

60. Arizona Department of Water Resources. Decision and Order No. 86-400619.0001 (City of Scottsdale Designation as Having and Assured Water Supply): Attachment A. 2010. Available online: https://infoshare.azwater.gov/docushare/dsweb/HomePage (accessed on 3 May 2017).

61. Arizona Department of Water Resources. Decision and Order No. 86-400679.0001 (City of Peoria Designation as Having and Assured Water Supply): Attachment A. 2010. Available online: https://infoshare.azwater.gov/docushare/dsweb/HomePage (accessed on 3 May 2017).

62. Arizona Department of Water Resources. Decision and Order No. 86-402208.0001 (Town of Gilbert Designation as Having and Assured Water Supply): Attachment A. 2010. Available online: https://infoshare.azwater.gov/docushare/dsweb/HomePage (accessed on 3 May 2017).

63. Arizona Department of Water Resources. Decision and Order No. 860002018.001 (City of Glendale Designation as Having an Assured Water Supply): Attachment A. 2010. Available online: https://infoshare.azwater.gov/docushare/dsweb/HomePage (accessed on 3 May 2017).

64. Lauver, L.; A Baker, L. Mass balance for wastewater nitrogen in the Central Arizona–Phoenix ecosystem. *Water Res.* **2000**, *34*, 2754–2760. [CrossRef]

65. Messerschmidt, M.; The Nature Conservancy of Arizona, Tucson, AZ, USA. Personal Communication—Preliminary Flow Data. 2016.

66. Gardner, R.; Ostrom, E.; Walker, J.M. The Nature of Common-Pool Resource Problems. *Ration. Soc.* **1990**, *2*, 335–358. [CrossRef]

67. Salt River Project. 2016 Runoff Season Falls Short of Promise. 2016. Available online: http://www.srpnet.com/newsroom/releases/080316.aspx (accessed on 3 May 2017).

68. Webinar by Clint Chandler, Assistant Director of Water Planning and Permitting, Central Arizona Project. Colorado River Shortage Briefing—Introduction. 18 May 2016. Available online: http://www.cap-az.com/documents/shortage/Introduction-Colorado-River-Briefing.pdf (accessed on 3 May 2017).

69. Arizona Department of Water Resources. Data Center. 2017. Available online: https://infoshare.azwater.gov/docushare/dsweb/HomePage (accessed on 3 May 2017).

70. Arizona Department of Water Resources. Long-Term Storage Account Summary. 2017. Available online: http://www.azwater.gov/azdwr/WaterManagement/Recharge/documents/2016LTSASummary05-31-17mro.pdf (accessed on 3 May 2017).

71. Arizona Department of Water Resources. Preliminary DRAFT 4MP Total GPCD Program. 2012. Available online: http://www.azwater.gov/AzDWR/WaterManagement/AMAs/documents/GUAC_4MP_GPCD_PROGRAM_PHXAMA.pdf (accessed on 3 May 2017).

72. Simonit, S.; Connors, J.P.; Yoo, J.; Kinzig, A.; Perrings, C. The Impact of Forest Thinning on the Reliability of Water Supply in Central Arizona. *PLoS ONE* **2015**, *10*, e0121596. [CrossRef] [PubMed]

73. Li, Y.P.; Huang, G.H.; Cui, L.; Liu, J. Mathematical Modeling for Identifying Cost-Effective Policy of Municipal Solid Waste Management under Uncertainty. *J. Environ. Inform.* **2019**, *34*, 55–67. [CrossRef]

74. Malagó, A.; Vigiak, O.; Bouraoui, F.; Pagliero, L.; Franchini, M. The hillslope length impact on SWAT streamflow prediction in large basins. *J. Environ. Inform.* **2018**, *32*, 82–97. [CrossRef]

75. Zhou, P.; Li, Z.; Snowling, S.; Goel, R.; Zhang, Q.; Solutions, I.H.E.S. Short-Term Wastewater Influent Prediction Based on Random Forests and Multi-Layer Perceptron. *J. Environ. Inform. Lett.* **2019**, *1*, 87–93. [CrossRef]

76. Wu, Q.; Xia, X. Response of Surface Water Quality in Urban and Non-urban Areas to Heavy Rainfall: Implications for the Impacts of Climate Change. *J. Environ. Inform. Lett.* **2019**, *1*, 27–36. [CrossRef]

77. Hogue, T.S.; Pincetl, S. Are you watering your lawn? *Science* **2015**, *348*, 1319–1320. [CrossRef]

78. Campbell, H.E.; Johnson, R.M.; Larson, E.H. Prices, Devices, People, or Rules: The Relative Effectiveness of Policy Instruments in Water Conservation1. *Rev. Policy Res.* **2004**, *21*, 637–662. [CrossRef]

79. McManamay, R.A.; Nair, S.S.; DeRolph, C.R.; Ruddell, B.L.; Morton, A.M.; Stewart, R.N.; Troia, M.J.; Tran, L.; Kim, H.; Bhaduri, B.L. US cities can manage national hydrology and biodiversity using local infrastructure policy. *Proc. Natl. Acad. Sci. USA* **2017**, *114*, 9581–9586. [CrossRef]
80. Scanlon, B.R.; Ruddell, B.L.; Reed, P.M.; Hook, R.I.; Zheng, C.; Tidwell, V.C.; Siebert, S. The food-energy-water nexus: Transforming science for society. *Water Resour. Res.* **2017**, *53*, 3550–3556. [CrossRef]
81. Brookshire, D.S.; Colby, B.; Ewers, M.; Ganderton, P.T. Market prices for water in the semiarid West of the United States. *Water Resour. Res.* **2004**, *40*. [CrossRef]
82. Kenney, D.S. Understanding Utility Disincentives to Water Conservation as a Means of Adapting to Climate Change Pressures. *J. Am. Water Works Assoc.* **2014**, *106*, 36–46. [CrossRef]

Article

Assessing Aquifer Water Level and Salinity for a Managed Artificial Recharge Site Using Reclaimed Water

Faten Jarraya Horriche and Sihem Benabdallah *

Centre of Water Research and Technologies, Geo-resources Laboratory, Techno-park Borj-Cedria, BP 273 Soliman, Tunisia; faten.horriche@topnet.tn
* Correspondence: sihem.benabdallah@certe.rnrt.tn; Tel.: +216-98-404-777

Received: 11 December 2019; Accepted: 20 January 2020; Published: 25 January 2020

Abstract: This study was carried out to examine the impact of an artificial recharge site on groundwater level and salinity using treated domestic wastewater for the Korba aquifer (north eastern Tunisia). The site is located in a semi-arid region affected by seawater intrusion, inducing an increase in groundwater salinity. Investigation of the subsurface enabled the identification of suitable areas for aquifer recharge mainly composed of sand formations. Groundwater flow and solute transport models (MODFLOW and MT3DMS) were then setup and calibrated for steady and transient states from 1971 to 2005 and used to assess the impact of the artificial recharge site. Results showed that artificial recharge, with a rate of 1500 m^3/day and a salinity of 3.3 g/L, could produce a recovery in groundwater level by up to 2.7 m and a reduction in groundwater salinity by as much as 5.7 g/L over an extended simulation period. Groundwater monitoring for 2007–2014, used for model validation, allowed one to confirm that the effective recharge, reaching the water table, is less than the planned values.

Keywords: artificial recharge; groundwater; treated wastewater

---

1. Introduction

Water is a critical resource in many Mediterranean countries because of its scarcity and uneven geographical, seasonal, and inter-annual distribution [1,2]. In a country with frequent water stress, several national strategies were established in Tunisia to optimize the management of water resources to meet growing freshwater needs and planning for climate change adaptation [3,4]. The use of reclaimed municipal wastewater was considered one of the main axes in the national water strategy, mainly for agriculture and groundwater artificial recharge [3]. Thus, treated wastewater (TWW) is used to improve groundwater storage and reduce seawater intrusion in coastal aquifers. Previous studies have demonstrated that such alternatives are reasonable when conventional freshwater sources become very limited [5–9]. Artificial recharge using conventional freshwater has already been acknowledged for inducing groundwater level rise and for improving water quality in several aquifers in Tunisia. The Teboulba coastal aquifer registered a rise of groundwater level up to 30 m following the artificial recharge in wells during six years [10]. The artificial recharge in the Zeroud riverbed of Kairouan aquifer led to an increase of water table between 0.2 and 5.25 m for a distance of up to 8 km [11]. A positive impact of the artificial recharge in El Khairat wadi was confirmed by the authors of [12], indicating an increase in water table between 0.4 and 2.6 m.

This study mainly focuses on the Korba aquifer, located in North-Eastern Tunisia. This aquifer registered groundwater decline and water salinization over the past three decades due to high groundwater pumping rates for agriculture uses. This serious threat has motivated many authors to study seawater intrusion by modelling [13–15] or by hydro-geochemical investigations in order to determine the spatial extension of the salinization or to identify the origins and the mechanisms governing its contamination [16–19]. The construction of the Korba–Elmida artificial recharge ponds, using treated municipal wastewater, was among the actions undertaken by the water authority for aquifer storage recovery to overcome groundwater level decline and salinity rise for the Korba aquifer [20].

Numerical modelling and hydro-geochemical investigations were often used by scientists to estimate aquifer storage or to carry out groundwater recovery based on exploratory simulations and scenarios development [8,9,21–24]. They can also help to estimate the potential benefits for constructed structures on hydrologic conditions under a range of management scenarios. Groundwater flow and solute transport models are applied as predictive tools to plan and quantify the impact on the local groundwater, and determine geochemical processes and the resulting recovery efficiency [25,26]. Basic descriptions of various physical and chemical equilibrium for solute transport models are given by the authors of [26], indicating that the development of geochemical transport models or hydrogeochemical models represents still new pursuit, although some mathematical flow models date back to the late 1960s. Thus, solute-transport models are inherently more complex in terms of conceptualization and governing equations, numerical methods, parameter estimation, and boundary conditions, as well as concerns about model complexity [27]. It should be noted that analytical geochemical methods and path way modelling were also used when geochemical data was available to explain groundwater salinity variations and to support groundwater management and preventing salinization [28,29]. Yet, numerical modelling was frequently used for coastal aquifers to simulate seawater intrusion in natural and anthropogenic conditions and to predict also climate change impact [13,26]. In this paper, we evaluate the impact of the constructed artificial recharge ponds on groundwater level and water salinity by using 2D flow and solute transport models. First, we give an overview for the hydrogeological characterization of the artificial recharge site, at a small scale. Secondly, we simulate the predicted impact of the artificial recharge site built on flow and solute transport calibrated models based on several scenarios. Sensitivity tests for the effective artificial recharge are then performed during the artificial recharge period between 2009 and 2014, by using a monitoring network for groundwater levels and salinities.

2. Materials and Methods

2.1. Study Area Description

The Korba aquifer is located in the Cap Bon Peninsula in Tunisia (Figure 1a). It covers an area of 500 km^2 and is bounded by the Mediterranean Sea to the south-east. The average annual rainfall is about 450 mm per year computed for the period 1959–2009 using the national rainfall network over the study area. The study area is characterized by its important agricultural activity, leading to high water demand and a groundwater overdraft [30]. The shallow aquifer is composed of the Plio-Quaternary formation, characterized mainly by sand surmounting the Miocene clay bedrock [30]. Groundwater levels and water quality have been monitored since the 1960s. Early measures showed that water salinity did not exceed 3 g/L [31]. Starting in 1970, the groundwater level declined and water quality degraded as well; an increase of the size of the affected areas was registered mainly in some overexploited localities where the groundwater level declined below the sea water level and the salinity exceeded 5 g/L reaching in some points 10 g/L [31]. This situation was a result of groundwater over abstraction, which generated a reverse hydraulic gradient and seawater intrusion [13–18]. In addition, other authors [18,30] attributed the groundwater salinization to the increase in irrigated agricultural areas, which induced soil leaching and migration of fertilizers to the aquifer.

In order to overcome the problem of groundwater level decline and water quality deterioration in the Korba aquifer, water authorities planned the construction of the Korba–Elmida artificial recharge site, using treated municipal wastewater. The site location was selected after a feasibility study that covered the Cap Bon region, considering several technical criteria related to geologic aspects, depth of water table, uses of groundwater, location of nearby wastewater treatment plant, and other economic constraints [20]. The selected site is located in the north-east of Tunisia, about 300 m north of the Korba wastewater treatment plant and 1.5 km from the coast (Figure 1a). The artificial recharge site, consisting of three recharge ponds with 1500 m² each, was designed for a recharge rate of 1500 m³/day. In fact, the monitoring of the actual recharge rate, which we realized between 2009 and 2014, ranged between 800 and 1745 m³/day for duration of 0 to 10 h per day [32]. The TWW was transferred from a nearby wastewater treatment plant (WWTP), which received an average inflow of 5000 m³/day with an average salinity 3.3 g/L in the year 2003 [20].

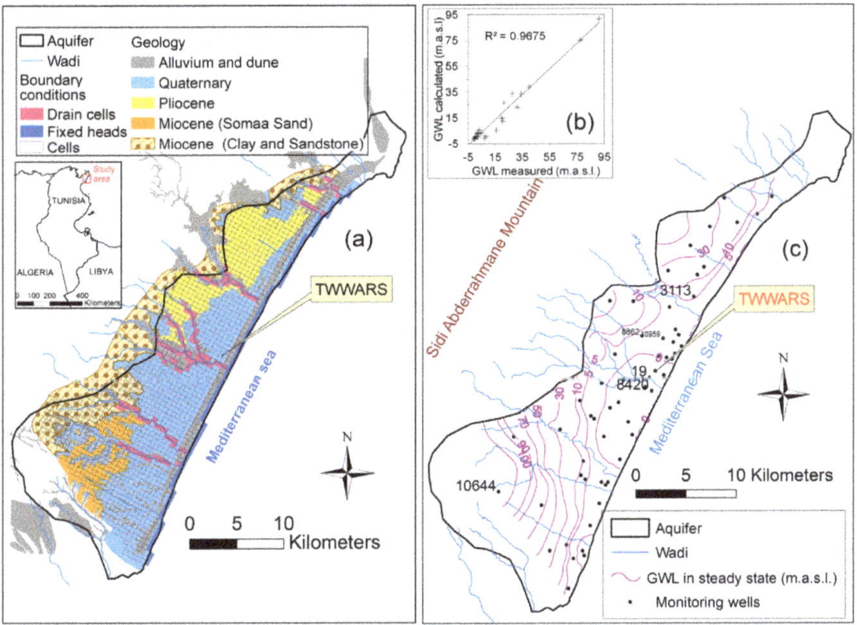

Figure 1. (**a**) Location map for the Korba aquifer with the treated wastewater artificial recharge site (TWWARS) and model setup structure and boundary conditions and simplified surface geology, (**b**) scatter diagram calibration of steady state groundwater flow modelling, (**c**) and groundwater level (GWL) map simulated for steady state (1971).

2.2. Hydrogeological Characterization

The efficiency of the artificial recharge depends on the aquifer hydrogeological characteristics [5,33,34]. The site characterization was based on the field hydrogeological investigations carried out before the project implementation and during artificial recharge period. Soil properties, which control the flow rate of infiltration and the downward percolation, are of special importance to this type of technique. Thus, subsurface formation was identified using well logs and test infiltration results.

A well (W), shown in Figure 2, was drilled up to a depth of 52 m (Figure 2a) and was used to investigate the subsurface formations and to estimate the horizontal transmissivity and the storage coefficient. In order to identify the unsaturated zone, eight complementary drills were performed to check the thickness of sandstone layers. Suitable sites for artificial recharge were identified using four surface infiltration tests (IF), followed by twelve infiltration tests in ditches (PM) (Figure 2b) to the depth of 1.8 m corresponding to the depth of the planned artificial recharge ponds.

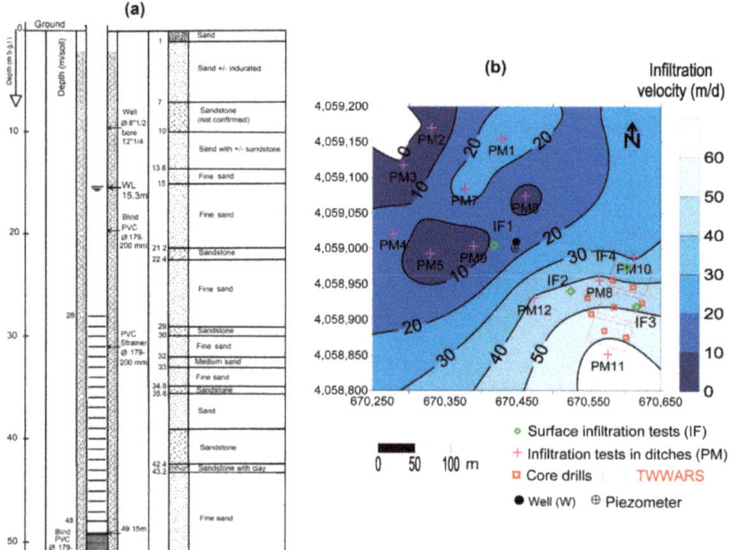

Figure 2. (**a**) Log for well (W) and (**b**) infiltration velocity map with location for test infiltration in subsurface (IF) and in ditches (PM).

2.3. Modelling Approach

The MODFLOW code was used to solve the 2D groundwater flow equation in saturated porous media by means of the finite-differences method [35], and the MT3DMS code was used to solve the solute transport equation [36]. Both codes were used within the Processing Modflow package [37]. The equations were resolved within a regular square grid of 25×10^4 m^2 cell size. The models were calibrated during steady and transient states for the period between 1971 and 1996, by using a manual process based on the trial and error method, referring to measured values in monitoring network. They were thereafter validated for the period 1997–2006. The model area was limited by the boundaries of the Plio-quaternary outcrops (Figure 1a). At the upstream, the Miocene formation outcrop was considered to be part of the modelling area. The bedrock is formed by Miocene clay formation with a thickness ranging between 100 m upstream and 1800 m in the center of the aquifer. The groundwater recharge was mainly assured by rain infiltration with about 7% of the average annual rainfall, as was estimated by the authors of [13,38]. The natural outlet represents the sea, the salt areas (marshes), and the downstream draining areas of the wadis. The coastal limit is presented by a fixed head condition of 0 m, and a fixed water salinity of 38 g/L, which is the average value for the Mediterranean Sea. Near the irrigated areas, salinity of recharged water was increased to consider irrigation return flow as justified by [11,16].

The calibrated models were used to predict the impact of the planned artificial recharge on groundwater level and salinity with a focus on the surrounded region. Several scenarios were used considering hypothesis for natural and artificial recharge and groundwater pumping rates during the period between 2007 and 2050. Model's results were validated during the period 2007–2014 using the actual recharge rates based on monitored values.

3. Results and Discussions

3.1. Characterization of the Artificial Recharge Site

Hydrogeological setting performed before and after the artificial recharge allowed for the characterization of the subsurface and surface zones. According to the drilled well (W) and the core drills (Figure 2), subsurface formations are mainly sandy with thin sandstone layers. The hydraulic conductivity of the sampled sand from the well, using a grain size analysis, was primarily calculated to 3×10^{-4} m/s. Tests performed for several depths in the core drills showed a range of hydraulic conductivity between 2×10^{-3} m/s for fine sand and 2×10^{-6} m/s for sandstone. Sampling analysis confirmed that unsaturated zone is mainly composed of sand; more than 95% have grain size less than 2×10^{-3} m. The pumping test carried out in well W produced an estimate horizontal transmissivity of 4×10^{-3} m^2/s and a storage coefficient varying between 4.5×10^{-4} and 6×10^{-4}.

The infiltration test results, applied in surface and ditches, gave a vertical infiltration velocity ranging between 1 and 65 m/day depending on the soil type. High values corresponded to sand, whereas low values were linked to sandstone and consolidated sand. The south-east zone indicated infiltration rates higher than 30 m/day, which can be considered as suitable values for artificial recharge, leading to the conclusion that the selected site was well adapted for the construction of the artificial recharge ponds.

3.2. Model Calibration and Validation

The calibration of the flow model during steady state allowed for the assessment of groundwater recharge ranging between 7% and 11% of annual average rainfall for the Quaternary outcrops. These values are close to those used by previous studies [13–15,38]. The calibration results were satisfactory by comparing the simulated groundwater levels with the observed values (Figure 1b). The coefficient of determination (R^2) is about 0.97. The groundwater renewable resources are evaluated as 1.31 m^3/s during 1971, as shown in Table 1. Aquifer recharge was provided mainly from recharge at outcrops, and wells' pumping represented the main outlet fluxes. During transient state, the calibrated porosity ranged between 0.05 and 0.35, which are close to the constant value of 0.12 used by [26], and those provided by [14] ranging between 0.04 and 0.25. The general trends of groundwater decline were reproduced by the calibrated model, with a maximum error of +/−1 m (Figure 3a). The results confirmed that the most affected areas were located between Korba, Diar Elhojjej, and Tafeloune agglomerations. The flow model was validated for the period 1997–2006. It reproduced close water level variation for several piezometers as shown in Figure 3a.

Table 1. Calculated water balance in steady state (1971).

Boundary Conditions	Input	Output
	(m^3/s)	
Fixed head (Sea)	0.033	0.091
Abstraction (wells)		0.892
Wadis' drainage	0.057	0.331
Recharge (from rainfall)		
Quaternary	0.307	
Tyrrenian dunes	0.045	
Pliocene	0.313	
Miocene	0.166	
Deep aquifer input (vertical leakance)	0.393	
Total	1.314	1.314

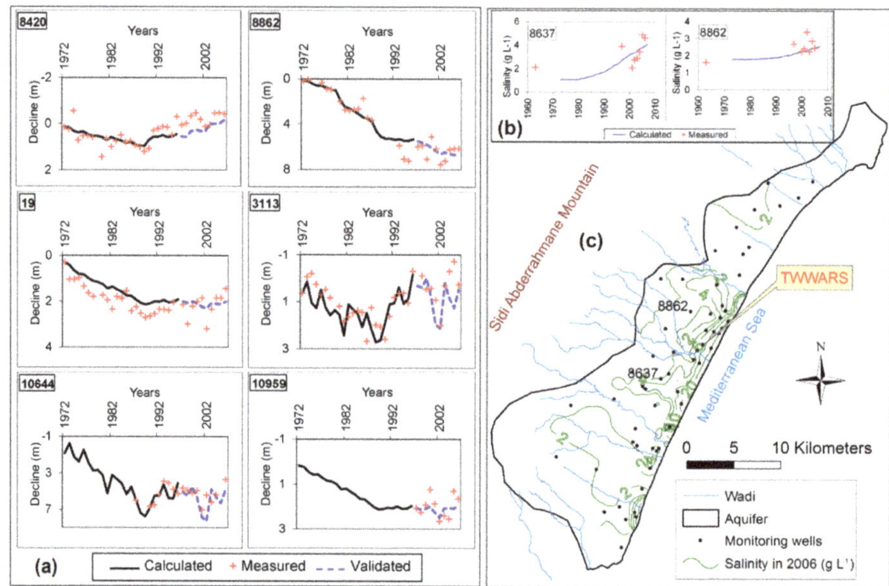

Figure 3. (**a**) Calibration and validation in monitoring wells during transient period (1972–2006). Calibration was established during the period 1972–1996, and validation was performed during the period 1997–2006; (**b**) simulated salinity during transient period (1972–2006); and (**c**) salinity map for 2006.

The calibration of the solute transport modelling was performed using the calibrated flow model for the period 1971–2006. The sea, represented by fixed head and fixed concentration conditions, contributed to groundwater salinization by convection and dispersion. Figure 3b presents simulated salinity results for the transient period. Calibration for salinity was not possible, in part due to the few available measurements but mainly due to the complexity of processes [27]. Simulated salinity for 2006 (Figure 3c) confirms the existence of seawater intrusion along the coastal area between the two water courses (Figure 1), spreading out in the aquifer up to 5 km.

3.3. Artificial Recharge Simulations

Two simulations (SIM1 and SIM2) were conducted during a forecast period from 2007 to 2050. In SIM1, we maintained the same boundary conditions throughout the simulation period, assuming the continuity of the current conditions of recharge and extraction rates as in 2006 and omitting the additional artificial recharge rate from the treated municipal wastewater. According to this scenario, the model predicted a groundwater decline of less than 2 m near the planned artificial recharge site by the year 2050 and an increase in salinity reaching 12 g/L along the coast.

In SIM2, we considered the same boundary conditions as in SIM1 and assumed an additional recharge flux of TWW of 1500 m³/day with a salinity of 3.3 g/L to be injected in the recharge pools, represented by one cell in the model. The comparison of the simulated groundwater level for 2050 according to SIM1 and SIM2 showed that an increase reaching a maximum of 2.7 m can be achieved (Figure 4a). The influenced area around the site, considering a minimum change of 0.1 m, is around 80 km². Moreover, the groundwater salinity can be reduced under recharge conditions by a maximum of 5.7 g/L near the site with a recovery area of around 26 km² (Figure 4c). It extends over a 6 km distance, including a 17 km² of recovery area with a salinity variation of less than 1.5 g/L.

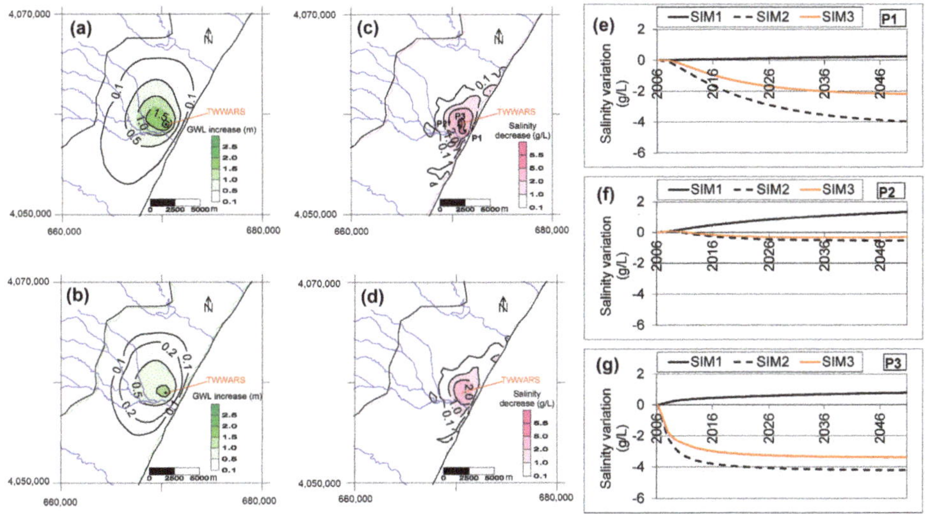

Figure 4. Impact of artificial recharge in 2050: groundwater level increase (**a**) SIM2—SIM1 and (**b**) SIM3—SIM1; groundwater salinity decrease: (**c**) SIM1—SIM2 and (**d**) SIM1—SIM3; and (**e–g**) groundwater salinity variations in wells P1, P2, and P3.

Results relatives to wells P1, P2 and P3, located, respectively, downstream, upstream, and close to the artificial recharge area as shown in Figure 4c, were used to compare the temporal evolution of groundwater salinity. Figure 4e–g also shows that artificial recharge may reduce salinity, especially near the ponds (P1 and P3). However, its impact was lesser away from the recharging ponds (P2). Salinity decrease reached 4.2, 1.9, and 5 g/L, respectively, for P1, P2, and P3. The salinity decrease occurred during the first ten years of the simulation by an annual decrease of 0.2 g/L (P1) and became insignificant thereafter due to the constant boundary conditions for the aquifer recharge and abstractions. The annual decrease was at a maximum in well P3, located close the ponds.

An additional scenario SIM3 was considered in which only 70% of the artificial recharge rate would reach the water table following the evaporation and losses in the unsaturated zone. This value would be justified for semi-arid regions, and would be close the rates calculated by [38] during the artificial recharge in the plain of Kairouan. As a consequence of this decline, we obtained a lower impact on groundwater level (Figure 4b,d). In fact, the positively influenced area was reduced to 58 km^2 and the maximum increase of groundwater levels reached 1.6 m. For the salinity, the influenced area was almost maintained, with a maximum decrease of 4.4 g/L near the recharge ponds.

3.4. Sensitivity Tests

In this section, we focus on the artificial recharge period between 2007 and 2014. In fact, the artificial recharge was practiced during 2009–2014 with a rate ranging between 0 and 1745 m^3/day. Given the uncertainties related to the clogging, the evaporation from the site, and from the unsaturated zone, we investigated whether the measured actual rates contributed effectively to the groundwater recharge by using P3 as an observation well. Two additional simulations (SIM4 and SIM5) were considered. In SIM4, the actual observed rates of the artificial recharge were used, while in SIM5, we considered losses by evaporation and clogging, assuming that only 70% of the measured rates was effective. In both simulations, we maintained an average salinity of 3.3 g/L for the TWW. In fact, the measured salinity of the TWW varied between 3 and 5 g/L. For the starting year of recharge, 2009, the variation of groundwater level for SIM4 is close to the observed values (Figure 5a). As of 2010, SIM5 fits better the observations. These results can be justified by a decrease in the efficiency of the artificial recharge pools due to clogging phenomena. In fact, we noticed that infiltration velocity decreased during this period. Before the artificial recharge operation, the infiltration velocity was more than 30 m/d. In June 2010, we measured an infiltration velocity of about 1.3 m/d. It was the result of the accumulation of suspended particles in subsurface soil, causing a decrease of the saturated hydraulic conductivity. This decrease of infiltration velocity is consistent with those reported by the authors of [39] during infiltration of TWW in a 3 m diameter column. The decrease of infiltration rates from 3.3 to 0.08 cm/h was due to the development of a clogging in the uppermost layer. The assumption of considering an effective rate of 70% of the recharge rate is thus justified.

For the period 2011–2014, we registered a trend for a general decrease of measured groundwater levels. Sensitivity tests were run in order to reproduce this decrease by reducing the infiltration rate linearly over the simulation period, starting with 70% in 2009. The best simulation, SIM 6 presented in Figure 5a, corresponds to an infiltration rate decreasing linearly with a negative slope of 0.12.

Regarding salinity, the observed variations do not match the simulated values of SIM 4, 5, and 6 (Figure 5b). However, they are much closer to the results given by SIM 2, which corresponds to the projected artificial recharge with a rate of 1500 m^3/day. Given that this volume is hard to achieve, we elaborated further sensitivity tests on the salinity of TWW by reducing the average salinity input value. The best simulation, SIM 7, corresponds to the same artificial recharge rates as for SIM 6 with a TWW salinity of 1 g/L. Simulated salinity results are much closer to the observed values. Given that the monitored salinity for the TWW varied between 3 and 5 g/L, we can assume that some of the salinity was retained by the unsaturated zone.

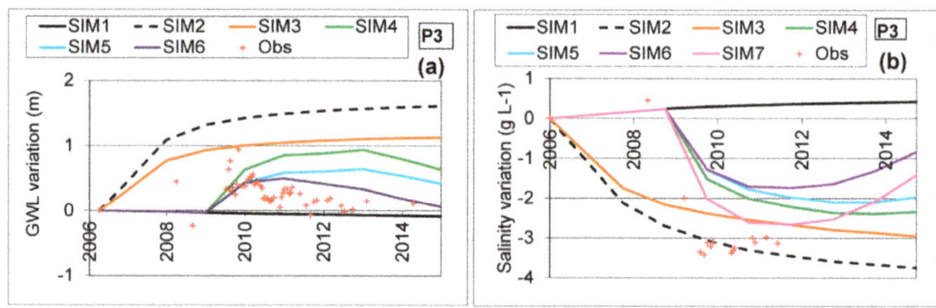

Figure 5. **Well P3** (**a**) Observed and simulated variations of groundwater level; (**b**) Observed and simulated variations of groundwater salinity.

4. Conclusions

The Korba–Elmida artificial recharge pond using TWW was a pilot site to investigate the recharge impact on groundwater level and salinity. The hydrogeological characterization allowed for the calculation of vertical infiltration velocity, which reached more than 30 m/day near the artificial recharge site. Sand formation was identified as the most suitable for infiltration. The planned recharge rate of 1500 m^3/day, and applied to an area of 3000 m^2 (two ponds), is equivalent to 0.5 m/day, which is very low compared to infiltration capacity of sand. Thus, there is still the potential to increase the rate of the artificial recharge to enhance the impact on groundwater level and salinity.

The groundwater flow and transport models allowed for the prediction of the artificial recharge impact on groundwater. The results of the simulations showed that, in the long-term, the artificial recharge with a rate of 1500 m^3/day would induce a maximum recovery of groundwater level of up to 2.7 m and a maximum relative decrease in salinity of 5.7 g/L. The influenced area encompassed up to 26 km^2 around the site, extending about 10 km belt along the coast. Hence, it was found that the implementation of an artificial recharge site would reduce groundwater decline, improve water salinity, and reduce seawater intrusion given that the water abstraction stays at the actual level. The results of the simulations may have been affected by predictive rainfall, artificial recharge, and abstraction, considered constant for both simulations. The sensitivity tests based on modifying the recharge rate and the reclaimed water salinity allowed for the calculation of different variabilities for groundwater level and water salinity.

The sensitivity of the models to the effective artificial recharge was performed during the period 2007–2014 using the affective artificial recharge rates and the observed groundwater level and salinity. We concluded that effective artificial recharge is less than the planned values. This can be justified by the evaporation and losses in the unsaturated zone and by the clogging phenomena in the ponds. Further, the impact of the artificial recharge is insignificant in terms of salinity.

Author Contributions: Conceptualization, F.J.H.; methodology, F.J.H. and S.B.; software, F.J.H.; validation, F.J.H. and S.B.; formal analysis, F.J.H.; investigation, F.J.H.; resources, F.J.H.; data curation, F.J.H.; writing—original draft preparation, F.J.H. and S.B.; writing—review and editing, F.J.H. and S.B. All authors have read and agreed to the published version of the manuscript.

Funding: This research received no external funding and the APC was funded by the Tunisian Scientific association "Association Eau et Dévelopement".

Acknowledgments: The authors would like to thank the staff of the water authority DGRE and its local representatives for their support and interest in this work.

Conflicts of Interest: The authors declare no conflict of interest.

References

1. Hamoda, M.F. Water strategies and potential of water reuse in the south Mediterranean countries. *Desalination* **2004**, *165*, 31–41. [CrossRef]
2. Blinda, M.; Thivet, G. Ressources et demandes en eau en Méditerranée: Situation et perspectives. *Sécheresse* **2009**, *20*, 9–16. [CrossRef]
3. DGRE. *Stratégie pour le développement des ressources en eau de la Tunisie au cours de la décennie 1991–2000*; National Report; DGRE: Tunis, Tunisia, 1990. (In French)
4. MARH; GTZ. *Stratégie nationale d'adaptation de l'agriculture tunisienne et des écosystèmes aux changements climatiques*; National Report; MARH>Z: Tunis, Tunisia, 2007. (In French)
5. Bouwer, H. Artificial recharge of groundwater: Hydrogeology and engineering. *Hydrogeol. J.* **2002**, *10*, 121–142. [CrossRef]
6. Sheng, Z. An aquifer storage and recovery system with reclaimed wastewater to preserve native groundwater resources in El Paso, Texas. *J. Environ. Manag.* **2005**, *75*, 367–377. [CrossRef] [PubMed]
7. Cazurra, T. Water reuse of south Barcelona's wastewater reclamation plant. *Desalination* **2008**, *218*, 43–51. [CrossRef]
8. Shammas, M.I. The effectiveness of artificial recharge in combating seawater intrusion in coastal aquifer Salalah, Oman. *Environ. Earth Sci.* **2008**, *55*, 191–204. [CrossRef]
9. Tamer, A.; Al-Assa, D.F.; Ahmad, A. Artificial groundwater recharge to a semi-arid basin: Case study of Mujib aquifer, Jordan. *Environ. Earth Sci.* **2010**, *60*, 845–859.
10. Bouri, S.; Ben Dhia, H. A thirty-year artificial recharge experiment in a coastal aquifer in an arid zone: The Teboulba aquifer system (Tunisian Sahel). *C. R. Geosci.* **2010**, *342*, 60–74. [CrossRef]
11. Salem, S.B.; Chkir, N.; Zouari, K.; Cognard-Plancq, A.L.; Valles, V.; Marc, V. Natural and artificial recharge investigation in the Zeroud Basin, Central Tunisia: Impact of Sidi Saad Dam storage. *Environ. Earth Sci.* **2012**, *66*, 1099–1110. [CrossRef]
12. Ketata, M.; Gueddari, M.; Bouhlila, R. Hydrodynamic and salinity evolution of groundwater during artificial recharge within semi-arid coastal aquifers: A case study of El Khairat aquifer system in Enfidha (Tunisian Sahel). *J. Afr. Earth Sci.* **2014**, *97*, 224–229. [CrossRef]
13. Paniconi, C.; Khlaifi, I.; Lecca, G.; Giacomelli, A.; Tarhouni, J. Modeling and analysis of seawater intrusion in the coastal aquifer of eastern Cap-Bon, Tunisia. *Transp. Porous Med.* **2001**, *43*, 3–28. [CrossRef]
14. Kerrou, J.; Renard, P.; Tarhouni, J. Status of the Korba groundwater resources (Tunisia): Observations and three-dimensional modelling of seawater intrusion. *Hydrogeol. J.* **2010**, *18*, 1173–1190. [CrossRef]
15. Zghibi, A.; Zouhri, L.; Tarhouni, J. Groundwater modelling and marine intrusion in the semi-arid systems (Cap-Bon, Tunisia). *Hydrol. Process.* **2011**, *25*, 1822–1836. [CrossRef]

16. Slama, F.; Bouhlila, R. Multivariate statistical analysis and hydrogeochemical modelling of seawater-freshwater mixing along selected flow paths: Case of Korba coastal aquifer Tunisia. *Estuar. Coast. Shelf Sci.* **2017**, *198*, 636–647. [CrossRef]

17. Kouzana, L.; Ben Mamou, A.; Felfoul, M.S. Seawater intrusion and associated processes: Case of the Korba aquifer (Cap Bon, Tunisia). *C. R. Geosci.* **2009**, *341*, 21–35. [CrossRef]

18. Hamouda, M.F.; Tarhouni, J.; Leduc, C.; Zouari, K. Understanding the origin of salinization of the Plio-Quatnary eastern coastal aquifer of Cap Bon (Tunisia) using geochemical and isotope investigations. *Environ. Earth Sci.* **2011**, *63*, 889–901. [CrossRef]

19. Cary, L.; Casanova, J.; Gaaloul, N.; Guerrot, C. Combining boron isotopes and carbamazepine to trace sewage in salinized groundwater: A case study in Cap Bon, Tunisia. *Appl. Geochem.* **2013**, *34*, 126–139. [CrossRef]

20. CRDAN. *Projet de recharge des nappes du Cap Bon à partir des eaux usées traitées. Etude de faisabilité, environnementale, d'avant projet et d'exécution*; Local Report; CRDAN: Nabeul, Tunisia, 2004. (In French)

21. Martinez-Santos, P.; Martinez-Alfaro, P.; Murillo, J.M. A method to estimate the artificial recharge capacity of the Crestatx aquifer (Majorca, Spain). *Environ. Earth Sci.* **2005**, *47*, 1155–1161. [CrossRef]

22. Izbicki, J.A.; Flint, A.L.; Stamos, C.L. Artificial recharge through a thick, heterogeneous unsaturated zone. *Ground Water* **2008**, *46*, 475–488. [CrossRef]

23. Vandenbohede, A.; Van Houte, E.; Lebbe, L. Groundwater flow in the vicinity of two artificial recharge ponds in the Belgian coastal dunes. *Hydrogeol. J.* **2008**, *16*, 1669–1681. [CrossRef]

24. Zhang, H.; Xu, Y.; Kanyerere, T. Site Assessment for MAR through GIS and Modeling in West Coast, South Africa. *Water* **2019**, *11*, 1–21. [CrossRef]

25. El Ayni, F.; Manoli, E.; Cherif, S.; Jrad, A.; Trabelsi-Ayadi, M.; Assimacopoulos, D. Deterioration of a Tunisian coastal aquifer due to agricultural activities and possible approaches for better water management. *Water Environ. J.* **2013**, *27*, 348–361. [CrossRef]

26. Elango, L.; Stagnitti, F.; Gnanasundar, D.; Rajmohan, N.; Salzman, S.; Leblanc, M.; Hill, J. A Review of recent solute transport models and a case study. *Environ. Sci. Environ. Comput.* **2004**.

27. Konikow, L.F. The secret to successful solute-transport modeling. *Groundwater* **2011**, *49*, 144–159. [CrossRef] [PubMed]

28. Critelli, T.; Vespasiano, G.; Apollaro, C.; Muto, F.; Marini, L.; De Rosa, R. Hydrogeochemical study of an ophiolitic aquifer: A case study of Lago (Southern Italy, Calabria). *Environ. Earth Sci.* **2015**, *74*, 533–543. [CrossRef]

29. Vespasiano, G.; Cianflone, G.; Romanazzi, A.; Apollaro, C.; Dominici, R.; Polemio, M.; De Rosa, R. A multidisciplinary approach for sustainable management of a complex coastal plain: The case of Sibari Plain (Southern Italy). *Mar. Pet. Geol.* **2019**, *109*, 740–759. [CrossRef]

30. Ennabli, M. Étude hydrogéologique des aquifères du Nord-Est de la Tunisie pour une gestion intégrée des ressources en eau. Ph.D. Thesis, University of Nice, Nice, France, 1980. (In French).

31. Jarraya Horriche, F.; Benabdallah, S.; Charef, A. Hydrogeological characterization of the Korba artificial recharge site. In Proceedings of the Tunisia-Japan Symposium on Regional Development and Water Resource. A New Vision for Sustainable Society, Tunis, Tunisia, 1–29 December 2010.

32. Jarraya Horriche, F.; Benabdallah, S.; Ayadi, M.; Triki, L.; Haggui, R. *Etude de l'impact de la recharge artificielle de la nappe de la Côte Orientale du Cap Bon par les eaux usées traitées (site de Korba-El Mida)*; Internal Report; CERTE&DGRE: Tunis, Tunisia, 2015; 45p.

33. Abu-Taleb, M.F. The use of infiltration field tests for groundwater artificial recharge. *Environ. Earth Sci.* **1999**, *37*, 64–71. [CrossRef]

34. Flint, A.L.; Ellett, K.M. The role of the unsaturated zone in artificial recharge at San Gorgonio Pass, California. *Vadose Zone J.* **2004**, *3*, 763–774. [CrossRef]

35. Harbaugh, A.W.; Banta, E.R.; Hill, M.C.; McDonald, M.G. *Modflow-2000, U.S. Geological Survey Modular Ground-Water Model: User Guide to Modularization Concepts and the Groundwater Flow Process. Report 00-92*; USGS: Reston, VA, USA, 2000; Volume 134.

36. Zheng, C.; Wang, P.P. MT3DMS: A Modular Three-Dimensional Multispecies Transport Model for Simulation of Advection, Dispersion and Chemical Reactions of Contaminants in Groundwater Systems. In *Documentation and User's Guide*; U.S. Army Corps of Engineers: Washington, DC, USA, 1999.

37. Chiang, W.H.; Kinzelbach, W. *3d-Groundwater Modelling with Pmwin, A Simulation System for Modelling Groundwater Flow and Pollution*; Springer: Berlin, Germany, 2001.

38. Nazoumou, Y.; Besbes, M. Simulation de la recharge artificielle de la nappe en oued par modèle à réservoirs. *Revue des Sciences de l'Eau* **2000**, *13*, 379–404. [CrossRef]
39. Ollivier, P.; Surdyk, N.; Azaroual, M.; Besnard, K.; Casanova, J.; Rampnoux, N. Linking water quality changes to geochemical processes occurring in a reactive soil column during TWW infiltration using a large-scale pilot experiment: Insights into Mn behavior. *Chem. Geol.* **2013**, *356*, 109–125. [CrossRef]

Article

Analyzing the Effectiveness of a Multi-Purpose Dam Using a System Dynamics Model

Sleemin Lee and Doosun Kang *

Department of Civil Engineering, Kyung Hee University, 1732 Deogyeong-daero, Giheung-gu, Yongin-si, Gyeonggi-do 17104, Korea; sleemin16@khu.ac.kr
* Correspondence: doosunkang@khu.ac.kr; Tel.: +82-31-201-2513

Received: 12 February 2020; Accepted: 7 April 2020; Published: 8 April 2020

Abstract: The increasing frequency of extreme droughts and flash floods in recent years due to climate change has increased the interest in sustainable water use and efficient water resource management. Because the water resource sector is closely related to human activities and affected by interactions between the humanities and social sciences, there is a need for interdisciplinary research that can consider various elements, such as society and the economy. This study elucidates relationships within the social and hydrological systems and quantitatively analyzes the effects of a multi-purpose dam on the target society using a system dynamics model. A causal loop was used to identify causal relationships between the social and hydrological components of the target area, and a simulation model was constructed using the system dynamics technique. Additionally, climate change and socio-economic scenarios were applied to analyze the future effects of the multi-purpose dam on population change, the regional economy, water use, and flood damage prevention in the target area. The model proved reliable in predicting socio-economic changes in the target area and can be used to make decisions about efficient water resource management and water-resource-related facility planning.

Keywords: climate change; multi-purpose dam; system dynamics; water management

---

1. Introduction

1.1. System Dynamics Approach for Water Management under Changing Environment

Water resource development studies originally focused on understanding natural phenomena and on securing water resources. However, economic and social developments have necessitated studies that consider not only engineering elements, but also their relationships with various human and social elements, such as the economy, environment, and ecosystems. In recent years, interest in securing and using sustainable water resources has increased due to the influence of climate change. The water-resource sector is closely related to human activities, and they interact in more complex forms because of diverse and complex changes in the humanities and social sciences. Therefore, the interdisciplinary study to quantify the mutual effects of water resources and socio-economic factors is potentially useful for establishing water resource policies and planning for large-scale water resource facilities.

During the past five decades, a systems approach has been applied in many practical and scientific fields, including management, ecology, economics, education, engineering, public health, and sociology [1]. Among the systems analysis techniques, the system dynamics (SD), feedback-based and object-oriented simulation approach, can define the complex relationships in water resources systems and understand the dynamic correlation between socio-economic and hydrological systems according to causality between the system components [2]. The SD scheme is popular because of its advantages to handle the complex interactions among system components [3]. It studies the dynamic, evolving, cause-effect interrelations, and information feedbacks that direct interactions in a system over time [4,5]. Using the SD approach, we can understand more accurately and fundamentally the problems surrounding the water resources environment in an interdisciplinary modeling framework [6].

1.2. Literature Review of Systems Approaches for Water Resources Management

A hydro-system contains disparate but interactive components, which function as a unit and should be handled as a whole [7]. The SD has been applied in the hydro-system modeling and water resource management for more than 20 years [8]. Some representative SD application literature in the water field are summarized as follows.

Ahmad and Simonovic [6] developed a system dynamics model for the simulation of reservoir operation for flood management. The reservoir operation rules were suggested for high flow/flood years for mitigation of flooding by changing the reservoir storage allocation, reservoir levels and outflows in the Shellmouth reservoir on the Assiniboine River in Canada. The developed model was beneficial to predict future system behavior and provide a decision for secured flood management.

Li and Simonovic [9] developed a model to simulate flood patterns by snowmelt under temperature change in the spring season. The model is composed of five tanks representing snow, interception, surface, subsurface and groundwater storage and capture a vertical water balance. They studied hydrological processes in North American prairie watersheds where floods are significantly contributed to by snowmelt. They found that snowpack accumulation and snowmelt are major importance on flood generation and the temperature is a critical factor to determine the snowmelt rate and the physical state of the soil.

Simonovic [10] developed a global water model (named, World Water Model), and conducted a macro-scale assessment of global water resources availability. The study showed the relationship between water resources and future industrial growth and revealed that water pollution is the most important water problem in the future.

Xu et al. [2] constructed a simulation model to evaluate the sustainability of the water resources system in the Yellow River basin. The model captured the dynamic characteristics of the main components affecting water demand and supply. Scenario analyses were conducted by predicting future water demand and supply conditions and evaluated future water sustainability in different sub-regions.

Stave [11] carried out a study to encourage the public to understand the value of water conservation in Las Vegas, Nevada. The study described the process of building a strategic-level system dynamics model for water management of the study area.

Simonovic and Rajasekaram [12] developed an Integrated Water Resources Management (IWRM) model, in which the dynamic interactions between the quantitative features of available water resources and the water use affected by socio-economic levels of development, population and physiological features were simulated. The study developed 12 scenarios to investigate policy options in wastewater treatment, economic and population growth, freshwater export, energy production and trade.

Neto et al. [13] developed a system dynamics model to analyze the complicated interrelationships among the agents affecting the Sepetibza Bay environment. The model simulated various hypotheses of economic growth and population increase on the watershed. The study explained how environmental problems should be managed when the industry and population are expected to grow rapidly. Furthermore, a model simulation through system dynamics can provide important information to the policy decision-makers and the public related to water resources management.

Madani and Mariño. [14] conducted an integrated study and suggested a model based on a causal loop to manage the complicated water system. The study showed that diverse options of demand management and population control can be effective in addressing the water crisis, increasing water storage capacity and controlling of groundwater withdrawal.

Khan et al. [15] developed a conceptual water balance model to simulate the hydrologic processes including percolation, surface runoff, actual evapotranspiration, and capillary rise. Using the model, the dynamic interactions between surface and groundwater in an irrigation area was simulated.

The model further applied to simulate responses of different irrigation management scenarios and reduce the cost of groundwater abstraction in lowland areas.

Davies and Simonovic [16] developed an integrated assessment model, ANEMI, which represents nonlinear feedbacks between water resources, socio-economic, and environmental systems. The model includes eight sectors, such as climate, carbon cycle, economy, land use, agriculture, population, natural hydrological cycle, water use and quality.

Gaupp et al. [17] examined how storage capacity can improve water security in large river basins using a water balance model. The model can simulate runoff, water use from surface and aquifer, evaporation and trans-boundary discharges in BCUs (basin country unit) scale. The study showed the balance between water for human use and water for the environment and helped to acknowledge the limitations of over-reliance on water storage.

Di Baldassarre et al. [18] analyzed the system dynamics between supply–demand cycles and reservoir effects using a causal loop diagram and demonstrated the case studies in Athens, Las Vegas and Melbourne. The reservoir effect showed that the construction of reservoirs can supply abundant water, but dependence on water infrastructure (e.g., reservoir) can also generate vulnerability and economic damage when water shortages occur. The supply–demand cycle explained that increasing supply enables agricultural, industrial or urban expansion, but it can also bring competition for limited water resources.

Wang et al. [19] suggested a system dynamics modeling framework for the Integrated Water Resources Management (IWRM). The model dealt with water demands, allocation, and uses under climate, population and economic scenarios. The modeling framework provided a comprehensive understanding of IWRM concepts and strategic trade-offs in efforts towards basin-scale water sustainability.

System dynamics approaches were also applied to sewage and water supply operation and maintenance problems. Park et al. [5] developed a system dynamics simulation model to predict future operational conditions of a sewerage system and identify the most efficient operation scheme. Using the model, the operating mechanism of the overall sewerage system was established in relation to pipe maintenance. Park et al. [20] identified the feedback loop mechanisms that are inherent in the management of water supply systems and showed the relationships between water supply rate, revenue water ratio, average unit water price and investment costs for the water supply system.

As summarized above, the system dynamics techniques have been widely applied in water resources management, flood control, climate change modeling, and policy-making decision support aiming to understand and simulate the interaction among systems with social, economic, hydrological, and environmental elements. However, little research has analyzed how water resource facilities affect human society or predicted plausible future changes in socio-economic factors caused by the construction of new facilities. Constructing a large water infrastructure (e.g., dam) can induce significant changes in social and hydrological components in the affecting area. Thus, it is important to analyze the quantitative effects of proposed water infrastructure on social, economic and hydrological systems. Quantification analyses can provide decision-making support for policymakers who are planning to construct water infrastructure.

1.3. Research Background and Purpose

In South Korea, most of the annual precipitation occurs in summer (between June and September), causing frequent summer floods. From autumn to spring, on the other hand, prolonged droughts occur. The concentrated precipitation can thus limit the stable use of water resources. Moreover, the recent increase in the frequency and intensity of droughts and floods caused by climate change is expected to make the sustainable water management more difficult. Therefore, efforts need to be made to establish the sustainable use of water resources and prepare for climate change. Multi-purpose dams are potential solutions to these problems because they enable water resource management through efficient operations. They prevent flood damage by securing storage space prior to the flood season, and the stored water can meet the water requirements for living, industry, and agriculture during non-flood periods, droughts, or dry seasons.

Multi-purpose dams also provide additional social and economic benefits to downstream areas. For example, the flood control by the dam minimizes damage to crops and human life in downstream areas, which can consequently reduce the number of emigrants that likely occur if the flood is not controlled. Flood control, population influx, and water storage can affect a regional economy in the long term.

In this study, we develop a system dynamics model to understand the cyclical relationship between the socio- and hydro-sectors of an area. We apply climate change and socio-economic scenarios and examine future changes in the socio- and hydro-sectors using the constructed model. In addition, we attempt to quantify the effectiveness of water management by constructing a multi-purpose dam in terms of changes in the residential population, gross regional domestic product (GRDP), and flood/drought damages. The developed model can reflect the characteristics of the target area because it is constructed based on an understanding of both the socio- and hydro-sectors. It can also simulate changes in the two sectors caused by climate and socio-economic changes and improve the reliability of long-term predictions by considering both sectors simultaneously.

In the methods section of this paper, we describe the causal loop structure used for system dynamics model construction, which reflects the human and hydrological characteristics of the target area. In the application and results section, we perform a model calibration by comparing the model results with statistical data. Possible climate change and socio-economic scenarios for the target area are simulated using the calibrated model. In this way, we quantitatively analyze the human and social effects of a multi-purpose dam. Finally, the results of the study are summarized, and future research directions are suggested.

2. Materials and Methods

2.1. Study Area

The Hoengseong multi-purpose dam (hereafter, H-dam), located in Gangwon-do, South Korea, was selected to demonstrate the developed model. Figure 1 shows the location of the dam, Seom-river and Hoengseong-gun and Wonju-si, which receive water from the dam. The construction of H-dam started in December 1993 and was completed in November 2000. Table 1 lists the specifications of the dam.

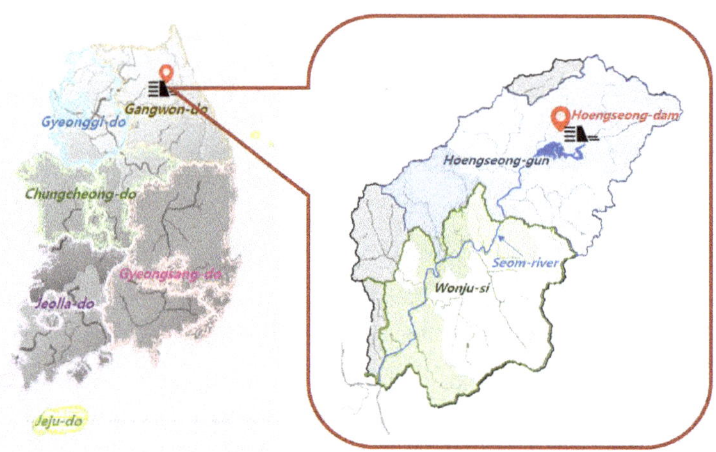

Figure 1. Location of the Hoengseong multi-purpose dam.

Table 1. Specifications of the Hoengseong multi-purpose dam.

Height (m)	48.5
Length (m)	205.0
Basin area $\left(km^2\right)$	209.0
Designed flood level $(EL.m)$	180.0
Restricted water level $(EL.m)$	178.2
Low water level $(EL.m)$	160.0
Total storage capacity $\left(million\ m^3\right)$	86.9
Effective storage capacity $\left(million\ m^3\right)$	73.4

Before the construction of the H-dam, the middle and downstream areas of the Seom-river suffered considerable flood damage in the wet season and serious water shortages in the dry season. To address those problems, the dam was constructed in the Seom-river basin to supply water to small and medium-sized cities in the downstream areas, such as Hoengseong-gun and Wonju-si. The dam plays its role in preventing flood damage by securing a flood control capacity of approximately 9.5 million m^3. The population of the area (Hoengseong-gun and Wonju-si) was approximately 370,000 in 2015 [21]. Agriculture and industry were developed in the area, and the dam supplies domestic, industrial, and agricultural water. Along with the construction of the H-dam, a water culture pavilion and a Hoengseong lake trail were created near the dam to provide local residents with recreational spaces. Additionally, the supply of maintenance water to the rivers in the downstream area improves the environment, water supply, and irrigation. The dam also reduces floods, thereby contributing to the stability and safety of residents.

Table 2 lists the statistic values of the study area in 2015, and the available data period. Note that the dam operation started in 2002 and the data related to the dam is only available after year 2002. The data used by the model were collected from the Water Supply Statistics [21], the Korean Statistical Information service [22], and the Water Management Information System [23].

Table 2. Statistics of the study area.

Element	Value in 2015	Available Data Period
Population	370,000	1965–2015
Residential land $\left(km^2\right)$	29.0	1991–2015
Industrial land $\left(km^2\right)$	2.6	1965–2015
Agricultural land $\left(km^2\right)$	121.0	1965–2015
GRDP *(billion KRW)*	8,600	1991–2015
Annual rainfall (mm)	717	1965–2015
Annual water use $\left(million\ m^3\right)$	59	1965–2015

GRDP = gross regional domestic product

2.2. System Dynamics Model Construction

2.2.1. System Dynamics

System dynamics is a research methodology developed by Forrester, an industrial engineering professor in the U.S., in 1961 [24]. This technique can be used to identify dynamic changes in an entire system with a focus on causal relationships and the feedback between system components. System dynamics has the following characteristics. First, it is not a single, one-way sequential line of events; instead, it reflects the interactive influence of feedback mechanisms [25]. Therefore, it is suitable for describing long-term changes in a target system, i.e., the future evolution and development processes of the system. Second, it finds the causes of changes in the system from the feedback structure. Because systems in modern society are complex, it is necessary to understand and analyze the feedback structure, which makes it possible to identify the causal relationships between components, rather than favoring one-way thinking [26].

System dynamics uses a causal loop diagram to describe the feedback structure. A causal loop diagram enables the visualization of causal relationships among the components that constitute the target system. A causal loop uses arrows to represent the relation and direction between elements, as well as (+) and (−) signs to indicate positive and negative correlations, respectively.

The modeling process uses system dynamics, which begins by defining the problem in the target system of interest. Next, various elements that constitute the target system are examined from a feedback perspective, and a causal loop is created to identify their causal relationships in the conceptualization phase [27]. A model can be constructed based on this phase, and long-term changes in the system are examined by simulating the target model through an appropriate scenario. In this study, we used Vensim DSS [28] to construct a model using the system dynamics technique. Vensim is the system dynamics simulation software developed by Ventana and is known to be useful in conceptualizing, designing, simulating, and analyzing complex systems [29].

2.2.2. Causal Loop

Figure 2 shows the causal relationships among the components of the developed model. An increase in population causes an increase in domestic water demand and urban land area (residential, commercial, and industrial land). The increase in urban land raises the demand for domestic and industrial water, but urban development in a limited land area causes a decrease in the agricultural area, thereby reducing agricultural water demand. An increase in industrial and decrease in agricultural land are related to the production per unit area, which affects GRDP. A growing GRDP reflects a growing economy, which attracts people to an urban area and increases the demand for urban land at the expense of agricultural land.

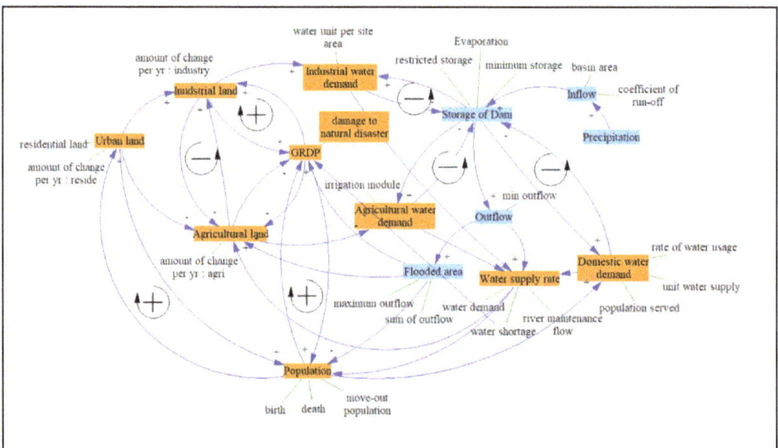

Figure 2. Schematic diagram of the causal loop in the proposed system dynamics model. The socio- and hydro-elements are represented in orange and blue, respectively.

Precipitation affects dam inflow and outflow. Thus, large precipitation causes an increase in dam inflow and outflow. When precipitation is small, the water stored in the dam is inevitably used due to water shortage, which again affects the storage and outflow of the dam. Prior to the flood season between June and September, the dam is required to maintain space for flood control. During normal periods, its storage and outflow must be adjusted to supply sufficient water by means of dam operation and management, such as securing storage by adjusting the water level of the dam and determining outflow to ensure a stable water supply during the flood and non-flood periods of each year, respectively. A large outflow in the flood season could cause inundation in the downstream area, resulting in damage to crops, and thereby affecting agricultural output and GRDP. In addition, a lack of available water in the dry season affects agricultural production, which also decreases GRDP. Changes in GRDP reflect a regional economy, which potentially affects the population migration of the area.

Based on these causal relationships, relational equations between the elements of each system can be determined. Here, the historical data of the area can be used to estimate the relationships. For the hydro-sector, the flooded area was included and dam operation was simulated using dam elements, such as the actual storage, restricted storage, minimum storage, inflow, outflow, and precipitation. For the socio-sector, a model was constructed using the population, residential land, industrial land, agricultural land, GRDP, and amount of damage to crops due to natural disasters. The water demand for domestic, industrial, and agricultural areas is also included in the model. The relationships among those elements can be seen in Figure 2.

2.2.3. Sub-Modules of the Socio-System

The components of the socio-system are population, land use (residential, industrial, and agricultural), water demand for the domestic, industrial, and agricultural sectors, GRDP, flood damage, and water supply rate (a ratio of water supply to demand), as shown by the orange shaded terms in Figure 2. Representative components reflecting the characteristics of the target area were selected to analyze the quantitative change in each component caused by dam construction in the target basin. The relational formulas between components were estimated using historical data and their causal relationships. For model construction, we first analyzed the historical data of the elements to identify the temporal pattern of each element and correlations among the elements. Based on the data analysis results, we sketched the causal loop diagram. The relational formulas were then developed and applied to the causal loop diagram to quantify the relations among the elements. Finally, the model calibration was conducted by adjusting the parameters of the relational formulas to fit the simulation results to the historical data. In the constructed model, the elements can affect each other because they are connected in a feedback relationship that allows each element to change the other elements.

The population was calculated by accumulating the annual population increment or decrement from the base year (Equation (1)). The increment or decrement was calculated considering births, deaths, and the number of people moving in and out. In this case, the number of people moving in and out was calculated by considering changes in land use (residential and industrial area), changes in GRDP, and move-out due to the damage from natural disasters (floods or droughts) in the target area. In Equation (1), $P(t_n)$ is the total population of year t_n; $P(t_0)$ is the initial population in the first year of the simulation period t_0, and $vp(t)$ is the net population increment or decrement *(person/yr)* in year *t*.

$$P(t_n) = P(t_0) + \int_{t_0}^{t_n} vp(t)dt. \tag{1}$$

Figure 3a shows the flooded area and move-out rate of the target area in the past (1971–2001) due to flood damage. Based on these results, the relationship between the two variables can be identified. Figure 3b shows the relationship between the water supply rate and the move-out rate of the area in the past (1971–2001), which was plotted to analyze the relationship between water shortages and move-out rates. Flood damage was judged by the flooded area in Figure 3a, and drought damage was judged by the water shortage (=1—water supply rate) in Figure 3b The move-out rate represents the ratio of the average number of people who moved out in the following year(s) to the number of people who moved out in the corresponding year. In Figure 3, positive move-out rates indicate that the number of people who moved out of the area increased, thereby reducing the population calculation. In other words, they indicate that when natural disasters, such as floods and droughts, occurred in the area, the number of people who moved to other areas increased due to the experience of inundation, damage to crops, and water shortages.

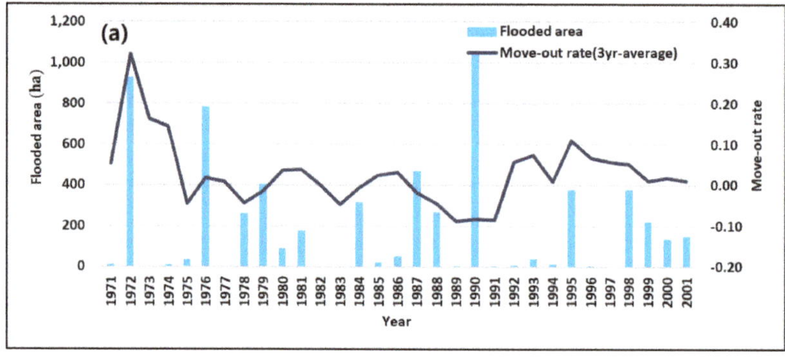

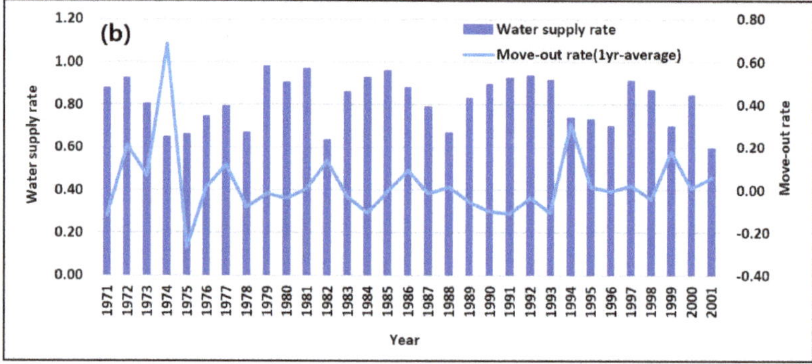

Figure 3. Relationship between natural disaster occurrence and the move-out rate (**a**) Flooded area, and (**b**) Water supply rate.

The move-out population for three consecutive years after a disaster was analyzed to identify the relationship between damage and move-out rates. In the case of flood damage, the move-out population increased for up to three years after considerable inundation damage. In the case of drought (i.e., water shortage), the move-out population increased for a year any time the water supply rate dropped below 0.7. Figure 3a shows the flooded areas and the three-year average move-out rates, and Figure 3b illustrates the water supply rates and the one-year average move-out rates. These results reveal that some residents of the target area evacuated because they no longer considered it safe after they had experienced damage from flooding and drought. Therefore, population change could be estimated reasonably by analyzing the number of people who moved out due to natural disasters.

The net population change was calculated using Equation (2), where $birth(t)$ and $death(t)$ are the number of births and deaths in year t, respectively. $I_i(t)$ is the industrial land change (km^2/yr) in year t, and $r_{p/i}$ is the population density in the industrial area (person/km^2). $I_r(t)$ is the residential land change (km^2/yr), and $r_{p/r}$ is the population density in the residential area (person/km^2). $I_g(t)$ is the GRDP change (KRW/yr) of the area, and $r_{p/g}$ is the population growth per GRDP increment (person/KRW). Note that 1 USD is equivalent to 1200 KRW. $t_p(t)$ is the number of people who moved out due to the occurrence of flooding and drought. Thus, the population change is assumed to be affected by changes in land use, GRDP, and disaster occurrence in the target area.

$$\int_{t_0}^{t_n} vp(t)dt = \int_{t_0}^{t_n} \Big(birth(t) - death(t) + I_i(t) \times r_{p/i} + I_r(t) \times r_{p/r} + I_g(t) \times r_{p/g} - t_p(t)\Big)dt. \tag{2}$$

Land use was classified into residential, industrial, and agricultural areas. The residential area was calculated using Equation (3), where, $A_r(t_n)$ and $A_r(t_0)$ are the residential land area (km^2) in year t_n and the first year of the simulation, respectively. $I_p(t)$ is the population change (person/km^2) in year t, and $r_{r/p}$ is the reciprocal of $r_{p/r}$ (km^2/person). $I_g(t)$ is the GRDP change (KRW/yr) in the area, and $r_{r/g}$ is the residential land change per GRDP change (km^2/KRW). $I_{A_a}(t)$ is the agricultural land change (km^2/yr), and r_{r/A_a} is the residential land change due to the agricultural land change (km^2/km^2). $I_{A_i}(t)$ is the industrial land change (km^2), and r_{r/A_i} is the residential land change due to the industrial land change (km^2/km^2). Thus, the changes in the residential area are affected by population, GRDP, and the land used for industry and agriculture.

$$A_r(t_n) = A_r(t_0) + \int_{t_0}^{t_n} \left(I_p(t) \times r_{r/p} + I_g(t) \times r_{r/g} + I_{A_a}(t) \times r_{r/A_a} + I_{A_i}(t) \times r_{r/A_i} \right) dt. \tag{3}$$

Similarly, the industrial area was calculated using Equation (4) and is affected by population, GRDP, and land use for residences and agriculture. $A_i(t_n)$ and $A_i(t_0)$ are the industrial land area (km^2) in year t_n and the first year of the simulation, respectively. $I_p(t)$ is the population change (person/yr), and $r_{i/p}$ is the reciprocal of $r_{p/i}$ (km^2/person). $I_g(t)$ is the GRDP change (KRW/yr) in the area, and $r_{i/g}$ is the industrial land change per GRDP change (km^2/KRW). $I_{A_r}(t)$ is the residential land change (km^2/yr), and r_{i/A_r} is the industrial land change due to the residential land change (km^2/km^2). $I_{A_a}(t)$ is the agricultural land change (km^2/yr), and r_{i/A_a} is the industrial land change due to the agricultural land change (km^2/km^2).

$$A_i(t_n) = A_i(t_0) + \int_{t_0}^{t_n} \left(I_p(t) \times r_{i/p} + I_g(t) \times r_{i/g} + I_{A_r}(t) \times r_{i/A_r} + I_{A_a}(t) \times r_{i/A_a} \right) dt. \tag{4}$$

Finally, agricultural land was calculated using Equation (5) and is affected by GRDP, land used for residences and industry, and the area damaged by flooding. $A_a(t_n)$ and $A_a(t_0)$ are the agricultural land area (km^2) in year t_n and the first year of the simulation, respectively. $I_g(t)$ is the GRDP change (KRW/yr), and $r_{a/g}$ is the agricultural land change per GRDP change (km^2/KRW). $I_{A_i}(t)$ is the industrial land change (km^2/yr), and r_{a/A_i} is the agricultural land change due to the industrial land change (km^2/km^2). $I_{A_r}(t)$ is the residential land change (km^2/yr), and r_{a/A_r} is the agricultural land change due to the residential land change (km^2/km^2). Moreover, $SA(t_n)$ is the flooding area (ha) in year t_n and is calculated using Equation (6). Flooding damage frequently occurs in the flood season. Therefore, the flooded area, $SA(t_n)$, was estimated using the total and maximum runoff in the flood season (June to September). A regression equation was estimated using historical runoff data and flood statistic values. Here, α and β are the parameters for calculating the flooded area according to the total and maximum runoff, respectively. Prior to dam construction, $os(t_n)$ and $op(t_n)$ represent the total streamflow (m^3) and maximum streamflow (m^3/month), respectively, during the flood season. After dam construction, $os(t_n)$ and $op(t_n)$ represent the total and maximum dam outflow (m^3), respectively, during the flood season.

$$A_a(t_n) = A_a(t_0) + \int_{t_0}^{t_n} \left(I_g(t) \times r_{a/g} + I_{A_i}(t) \times r_{a/A_i} + I_{A_r}(t) \times r_{a/A_r} \right) dt - SA(t_n) \times 10^{-2} \tag{5}$$

$$SA(t_n) = \alpha \times os(t_n) + \beta \times op(t_n). \tag{6}$$

GRDP is an economic index representing the value created by economic activities. It is affected by the population, industrial and agricultural production, and damage caused by flooding or drought and is calculated using Equation (7). $G(t_n)$ is the GRDP (KRW) in year t_n; $r_{g/p}$ is the production per capita (KRW/person); r_{g/A_i} is the production per industrial land area (KRW/km^2), and r_{g/A_a} is the production per agricultural land area (KRW/km^2). Additionally, $Shortage(t_n)$ is the water shortage amount (m^3) in year t_n, and u_a is the amount of water required per unit agricultural area (m^3/km^2).

Thus, the fourth term in Equation (7) indicates the economic loss caused by water shortages in the agricultural area. $FD(t_n)$ indicates the flood damage amount and is estimated as the sum of runoff, maximum runoff, and flooded area during the flood season, as expressed in Equation (8). Here δ, ϵ, and ζ are the parameters for calculating the flood damage according to the total runoff, maximum runoff, and flooded area, respectively.

$$G(t_n) = P(t_n) \times r_{g/p} + A_i(t_n) \times r_{g/A_i} + A_a(t_n) \times r_{g/A_a} - \left(Shortage(t_n) \times \frac{1}{u_a} \times r_{g/A_a} \right) - FD(t_n) \qquad (7)$$

$$FD(t_n) = \delta \times os(t_n) + \epsilon \times op(t_n) + \zeta \times SA(t_n). \qquad (8)$$

The water supply rate is quantitatively calculated to determine the water supply availability in the target area. Total water use includes domestic, industrial, and agricultural water consumption and river maintenance flow, and the water supply capacity includes the streamflow (before dam construction) or dam outflow (after dam construction). Therefore, the water supply rate is calculated as the ratio of the water supply capacity to the total water use, as expressed in Equation (9), where $WSR(t_n)$ is the water supply rate; $TW(t_n)$ is the total water demand (m^3), and $PO(t_n)$ is the water supply capacity in year t_n. For agricultural water demand, it was assumed that 80% of the total demand would be evenly consumed during the irrigation period (April–September), with the remaining 20% evenly allocated to the non-irrigation period (October–March). The annual domestic and industrial water demand is evenly distributed over the 12 months regardless of the season.

$$WSR(t_n) = PO(t_n)/TW(t_n). \qquad (9)$$

$Shortage(t_n)$ is the water shortage (m^3) in year t_n, and is expressed as shown in Equation (10).

$$Shortage(t_n) = TW(t_n) - PO(t_n). \qquad (10)$$

2.2.4. Sub-Modules of the Hydro-System

The blue blocks in Figure 2 indicate the components of the hydro-system: precipitation, dam (inflow, storage, outflow, evaporation), and river maintenance flow. The dam operation model for the simulation was constructed on a monthly basis. The monthly inflow into the dam was determined by the monthly precipitation and watershed information, and the dam storage and outflow were determined using the dam operation model.

The dam inflow was calculated using the monthly precipitation, dam basin area, and runoff coefficient, as expressed in Equation (11). $I(t_m)$ is the dam inflow (m^3/month) in month t_m; $P(t_m)$ is the monthly rainfall (mm/month); BA is the basin area (km^2), and C is the runoff coefficient. Here, the dam outflow is adjusted according to the inflow and storage, as obtained through the monthly operation simulation.

$$I(t_m) = P(t_m) \times BA \times C \times 10^{-3}. \qquad (11)$$

In the dam operation model, the restricted storage and minimum storage were set considering the specifications of the dam. Minimum monthly outflow was set by analyzing the water demand and downstream river maintenance flow in the past. In particular, when the dam storage was insufficient in the dry season or when restricted water supply was required, the minimum outflow was adjusted according to the dam operation rule.

Equation (12) is a water-budget equation for a dam (Figure 4). Here, $S(t_{m+1})$ and $S(t_m)$ are the water storage (m^3) of the dam in months t_{m+1} and t_m, respectively. $E(t_m)$ is the evaporation (m^3), and $O(t_m)$ is the outflow (m^3) from the dam in month t_m.

$$S(t_{m+1}) = S(t_m) + I(t_m) - E(t_m) - O(t_m). \qquad (12)$$

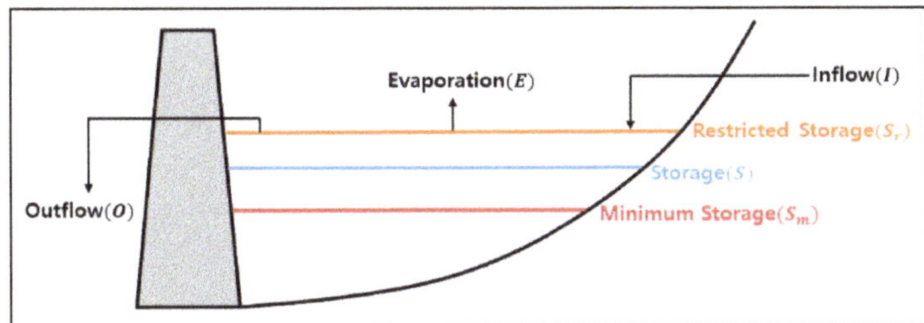

Figure 4. Dam water-budget analysis.

The outflow can be calculated for two cases. In the first case, when the inflow exceeds the restricted storage, the outflow is calculated using Equation (13). That is, the maximum value between the excess of the restricted storage and the minimum outflow is determined to be the outflow. Here, $S_R(t_m)$ is the restricted storage (m^3) in month t_m, and $\min O(t_m)$ is the predefined minimum outflow (m^3).

$$O(t_m) = \max[S(t_m) + I(t_m) - E(t_m) - S_R(t_m),\; \min O(t_m)]. \tag{13}$$

In the second case, when the inflow fails to reach the restricted storage, the outflow is calculated using Equation (14). That is, the maximum value between 50% of the water quantity obtained by subtracting the evaporation from the inflow and the minimum outflow is determined to be the outflow.

$$O(t_m) = \max[(I(t_m) - E(t_m)) \times 0.5,\; \min O(t_m)]. \tag{14}$$

Note that the minimum dam storage is specified, and the dam operation is simulated within the range of [minimum storage, restricted storage]. If downstream suffers flooding damages, the operators would increase the storage capacity to store more water in the dam; on the other hand, under severe droughts, the minimum outflow would be adjusted temporarily to secure water supply. In addition, if the downstream area requires more water due to urbanization, the minimum outflow would be increased (by changing the operation rules permanently) to supply the increased demand. In this way, the feedback from downstream can be reflected in the dam outflow in the model.

2.3. Scenario Development

In this study, we developed and simulated three plausible future scenarios, namely the baseline scenario, the extreme climate scenario, and the rapid urbanization scenario.

2.3.1. Baseline (Scenario 1)

The Representative Concentration Pathways (RCPs) describe four different 21st-century pathways of greenhouse gases and atmospheric concentrations, air pollutant emissions, and land use [30,31]. Comparing four pathways, we chose the RCP8.5 as a baseline because this scenario seems to represent the current conditions of the study site and plausible future condition. Note that the RCP8.5 indicates no specific climate change mitigation target in the area [32]. The projected annual rainfall data were obtained from the Korea Meteorological Administration, generated by multi-model ensemble means from the CMIP5 (e.g., more than 20 models) under the RCP 8.5 scenario [33].

2.3.2. Extreme Climate (Scenario 2)

The extreme climate scenario, which is a modification of the Baseline scenario (Scenario 1), hypothetically generates drought and flood years by manipulating precipitation to observe changes in the social and hydrological components of the target area. Figure 5 shows the annual rainfall for the next 30 years in the extreme climate scenario compared with scenario 1. For years of 2017, 2018, 2032, and 2040 in blue circles, which correspond to flood years, the annual rainfalls are the same as scenario

1, but the monthly rainfalls in the flood season (June to September) were purposely increased to induce flood events. On the other hand, for years of 2021, 2028–2029, and 2042–2043 in brown circles, which correspond to drought years, the monthly rainfalls of the past drought events (2014–2015 drought) in the target area were applied. Note that the 2014–2015 drought was one of the worst droughts occurred in the target area.

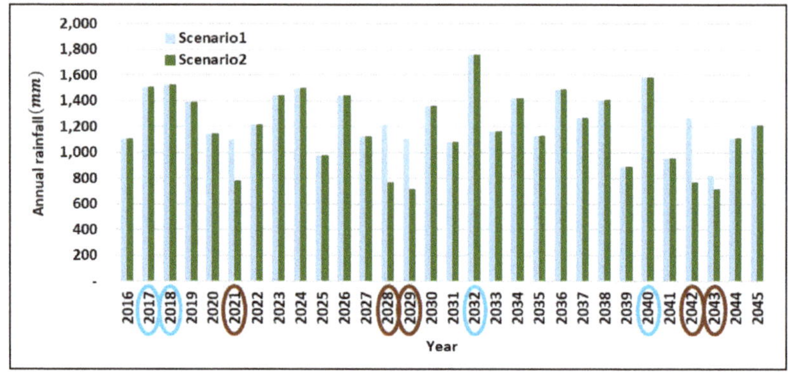

Figure 5. Annual rainfall with scenarios 1 and 2 (blue circle indicates flood years and brown circle indicates drought years).

2.3.3. Rapid Urbanization (Scenario 3)

For simulation of rapid urbanization in the target area, the following assumptions were made for the next 30 years: (1) higher-than-expected births, (2) an increase in water consumption per capita due to improved living standards, and (3) an increase in production per unit of industrial land due to technological developments. This simulation was conducted using the annual rainfall of the Baseline scenario with dam operation.

3. Results and Discussion

3.1. Model Calibration

To determine the applicability of the developed model for future scenarios, we first performed model calibration by comparing the statistical data with the simulation results of the model. The construction of the H-dam was completed in November 2000, and operations began in 2002. Therefore, the periods before and after dam construction were divided based on the year 2002. Table 3 summarizes the list of the parameters used for model calibration. The calibration was conducted using a trial-and-error, until the simulation results follow the trend of the historical data (visual inspection) and acceptable error is achieved by adjusting the parameters. Table 4 summarizes the overall root mean square error (RMSE) between the historical data and the simulation results.

Table 3. Model calibration parameters.

Element	Parameter	Description	Unit
	$r_{p/i}$	population density in the industrial area	person/km^2
Population	$r_{p/r}$	population density in the residential area	person/km^2
	$r_{p/g}$	population growth per GRDP increment	person/KRW
	$r_{r/p}$	residential land change per population change	km^2/person
Residential area	$r_{r/g}$	residential land change per GRDP increment	km^2/KRW
	r_{r/A_a}	residential land change due to the agricultural land change	km^2/km^2
	r_{r/A_i}	residential land change due to the industrial land change	km^2/km^2

Table 3. *Cont.*

Element	Parameter	Description	Unit
Industrial area	$r_{i/p}$	industrial land change per population change	km^2/person
	$r_{i/g}$	industrial land change per GRDP increment	km^2/KRW
	r_{i/A_r}	industrial land change due to the residential land change	km^2/km^2
	r_{i/A_a}	industrial land change due to the agricultural land change	km^2/km^2
Agricultural area	$r_{a/g}$	agricultural land change per GRDP increment	km^2/KRW
	r_{a/A_i}	agricultural land change due to the industrial land change	km^2/km^2
	r_{a/A_r}	agricultural land change due to the residential land change	km^2/km^2
Flooded area	α	regression parameter for total runoff	-
	β	regression parameter for maximum runoff	-
GRDP	$r_{g/p}$	production per capita	KRW/person
	r_{g/A_i}	production per industrial land area	KRW/km^2
	r_{g/A_a}	production per agricultural land area	KRW/km^2
Dam operation	C	Runoff coefficient	-
	E	Evaporation	m^3/month

Table 4. Calibration error root mean square error (RMSE).

Element	RMSE	Unit
Population	2095	person
Residential land	2.24	km^2
Industrial land	0.18	km^2
Agricultural land	1.30	km^2
Flooded area	0.18	km^2
Dam outflow	4.4	million m^3

3.1.1. Population

The calibration period was from 1965 to 2015, and the results are shown in Figure 6. The population of the target area in 1965 was approximately 240,000, which increased to approximately 370,000 by 2015. The population did not change significantly until the early 1990s and has slowly increased since the mid-1990s. It has increased continuously since the construction of the dam. Throughout the simulation period, the simulated values closely followed the statistical data.

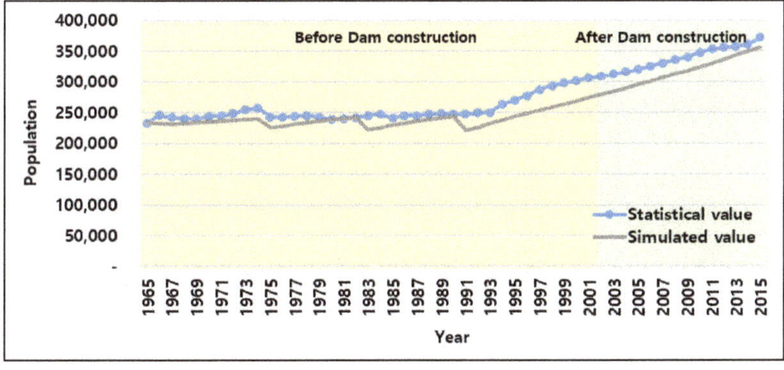

Figure 6. Calibration results—population.

3.1.2. Land Use

The land use of the target area was divided into residential, industrial, and agricultural land. The calibration period for residential land was from 1991 to 2015 due to a lack of data, and that for industrial and agricultural land was from 1965 to 2015. Figure 7 shows the simulation results for each land-use type. For residential land (Figure 7a), the area has increased continuously since the 1990s, which is similar to the population change results. For industrial land (Figure 7b), the area sharply increased in the late 1980s due to the industrial development and a population influx. The developed model could not accurately reflect this sharp, rapid increase, but it did simulate the overall increasing trend. For agricultural land (Figure 7c), the area has decreased continuously, likely because of the conversion of farmland into residential and industrial land following urbanization and industrialization. The model accurately simulated this decreasing trend.

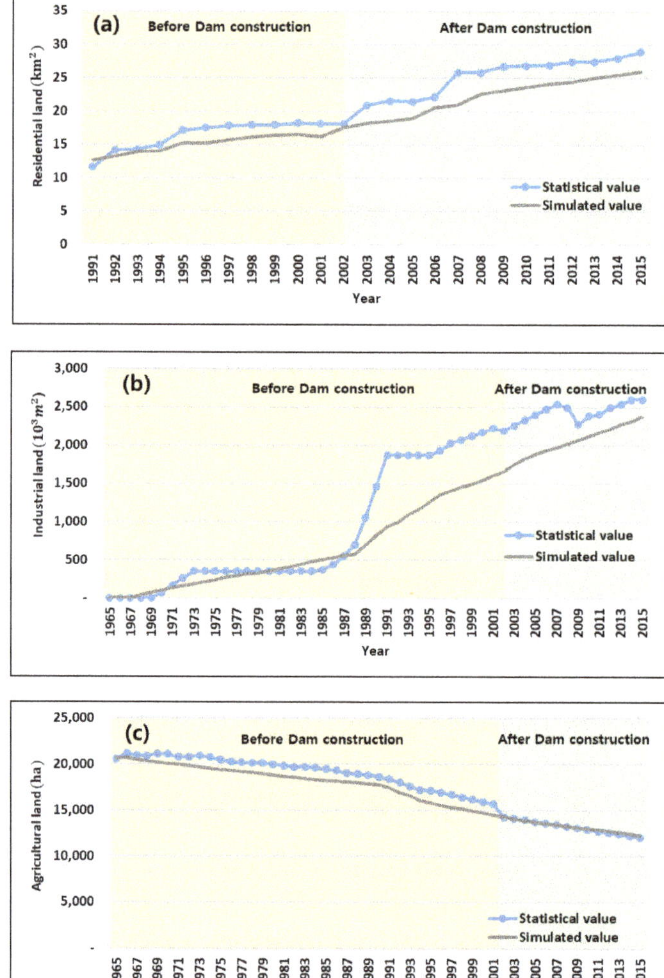

Figure 7. Calibration results—land use (**a**) residential land, (**b**) industrial land, and (**c**) agricultural land.

3.1.3. Flooded Area

The calibration period for the flooded area was 1971 to 2015, and the results are shown in Figure 8. As shown in the figure, the flooded area decreased significantly after dam construction, which directly exhibits the flood reduction effect of the dam. Although relatively large rainfall events occurred after dam construction, there was no large inundation damage in the area downstream of the dam due to its storage and outflow control. The dam simulation results also reflect this flood damage reduction effect.

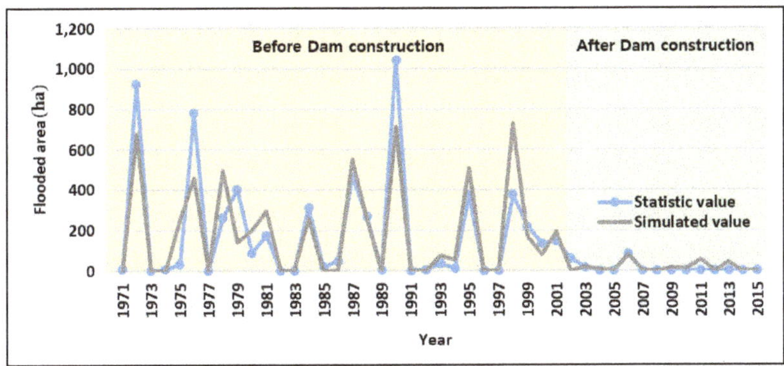

Figure 8. Calibration results—Flooded area.

3.1.4. Dam Outflow

To calibrate the dam operation results, actual dam outflow data were compared with the simulated outflows. The calibration period was 2002 to 2015, and the results are shown in Figure 9. In the target basin, there were large differences in annual outflows between the years with large floods and those with severe drought. Overall, the simulated values are similar to the statistical data.

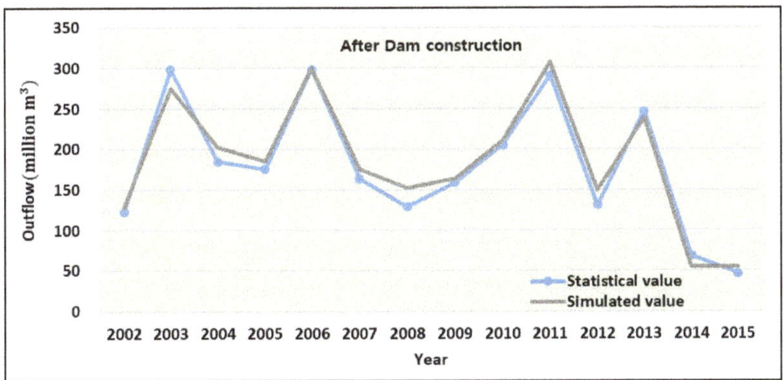

Figure 9. Calibration results—dam outflow.

3.2. Scenario Analysis

3.2.1. Baseline (Scenario 1)

This scenario is a base scenario. Two simulations were performed, one on the assumption that the dam was constructed as it is now (With DAM) and the other that it was not constructed (Without DAM), and the results were compared and analyzed. From the comparison, the effects of dam construction on the target area were quantitatively estimated for the following four elements: population change,

GRDP, inundation damage, and water supply rate. The future simulation period for the scenario was set to 30 years (2016–2045).

Population

In the simulation, the population was found to increase continuously following the previous trend. As can be seen in Figure 10, the case with dam construction (With DAM) exhibited a faster population increase than the case without dam construction (Without DAM). As of 2045, the population was estimated to be approximately 460,000 and 380,000 for the With DAM(S1) and Without DAM(S1) cases, respectively, resulting in a difference of approximately 80,000 people. Meanwhile, when the number of people who moved out during 30 years (2016–2045) was compared, With DAM(S1) exhibited approximately 10,000 fewer people than Without DAM(S1), apparently because efficient water resource management through the construction of the multi-purpose dam reduced the number of people who moved out of the target area by mitigating the damage caused by floods and droughts and inducing the activation of the regional economy. In the case without dam construction (Without DAM), the population increase in the target area was found to be slow due to an increase in the number of people moving out and slow regional development because of the future flooding and drought damages.

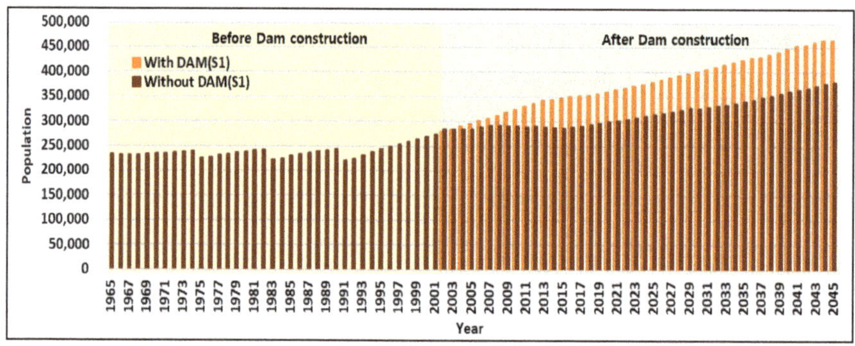

Figure 10. Population comparison between the With DAM and Without DAM simulations for Scenario 1.

Gross Regional Domestic Product (GRDP)

Figure 11 shows the GRDP change patterns of the target area from 1991 to 2045. In the With DAM(S1) case, it was predicted that GRDP would consistently increase by approximately 150 billion KRW each year (on average) from approximately 7 trillion KRW in 2016 to 12 trillion KRW in 2045. In the Without DAM(S1) case, on the other hand, GRDP decreased in 2006 and 2011 when large flood damage occurred in the target area and again after 2014 and 2015 when severe drought damage occurred. From 2016, the GRDP of Without DAM(S1) is expected to increase by an average of approximately 60 billion KRW each year for 30 years. Thus, the average GRDP increment of With DAM(S1) is approximately 2.5 times larger than that of Without DAM(S1). By 2045, the GRDP of the Without DAM(S1) case is expected to be approximately 8 trillion KRW, about 4 trillion KRW lower than that in the With DAM(S1) case. In the developed model, GRDP was affected by population, industrial land, agricultural land, and damage by natural disasters. Therefore, the presence of the dam affected those variables, and they, in turn, affected GRDP. If the developed model is used in planning for a new dam, it will be possible to quantitatively analyze the future effects of dam construction on the economy of the target area.

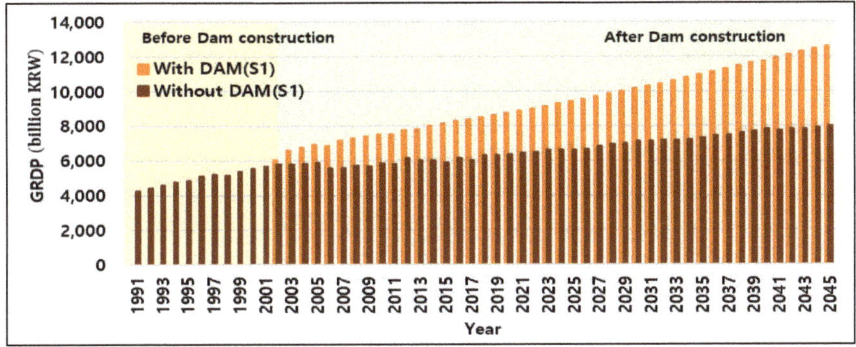

Figure 11. Gross regional domestic product (GRDP) comparison between the With DAM and Without DAM simulations for Scenario 1.

Flooded Area

The flooded area clearly shows the difference between the With DAM and Without DAM cases. As shown in Figure 12, approximately 4,500 ha of total inundation damage is expected to occur in the Without DAM(S1) case between 2016 and 2045, which corresponds to approximately 90 billion KRW when converted into a monetary value. For the With DAM(S1) case, however, almost no inundation damage (approximately 15 ha) is expected. The future inundation damage reduction through dam construction is expected to improve social values, such as economic vitality, by encouraging a population influx into the area and securing the land.

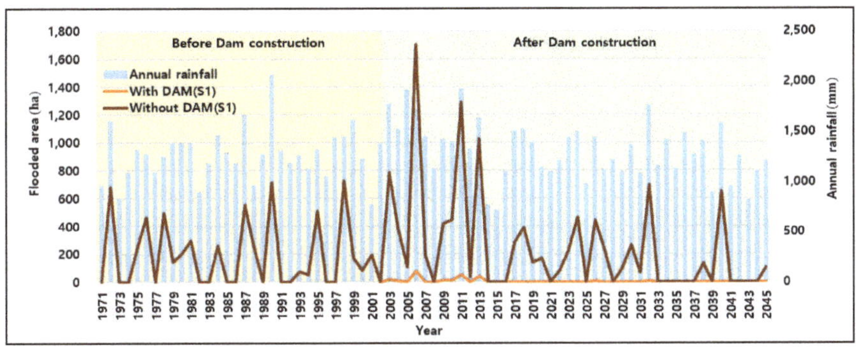

Figure 12. Flooded area comparison between the With DAM and Without DAM simulations for Scenario 1.

Water Supply Rate

Figure 13 shows the water supply rate for the target area from 1965 to 2045. The lowest water supply rate before dam construction was 0.61 in 2001, followed by 0.64 in 1982. The annual rainfall in 2001 was 780 mm, and that in 1982 was 900 mm, both significantly lower than the average annual rainfall for the area. Moreover, when the situations in 2014 and 2015, in which severe drought occurred in South Korea for two consecutive years, were simulated under the no-dam assumption, the water supply rates were calculated to be 0.56 and 0.51, and the annual rainfall amounts were found to be 765 and 717 mm, respectively. In the With DAM(S1) case during the same period, the actual water supply rates were 0.93 and 0.92, indicating that a stable water supply was possible.

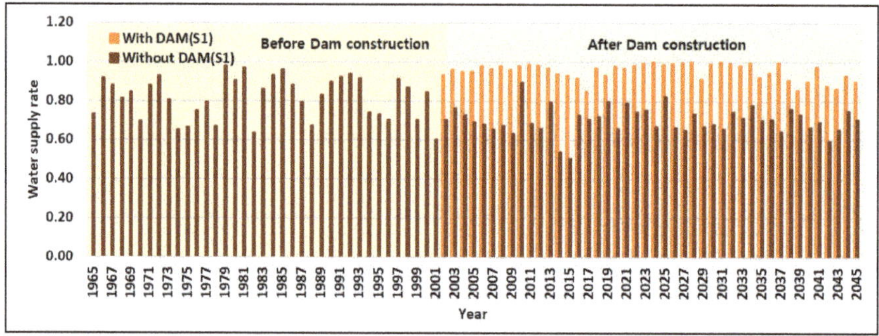

Figure 13. Water supply rate comparison between the With DAM and Without DAM simulations for Scenario 1.

The average annual water supply rate for 30 years (2016 to 2045) was calculated to be approximately 0.95 for With DAM(S1) and 0.71 for Without DAM(S1), indicating that water supply will be approximately 1.4 times more stable due to the dam construction.

3.2.2. Extreme Climate (Scenario 2)

Figure 14 shows the results of the simulation performed for the target area with the dam for the next 30 years (2016–2045) using the RCP8.5 scenario (Scenario 1) and the extreme climate scenario (Scenario 2). The population simulation results in Figure 14a show that the population in 2045 will be approximately 460,000 in Scenario 1 and 440,000 in Scenario 2, for a difference of approximately 20,000 people. When the number of people who moved out during 30 years was compared, Scenario 2 had approximately 2000 more than Scenario 1. In Figure 14b, the GRDP simulation results exhibit patterns similar to those for population change. The GRDP in 2045 is expected to be approximately 12 and 11 trillion KRW for Scenarios 1 and 2, respectively, for a difference of approximately one trillion KRW. Figure 14c shows the flooded area simulation results. The total flooded area expected for the next 30 years was approximately 15 ha for Scenario 1 and 50 ha for Scenario 2, an insignificant difference. In other words, no large inundation damage is expected to occur because of the dam operation, even if large floods similar to those observed in the past occur in the target area in the future. Figure 14d shows the simulation results for the water supply rate during the next 30 years. The 30-year average water supply rate for Scenario 1 was approximately 0.95, and that for Scenario 2 was approximately 0.91, indicating that the water supply rate could be reduced by approximately 4% by extreme drought events. In the case of Scenario 2, the water supply rate for the three years from 2042 to 2044 dropped below 0.8 due to the occurrence of extreme drought for the two consecutive years of 2042 and 2043. These simulation results indicate that water shortages might occur if extreme drought occurs consecutively in the target area for a certain period, and therefore adequate preparation is required.

Based on the comparison of Scenarios 1 and 2, it is expected that future extreme climate events in the target area are unlikely to cause significant changes in the social and hydrological elements. In other words, massive damage caused by flooding or drought will probably not occur in the area if the dam is properly operated, and the population and GRDP is also expected to increase consistently.

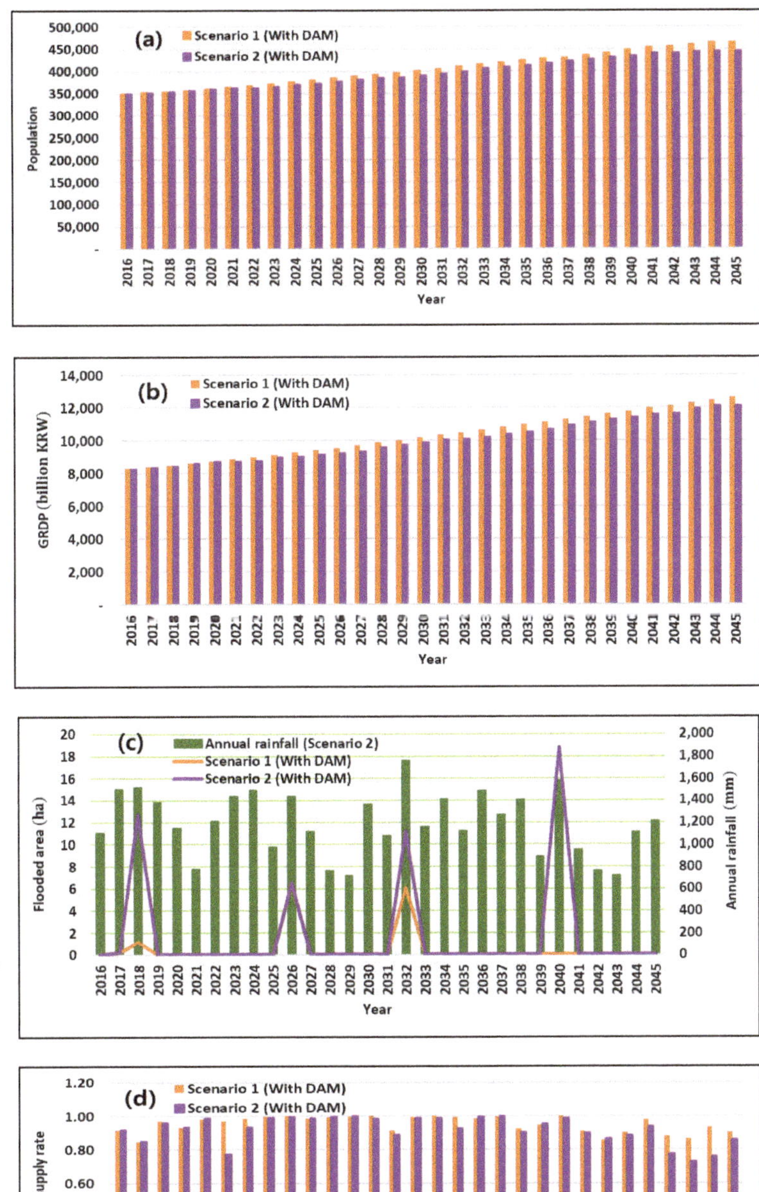

Figure 14. Comparison of Scenarios 1 and 2 for (**a**) population, (**b**) gross regional domestic product (GRDP), (**c**) flooded area, and (**d**) water supply rate.

3.2.3. Rapid Urbanization (Scenario 3)

In Figure 15, the simulation results for total water use, which combines domestic, industrial, agricultural, and river maintenance flow, and the water supply rate are compared between Scenarios 1 and 3. The simulation results for Scenario 3 show that the GRDP increased higher than that of Scenario 1. The increased GRDP then accelerated the population influx into the area and caused the expansion of urban areas (residential and industrial), with a consequent reduction in agricultural land. Thus, the demand for domestic and industrial water soared, but the demand for agricultural water decreased. The target area receives most of its domestic and industrial water from the H-dam, whereas agricultural water is supplied from small-scale agricultural reservoirs scattered in the basin. Therefore, irrigation dependence on the H-dam is generally low.

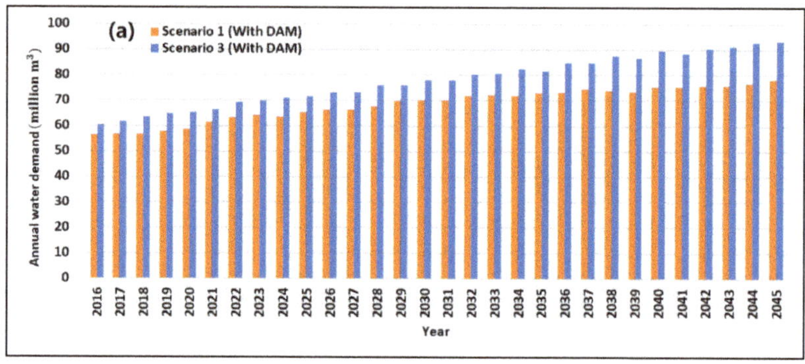

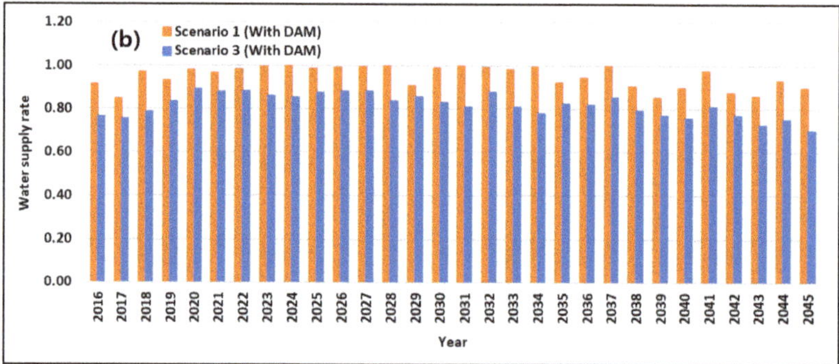

Figure 15. Comparison of Scenarios 1 and 3 for (**a**) Annual water demand and (**b**) Water supply rate.

As shown in Figure 15a, the total water uses in 2045 is expected to be approximately 78 million m³ in Scenario 1 and 95 million m³ in Scenario 3, for a difference of approximately 17 million m³. In Figure 15b, the average water supply rate in Scenario 3 is approximately 0.81, which is significantly lower than that for Scenario 1 (0.95). Thus, water supply shortages could occur if urbanization and industrialization progress faster than expected in the target area. Therefore, sustainable water management plans (e.g., water demand management, water reuse) should be established in conjunction with H-dam operation.

4. Conclusions

For multi-purpose dams, which are representative of large-scale water resource structures built for flood control and water supply, there is insufficient quantitative research on their effects on downstream areas. In this study, we developed a system dynamics model and used it under different scenarios to quantitatively analyze the effects of a multi-purpose dam on the population, economic change, response to disaster, and water supply in the downstream area. To apply the developed model, we selected the Hoengseong multi-purpose dam in Gangwon-do, South Korea. Prediction simulations were performed for the next 30 years (2016–2045), and the following conclusions were derived.

1. For Scenario 1, two case simulations were performed, one assuming dam construction (With DAM) and the other assuming no dam construction (Without DAM), and the results were compared to analyze the effects of the Hoengseong dam on the downstream area. When changes for the next 30 years were simulated, the population and GRDP were predicted to increase by approximately 80,000 and four trillion KRW, respectively, as of 2045 due to the 2002 completion of the Hoengseong dam. Furthermore, the flooded area will decrease by approximately 4480 ha, and the water supply rate will increase by approximately 1.4 times.

2. Scenario 2 simulated flood and drought years to analyze the effects of future climate changes in the target area. When Scenarios 1 and 2 were compared and analyzed from 2016 to 2045, it was concluded that future extreme climate events in the target area would not cause significant changes to social and hydrological elements. It appeared that no massive damage caused by flooding or drought would occur in the area, and the population and GRDP were expected to increase consistently.

3. Scenario 3 assumed increases in births, water consumption per capita, and production per unit of industrial land in the target area. The total water use, including domestic and industrial water, was expected to increase due to urbanization and economic revitalization. The 30-year average water supply rate dropped significantly; thus, water security plans would be required in conjunction with efficient dam operations.

In this study, we developed a system dynamics model to examine the effects of a multi-purpose dam on socio-economic factors of the downstream area. The model can be used as a decision-making tool to present grounds for engineering decisions and assist policymakers in planning and constructing a new multi-purpose dam. We also expect that the model can be adapted to perform effectiveness analyses and planning for other water resource facilities in the future.

Author Contributions: Conceptualization, D.K.; data curation, S.L.; formal analysis, S.L.; methodology, D.K. and S.L.; software, S.L.; supervision, D.K.; writing—original draft, S.L.; writing—review and editing, D.K. All authors have read and agreed to the published version of the manuscript.

Funding: This work was financially supported by the Korea Ministry of Environment (MOE) as a "Graduate School specialized in Climate Change".

Conflicts of Interest: The authors have no conflicts of interest to declare.

References

1. Sterman, J.D. *Business Dynamics, Systems Thinking and Modeling for a Complex World*; McGraw-Hill: Boston, MA, USA, 2000.
2. Xu, Z.X.; Takeuchi, K.; Ishidaira, H.; Zhang, X.W. Sustainability analysis for Yellow River water resources using the system dynamics approach. *Water Resour. Manag.* **2002**, *16*, 239–261. [CrossRef]
3. Zomorodian, M.; Lai, S.H.; Homayounfar, M.; Ibrahim, S.; Fatemi, S.E.; El-Shafie, A. The state-of-the-art system dynamics application in integrated water resources modeling. *J. Environ. Manag.* **2018**, *227*, 294–304. [CrossRef] [PubMed]
4. Wei, S.; Yang, H.; Song, J.; Abbaspour, K.C.; Xu, Z. System dynamics simulation model for assessing socio-economic impacts of different levels of environmental flow allocation in the Weihe River Basin, China. *Eur. J. Oper. Res.* **2012**, *221*, 248–262. [CrossRef]
5. Park, S.W.; Lee, T.G.; Kim, B.J.; Kim, T.Y. Development of a System Dynamics Model for the Efficient Operation and Maintenance of Sewerage Systems. *J. Korea Water Resour. Assoc.* **2012**, *45*, 101–111. [CrossRef]
6. Ahmad, S.; Simonovic, S.P. System dynamics modeling of reservoir operations for flood management. *J. Comput. Civil Eng.* **2000**, *14*, 190–198. [CrossRef]

7.   Simonovic, S.P. *Managing Water Resources: Methods and Tools for a Systems Approach*; UNESCO: Paris, France; Earthscan James & James: London, UK, 2009.

8.   Kim, Y.J.; Jeong, E.S. Socio-hydrology or Hydro-sociology: Study on the co-evolution between human and the water cycle. *Water Future* **2015**, *48*, 34–43.

9.   Li, L.; Simonovic, S.P. System dynamics model for predicting floods from snowmelt in North American prairie watersheds. *Hydrol. Process.* **2002**, *16*, 2645–2666. [CrossRef]

10. Simonovic, S.P. World water dynamics: Global modeling of water resources. *J. Environ. Manag.* **2002**, *66*, 249–267. [CrossRef]

11. Stave, K.A. A system dynamics model to facilitate public understanding of water management options in Las Vegas, Nevada. *J. Environ. Manag.* **2003**, *67*, 303–313. [CrossRef]

12. Simonovic, S.P.; Rajasekaram, V. Integrated analyses of Canada's water resources: A system dynamics approach. *Can. Water Resour. J./Revue Can. des Ressour. Hydr.* **2004**, *29*, 223–250. [CrossRef]

13. Neto, A.D.C.L.; Legey, L.F.; González-Araya, M.C.; Jablonski, S. A system dynamics model for the environmental management of the Sepetiba Bay watershed, Brazil. *Environ. Manag.* **2006**, *38*, 879–888. [CrossRef] [PubMed]

14. Madani, K.; Mariño, M.A. System dynamics analysis for managing Iran's Zayandeh-Rud river basin. *Water Resour. Manag.* **2009**, *23*, 2163–2187. [CrossRef]

15. Khan, S.; Yufeng, L.; Ahmad, A. Analysing complex behaviour of hydrological systems through a system dynamics approach. *Environ. Model. Softw.* **2009**, *24*, 1363–1372. [CrossRef]

16. Davies, E.G.; Simonovic, S.P. Global water resources modeling with an integrated model of the social–economic–environmental system. *Adv. Water Resour.* **2011**, *34*, 684–700. [CrossRef]

17. Gaupp, F.; Hall, J.; Dadson, S. The role of storage capacity in coping with intra-and inter-annual water variability in large river basins. *Environ. Res. Lett.* **2015**, *10*, 125001. [CrossRef]

18. Di Baldassarre, G.; Wanders, N.; AghaKouchak, A.; Kuil, L.; Rangecroft, S.; Veldkamp, T.I.; Van Loon, A.F. Water shortages worsened by reservoir effects. *Nat. Sustain.* **2018**, *1*, 617–622. [CrossRef]

19. Wang, K.; Davies, E.G.; Liu, J. Integrated water resources management and modeling: A case study of Bow river basin, Canada. *J. Clean. Prod.* **2019**, *240*, 118242. [CrossRef]

20. Park, S.W.; Kim, K.L.; Kim, B.J.; Lim, K.Y. Development of a System Dynamics Model to Support the Decision Making Processes in the Operation and Management of Water Supply Systems. *J. Korea Water Resour. Assoc.* **2010**, *43*, 609–623. [CrossRef]

21. Ministry of Environment. *2016 Water Supply Statistics*; Ministry of Environment: Sejong-si, Korea, 2018.

22. Korean Statistical Information Service (KOSIS). Available online: http://kosis.kr/ (accessed on 8 April 2020).

23. Water Management Information System (WAMIS). Available online: http://wamis.go.kr/ (accessed on 8 April 2020).

24. Forrester, J. *Industrial Dynamics*; The MIT Press: Cambridge, MA, USA; Wiley: New York, NY, USA, 1961.

25. Hjorth, P.; Bagheri, A. Navigating towards sustainable development: A system dynamics approach. *Futures* **2006**, *38*, 74–92. [CrossRef]

26. Jung, S.H.; Joo, Y.J. A Study on the Analysis of Policy Effects for System Dynamics Methodology: Focusing on the sex trade special law. *Korean Assoc. Public Adm.* **2005**, *39*, 219–236.

27. Jeong, H.S. Modeling Socio-Hydrological Systems for Wastewater Reused Watersheds. Ph.D. Thesis, The Graduate School Seoul National University, Seoul, Korea, 2014.

28. Ventana Systems. *Vensim User's Guide Version 6*; VENTANA Systems Inc.: Harvard, MA, USA, 2013.

29. Kim, K.C.; Jung, K.Y.; Kim, S.W. *SYSTEM DYNAMICS Using Vensim*; Seoul Economic Management Publisher: Seoul, Korea, 2014.

30. IPCC. Summary for Policy makers. In *Climate Change 2013: The Physical Science Basis. Contribution of Working Group I to the Fifth Assessment Report of the Intergovernmental Panel on Climate Change*; Cambridge University Press: Cambridge, UK; New York, NY, USA, 2013.

31. Pachauri, R.K.; Allen, M.R.; Barros, V.R.; Broome, J.; Cramer, W.; Christ, R.; Dubash, N.K. *Climate Change 2014: Synthesis Report. Contribution of Working Groups I, II and III to the Fifth Assessment Report of the Intergovernmental Panel on Climate Change (p. 151)*; IPCC: Geneva, Switzerland, 2014.

32. Riahi, K.; Rao, S.; Krey, V.; Cho, C.; Chirkov, V.; Fischer, G.; Rafaj, P. RCP 8.5—A scenario of comparatively high greenhouse gas emissions. *Clim. Chang.* **2011**, *109*, 33. [CrossRef]
33. Korea Meteorological Administration (KMA). Available online: https://www.weather.go.kr/ (accessed on 8 April 2020).

 *water*

Article

Applying the Systems Approach to Decompose the SuDS Decision-Making Process for Appropriate Hydrologic Model Selection

Mohamad H. El Hattab [1,2,*], Georgios Theodoropoulos [1], Xin Rong [1] and Ana Mijic [1]

[1] Department of Civil and Environmental Engineering, Imperial College London, London SW7 2AZ, UK; georgios.theodoropoulos16@imperial.ac.uk (G.T.); xin.rong17@imperial.ac.uk (X.R.); ana.mijic@imperial.ac.uk (A.M.)

[2] Centre for Doctoral Training in Sustainable Civil Engineering, Imperial College London, London SW7 2AZ, UK

* Correspondence: mohamad.el-hattab13@imperial.ac.uk

Received: 5 January 2020; Accepted: 22 February 2020; Published: 26 February 2020

Abstract: Sustainable Urban Drainage Systems (SuDS) have gained popularity over the last few decades as an effective and optimal solution for urban drainage systems to cope with continuous population growth and urban sprawl. A SuDS provides not only resilience to pluvial flooding but also multiple other benefits, ranging from amenity improvement to enhanced ecological and social well-being. SuDS modelling is used as a tool to understand these complex interactions and to inform decision makers. Major developments in SuDS modelling techniques have occurred in the last decade, with advancement from simple lumped or conceptual models to very complex fully distributed tools. Several software packages have been developed specifically to support planning and implementation of SuDS. These often require extensive amounts of data and calibration to reach an acceptable level of accuracy. However, in many cases, simple models may fulfil the aims of a stakeholder if its priorities are well understood. This work implements the soft system engineering and Analytic Network Process (ANP) approaches in a methodological framework to improve the understanding of the stakeholders within the SuDS system and their key priorities, which leads to selecting the appropriate modelling technique according to the end-use application.

Keywords: SuDS; decision-making; Soft Systems; ANP; modelling; stakeholder

---

1. Introduction

In 2018, the percentage of the world's population living in cities reached 55% and the level of urbanisation in Europe reached 74% [1]. The urbanization process inevitably diminishes the porous green spaces of cities. For instance, London loses the equivalent of 2.5 Hyde Parks of green space annually. This urbanisation and population growth, accompanied by changes in rainfall patterns due to climate change and the insufficient capacity of current sewer systems, is leading to increased urban flooding. The increase in flooding has reached a level that triggers global concern because it not only poses direct threats to human wellbeing and property safety but also has knock-on effects on economic and social development. Therefore, to mitigate the risk of flooding in a sustainable manner, Sustainable Urban Drainage Systems (SuDS), which fall within the context of Blue Green Infrastructure (BGI) practices, have been proposed as one of the first strategies to pursue.

Traditional management of urban water systems considers all components independently in a fragmented manner [2]. However, with new factors such as rapid urban growth driving water system development as well as burgeoning needs for infrastructure rehabilitation and climate change adaptation, integrated urban water modelling is currently expanding in new directions, stimulated by improvements in computational efficiency. Integrated 1D–2D modelling of the interactions between urban drainage systems and urban landscapes during large pluvial flooding incidents has become possible in the last decade (e.g., [3–5]). Modelling that extends beyond the strictly technical and biophysical domain of the water system (e.g., social and economic domains) has also attracted increasing interest in recent years [6].

Integrated Urban Drainage Models (IUDM) have been one of the most well-known and recognised forms of integrated models. These typically consist of studying upgrade options for a local Wastewater Treatment Plant, assessing ways of reducing Combined Sewer Overflow emissions or showing the combined impact of different parts of the drainage system on receiving waters [7]. They recognise both combined and separate drainage systems and can simulate real-time control (RTC) strategies for optimisation of new and existing complex systems [8].

The advancements in modelling capabilities have increased the need for a universal approach to select and develop urban drainage models, which are usually iteratively refined during model development by finding a balance between the study objectives, model structure, data requirements and availability and computational power efficiency [7]. Given the array of benefits provided by SuDS, spanning several disciplines to maintain multiple ecosystem services [9], it is very important to manage the level of modelling required to identify and quantify such benefits in the planning phase. The whole process for developing a SuDS model, from building to testing, should follow a systematic approach, starting with clearly defined aims and objectives and an initial assessment of data availability before selecting the potential model features. Hence, the first and most important step is to understand the aims and objectives of the modelling exercise.

This step is linked to understanding stakeholders' priorities; in turn, that understanding is linked to the whole decision-making process. In general, there are four main integrating approaches for decision-making: Cost-Benefit Analysis (CBA), Multi-Criteria Analysis (MCA), Triple Bottom Line (TBL) and Integrated Assessment (IA) [10].

CBA is one of the simplest forms of integrative approaches. It is usually applied as a pragmatic tool for aiding decision making rather than as a framework. It has proved to be useful because of the one single aggregated result obtained that helps to clarify and provide information about the costs and benefits of alternatives [11]. CBA has been widely adopted in water engineering applications traditionally because of its simple monetising approach [12]. Arrow et al. [13] provided a good philosophical foundation for the role of CBA in the management of natural resources such as water. CBA was used by Ossa-Moreno et al. [14] and Liu et al. [15] to capture the broader benefits of SuDS and not focus only on the system's performance regarding flood management and water quality improvement. In those authors' studies, they tried to compare the different technologies of SuDS within the framework of a cost benefit analysis.

MCA is a structured approach for supporting decision-making when dealing with more than a single criterion and allows relative importance to be placed upon each criterion by the user [16]. It is generally used as an analytical tool, but it can also be applied as an integrated framework by coupling with appropriate problem structuring methods (PSMs) [17]. In the water resource planning and management sector, MCA is heavily used for water policy evaluation, strategic planning and infrastructure selection [18]. MCA was used by Ellis et al. [19] to assess the best choice of SuDS with the aim of quantifying different and wider benefits of SuDS implementation. For the MCA evaluation, the benefits were separated using primary and secondary criteria and indicators.

TBL is itself not a truly integrative approach, but it is included because it can be used as a decision-making framework for guiding selection of indicators for measuring performances. It extends corporate social responsibility from the concept of sustainability, motivating organisations to address sustainability issues in a more integrated way [20]. A study by Viavattene and Ellis [21] combined the TBL and MCA approaches. The TBL was applied to account for the economic, social and environmental aspects of SuDS. MCA was used to enable the stakeholders to make a judgment among 16 different criteria that included the wider benefits of SuDS, such as amenity and aesthetic benefits. Moreover, the user had the opportunity to rank different SuDS technologies regarding the criteria that they had established.

IA is an emerging discipline that uses scenario management and stakeholder engagement while emphasising on the process to bring together a broad set of disciplines characteristic of the decision problem through [22]. Brouwer et al. [23] provided a comprehensive review of the IA concept and methods for water and wetland management. Some of the integrative approaches adopted by IA are system dynamics, Bayesian networks, agent-based models and expert systems.

Each of the above listed approaches has one or more limitation when dealing with SuDS decision-making problems. CBA generally strives to identify the gains of the winners and thereby unintentionally disregards the loss of the losers. MCA has three main problems: the first one is that it assumes independence between the criteria, the second is that there is a possibility of double counting and the third is the lack of transparency of the methods and the results. TBL suffices as a corporate reporting and communication tool; therefore, the impact of secondary stakeholders such as residents in the decision-making process is not fully accounted for. IA is an iterative adaptive approach, which makes it qualitative in nature without a robust model.

These limitations in the previous approaches can be solved by applying a framework based on systems thinking. The proposed framework is an application of the soft system methodologies to map the stakeholders and their interactions; then, analyse the stakeholders' priorities by applying the Analytical Network Process (ANP) method. As mentioned by Saaty [24], "the ANP overcomes the limitation of linear hierarchic structures and their mathematical consequences". Furthermore, by mapping and clustering the stakeholders, the assessment would be more robust and the procedure more transparent. By assigning weights to the stakeholders according to their role in the decision-making process, the ANP overcomes the limitations of CBA and TBL. In addition, by establishing links between criteria, ANP solves the problems of MCA. Finally, the quantitative nature of the ANP where all entries are assigned values overcomes the qualitative nature of the IA.

The ANP method has been successfully applied as a systematic selection process to guide decision-making in various industries, such as information system technologies [25], commodities [26], water and wastewater treatment [27,28], urban design [29] and renewable energy [30]. However, a review of the state-of-the-art of ANP application studies revealed that, to date, no attempts have been made to apply the ANP method in urban surface water management. Here, we seek to make an important contribution by explicitly applying the systems thinking approach to urban water management.

This study aims to introduce a methodological framework to manage SuDS modelling efforts when multiple stakeholders are involved in a project, such as retrofitting Sustainable Urban Drainage Systems (SuDS) in the UK to an existing urban environment, to deliver the maximum out of the available resources and to suppress the tendency towards seeking complex models. That is achieved by first identifying the key stakeholders and then analysing and understanding their priorities in relation to the wider benefits of SuDS and using the ANP approach to determine the desired outputs required from the modelling exercise.

2. Materials and Methods

2.1. Outline of the Proposed Framework

Hydrological Assessment and Management of green Infrastructure to Enhance Decision-making (HAMIED) is a framework that allows systematic management of the modelling needs of the different agents within the SuDS system of stakeholders in a series of steps, as outlined in Figure 1. The analysis formulation presented in the figure allows for defining the usage of available SuDS modelling techniques based on stakeholder needs.

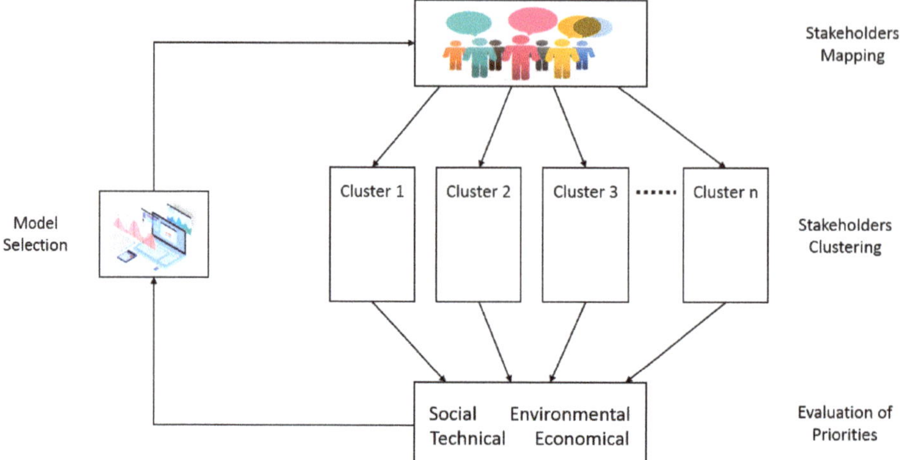

Figure 1. Hydrological Assessment and Management of green Infrastructure to Enhance Decision-making (HAMIED): A framework to adapt the optimal Sustainable Urban Drainage Systems (SuDS) modelling approach according to stakeholders' priorities.

In the first step of the overall selection framework, a list of potential stakeholders involved in SuDS projects is formulated. These stakeholders are then evaluated individually, and the interdependencies between them are mapped. Then, the stakeholders are clustered based on their roles within the decision-making process. In the next step, the priorities of stakeholders are evaluated with respect to different aspects of the SuDS' functions. Lastly, the appropriate modelling complexity is assigned to each group of stakeholders according to their interest in SuDS. The detailed steps for implementing the framework presented in Figure 1 are applied here to a full-scale UK case study, as explained in the following sections.

2.2. Classification of Model Complexity

Hydrological models are often classified based on their spatial and temporal resolution. Models vary in their temporal resolution, with daily models increasingly being replaced by models running at sub-daily time steps (e.g., 1 or 5 min). Such fine resolutions are necessary to represent the fast response typical of urban catchments. Spatial resolution of models ranges from single cell (of few square meters in size) to large catchments (covering several square kilometres). The larger the spatial scale is the smaller the confidence level in the model outputs is and the lighter the computational capabilities needed. A model's level of complexity is therefore defined based on the effort and cost required to validate a model (data requirements, model setup, calibration, simulations and uncertainty analysis) and the appropriateness of purpose and operational efficiency.

Urban hydrologic models can be grouped into five categories: (1) Conceptual (e.g., mass balance of a rainfall-runoff model); (2) Lumped (e.g., rainfall-runoff model where the catchment is described as a single entity); (3) Semi-distributed (e.g., rainfall-runoff model where the catchment is described as small individual units); (4) Hydrologic Response Unit (HRU) based, where an HRU represents an area with the same soil and land use type (e.g., rainfall-runoff model where similar hydrological behaviour in each unit is represented); and (5) Grid-based spatially distributed (e.g., rainfall-runoff model where the catchment is sub-divided into a raster system for representing the spatial variability of different attributes). Each of these can be constructed with different spatial and temporal resolutions, requiring various amounts of accurate data. A greater data requirement increases the computational resources needed, so we propose a scale of model complexity from light, for models requiring the least amount of data and effort, to very complex, for example, for fully distributed models (Table 1).

Table 1. Proposed scaling for assessing urban hydrologic models' level of complexity.

Type of Model	Spatial Scale	Time Scale	Level of Complexity
Conceptual	Site scale	Event based	Light
Lumped	Site scale	Event based/continuous simulation	Moderate
HRU	Catchment/regional	Event based/continuous simulation	Complex
Semi-distributed	Catchment scale	Event based/continuous simulation	Complex
Fully-distributed	Catchment scale	Event based/continuous simulation	Very Complex

Note: HRU, Hydrologic Response Unit.

2.3. Case Study

Counters Creek (CC) is one of the lost rivers of London and is situated on the boundary of the London Borough of Hammersmith and Fulham (LBHF) and the Royal Borough of Kensington and Chelsea (RBKC). This former river and its catchment (Figure 2) are now part of the sewerage system, draining surface water from buildings and roads as well as wastewater from toilets, bathrooms and kitchens. There are over 1700 properties across parts of LBHF and RBKC reported to be at risk of sewer flooding [31]. One of the proposed solutions to address the problem is to retrofit SuDS in these hotspots. Three streets in London, each with a nearby control street, were chosen to implement SuDS retrofitting technologies. The SuDS featured in this case study included rain gardens, permeable pavement and porous asphalt [32].

Figure 2. Counters Creek Catchment (in green) within the London Borough of Hammersmith (LBHF) and The Royal Borough of Kensington and Chelsea (RBKC), London, UK.

Multiple stakeholders were involved in this pilot project, including the water company, the local councils, the residents, the designers and a research institute, among others. Hence, the results produced on the performance of different SuDS options differ in terms of the technical level needed for each stakeholder, whose priorities with respect to the impact of SuDS on the urban system have to be addressed.

2.4. Identification of Key Stakeholders

The multi-functionality of SuDS, spanning from hydrological [14] to environmental [33] and socio-economic [34] functions, results in a wide range of benefits for each of the relevant sectors. Due to this multifunctional nature of SuDS, the relationships between stakeholders are very complex. Eventually, most SuDS functions involve multiple stakeholders that have different backgrounds and purposes, which adds to the complexity of the problem (e.g., [35,36]).

To analyse the complex system of SuDS stakeholders in the UK, the Soft System Methodology was applied [37]. A map of all stakeholders and their interdependencies was created and then analysed using the N2 method [38] to decompose the complex stakeholder system into its independent components. The analysis of the stakeholder system with the N2 method assesses the degree of binding and coupling in a system and therefore enables holistic stakeholder management and eventually improved system efficiency. It also aids in understanding the behaviours of stakeholders by identifying interconnectivity in the system [39].

The proposed algorithm is based on graph theory techniques [40] where a diagram in the shape of a matrix is formed to record the interconnections between the system elements. It is used to assess the degree of binding and coupling in a system and thereby determine the candidate architecture based on the natural structure of the system. It is also used to identify and document the interconnectivity in a system to help understand observed behaviour and to provide guidance for improvement. First, all system elements or functions are listed along the diagonal of the N2 chart and changed to numbers for convenience of grouping. Second, all of the isolated elements, the source and destination elements and the critical element (the most influential element that has the most connections to other elements) are identified from the system. Feedback loops where two elements are tightly bound to cascade flows between elements are also recognised, as they determine the partition result. The N2 chart is then clustered step-by-step to a state where no more clusters are allowed. The resulting clusters indicate how stakeholder management should be designed and improved.

2.5. Understanding Stakeholders' Priorities

2.5.1. Questionnaire Survey

A questionnaire survey is an effective and common way to study the perceptions, attitudes and behaviours related to a certain activity [41]. To obtain stakeholders' feedback on different aspects of a SuDS project in the UK, an 8-question survey (can be found in the Supplementary Materials) was developed for all stakeholders involved in the project. They were asked to provide their opinions on multiple aspects related to technical, economic, environmental and social functions linked to SuDS implementation. The survey was developed using Qualtrics and distributed through email and personal communication.

The questionnaire started by providing the survey aim, an introduction to SuDS and the voluntary and acceptance terms. The survey was designed to gather directly involved stakeholders' perceptions about SuDS and the importance of each SuDS function to them. The stakeholders where asked to answer seven questions related to the wide array of SuDS benefits which were clustered in four groups based on their aspects: (i) Economical, (ii) Technical, (iii) Social and (iv) Environmental and others. In total, seven out of fourteen stakeholders responded to the questionnaire. Although a higher response rate would have been more favourable and provided a wider spectrum of the population, the stakeholders who responded were among the most influential ones in the decision-making hierarchy, which provides the necessary robustness to the analysis conducted afterward.

2.5.2. Analytic Network Process (ANP) for Understanding Decision-Making in SuDS

The system of stakeholders mapped by the N2 method and the links between them increase the complexity of selecting a modelling tool; therefore, a processing method is needed to disentangle these interdependencies and assign the proper modelling tool for each group of stakeholders.

The Analytical Network Process (ANP), first proposed by Saaty [42], is a measurement theory based on multiple criteria that is used to derive relative priority scales of absolute numbers from a series of individual judgments that also belong to a fundamental scale of absolute numbers. These judgments represent the relative influence of each of two elements with respect to an underlying control criterion in a pairwise comparison process. The ANP synthesises the outcome of dependence and feedback within and between clusters of elements through its supermatrix, whose entries are themselves matrices of column priorities.

In the case of SuDS projects that could provide a wide range of benefits for multiple stakeholders, because of the nature of the links between the criteria and the environments analysed, the ANP is the methodology that allows the best benchmarking and that provides accurate results compared to other techniques, such as MCA [43].

For the evaluation of the ANP method, the open-source commercial program Super Decisions was used [44]. Super Decisions is a free educational decision support software program that has the necessary tools to create and manage the ANP models [26].

Each type of SuDS has different attributes and benefits, which result in different performance values for a criterion defined within the four selected aspects of SuDS functions. A scoring system in which each SuDS intervention could take a score that was scaled from zero (worst performance) to one (best performance) was used in order to define the performance for each criterion. Additionally, the performance of each SuDS was scored relative to the performance of all other selected SuDS technologies. For that reason, it was decided to keep the reference from one source, and the UK SuDS manual [27] was used for this purpose. The criteria were divided into two main categories, (i) Quantitative and (ii) Qualitative, as shown in the Supplementary Materials (Table S1). The criteria were directly linked to the questionnaire survey mentioned previously.

The first step in the process was to create the structure of the ANP network based on the criteria, and the connections based on the relationships between the criteria. The problem was decomposed into a rational system of network type. As shown in Figure 3 for a certain stakeholder, it consisted of seven clusters in total, each of which contained nodes that represent the selected criteria.

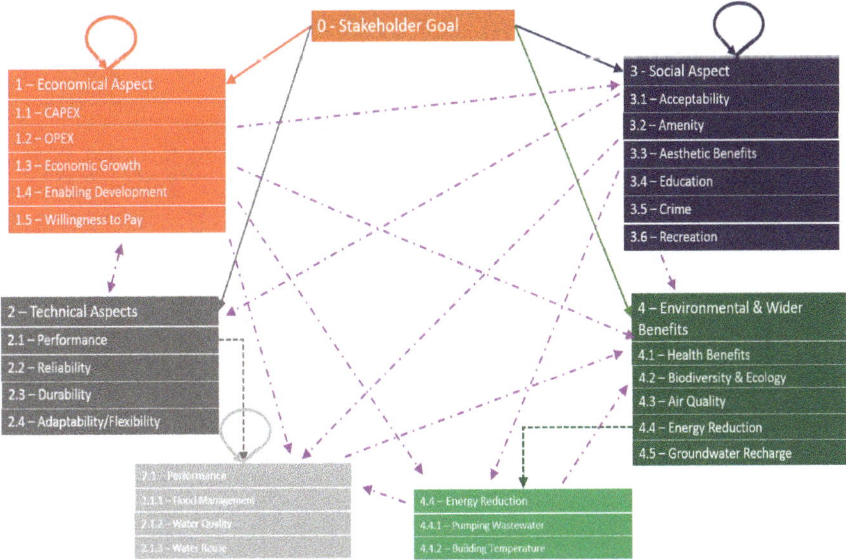

Figure 3. Example ANP network structure for SuDS stakeholders' priorities assessment. The connections with the stakeholders 'goals are represented by a continuous line; the connections of the main criteria with the sub-criteria are represented by a dashed line; and the interactions among the criteria are represented by a dashed dot line. The loops indicate inner dependence among the elements in the cluster. ANP, Analytic Network Process; CAPEX, CAPital Expenditure; OPEX, OPerating EXpense.

The next step was to determine the relative importance between the criteria. Pairs of decision-elements for each cluster were compared with respect to their importance to their control policies [45]. In the ANP, the judgment and determination of relative importance was made by answering a pairwise comparison [10].

The same network structure was created for each of the stakeholders who participated in the survey, and these were linked together according to the weight of the respective stakeholder within the decision-making system. The weights were assigned based on the results of the stakeholder screening from the N2, where the easily groupable stakeholders have the least weight and the self-acting stakeholders have the strongest weight. For instance, the local authority is a major stakeholder in the decision-making of SuDS; hence, it has a high weight compared to hospitals for example.

The next step was to make the supermatrix. It shows the intensity of the links between the criteria for each stakeholder (e.g., how much technical performance aspects are correlated with the economic aspects). This is a process similar to the creation of a Markov chain [45]. To achieve global priorities in a system with interdependent influences, vectors of local priorities are incorporated in the corresponding columns of a matrix. This supermatrix is actually a partitioned matrix in which each segment represents a relationship between two clusters in a system. The software produces the unweighted matrix, which shows the relative weights between the criteria based on the relative importance obtained from the stakeholder.

In the final step, a stochastic weighted supermatrix is created. The software produces the limit matrix, which is the long-run or limit priority of influence of each element on every other element. More specifically, the weighted matrix is raised in a high power in order for the limit matrix to be identified [42]. The values, which represent the stakeholders' priorities, are then extracted from the limit matrix.

3. Results and Discussion

3.1. Stakeholder Mapping

The stakeholder system map representing all UK stakeholders and their interdependencies for the SuDS system is shown in Figure 3. According to the primary role of each stakeholder, the system is divided into three main groups: Institutional, Design and Build as well as Adoption and Maintenance.

The Institutional group represents the government administration with approval power over the water system. It consists of all stakeholders within the government (marked blue in Figure 4). The Design and Build group (marked in orange) represents the consultants and contractors, which are both employed by the developers. Lastly, the Adoption and Maintenance group (marked in purple) represents an agreement system for adoption and maintenance consisting of adopters and maintenance companies.

It is worth mentioning that interrelations still exist among the three groups due to the complexity of the situation. Developers coordinate SuDS projects and have linkages to all three groups and all users. Users/citizens utilize the results of SuDS projects, and they give feedbacks to designers in the form of consultation. They can also give advice to the Institutional group as a non-statutory consultee.

The N2 method was then applied to the system to analyse the interdependencies between stakeholders. In the UK system analysed, there are 14 stakeholders and 37 interdependencies in total. As developers have connections to all other groups and have the most interactions among all stakeholders, they were identified as the critical element. The users are the source element and consultants and contractors are connecting elements. The un-clustered matrix created according to the stakeholder system map is provided in Supplementary Materials (Figure S1). Each stakeholder was then assigned a number for convenience of clustering (Supplementary Materials, Table S2).

The result of the grouping is shown in Figure 5, where all the government authorities, managing companies, developers, owners and acting parties are grouped together. In other words, the Institutional and Adoption and Maintenance groups should be managed together due to mutual interdependencies through information flows. As there are no strong feedbacks from consultancies and construction companies, they cannot be grouped, and hence they have been identified as isolated elements in the system. This is consistent with the fact that in the UK, consultancies and contractors only form a contract with developers and do not take part in the decision-making process. Users were defined as a separate component of a system as well, because their acceptance and behaviour are drivers for the whole SuDS system (more details of clustering are given in the Supplementary Materials, Table S2).

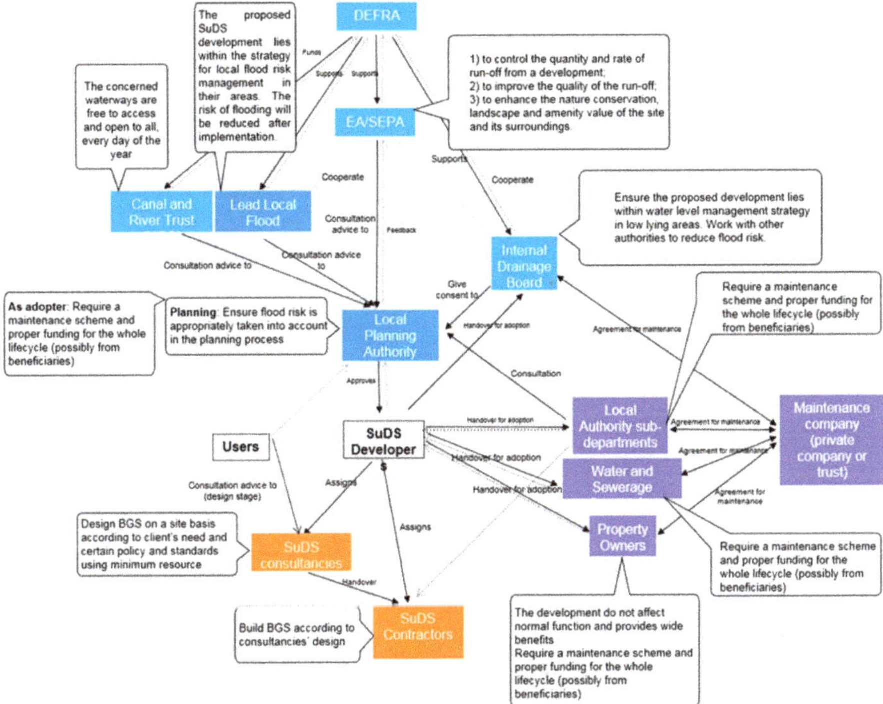

Figure 4. Stakeholders system map for the UK SuDS system (blue colour for the Institutional group, orange colour for the Design and Build group, purple colour for the Adoption and Maintenance group; dotted arrows represent a feedback link). DEFRA, Department for Environment, Food and Rural Affairs; EA, Environment Agency; SEPA, Scottish Environment Protection Agency; BGS, Blue Green Solution.

*S12 11 10 7654321	1	1	
	S8	1	
		S9	
1	1		S14

Figure 5. Results of the N2 method (orange colour for critical elements in the decision-making, i.e., Institutional group and Adoption and Maintenance group, and blue colour for source, i.e., users). S8 and S9 correspond to consultancies that do the designs and the contractors that do the construction. Stakeholder numbers (S1–14) correspond to those defined in Table S2.

3.2. Stakeholder Priorities

The stakeholders selected for the survey were drawn from the results of the N2 method; at least one stakeholder from each group was included in the ANP network (Table 2) to account for the key elements of the SuDS.

Table 2. List of stakeholders who participated in the survey and their associated group according to the N2 results.

Stakeholder No	Position	N2-No	N2-Group
1	Local Authority Engineer	S5	Institutional
2	Economist	S4	Institutional
3	Public Realm Manager	S2	Institutional
4	Head of Maintenance	S13	Adoption and Maintenance
5	Designer	S8	Design and Build
6	Water Utility	S11	Adoption and Maintenance
7	Resident	S14	Source

The priorities of each stakeholder were calculated using the ANP method facilitated by the Superdecision software. Figure 6 illustrates the priorities of each stakeholder as well as the average value, which can be seen as a proxy for a shared perception of SuDS. For a majority of the stakeholders, the most important criteria were economic and technical, while the other criteria such as social ones had lower priority. For all of the stakeholders apart from Stakeholder 7, the most important criterion was the Operating Expenditure (OPEX). This was reasonable considering that the criterion of Durability, which scores high as well, was related directly to the OPEX in the ANP network. The exceptions to this pattern were Stakeholders 6 and 7. More specifically, for Stakeholder 6, the water company, Groundwater Recharge was the second most important criterion since it is a major source of water for them. For Stakeholders 7 and 3, who represent the residents and the environmentalists, respectively, the most important criteria were social, environmental and wider benefits, with the most important criterion being Biodiversity and Ecology.

The results show that not a single criterion was excluded from the stakeholders' scoring, which means that all of them are relevant but at different intensities. The mix of stakeholders surveyed secures a balance between the trend of prioritising the physical benefits of SuDS (such as OPEX) and the social and environmental benefits, especially as Stakeholder 7 is an insolvent stakeholder in the N2 method (S14) with a high weight. This strengthens the generality of the results and increases their credibility as a solid basis on which to build conclusions.

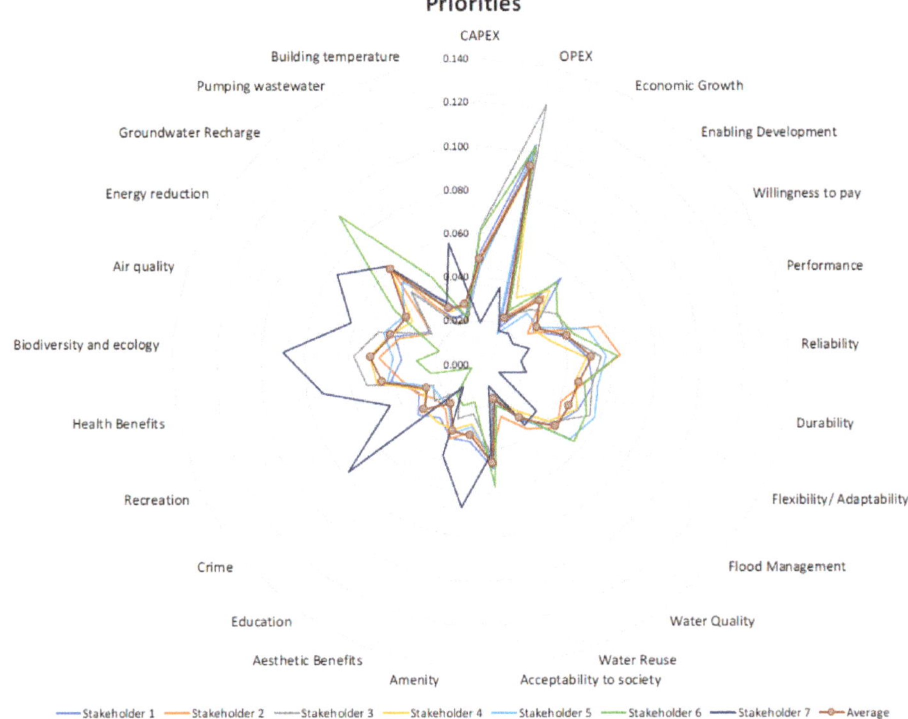

Figure 6. Priorities of each stakeholder and the average value using ANP for the UK SuDS case study. ANP, Analytic Network Process; CAPEX, CAPital Expenditure; OPEX, OPerating EXpense.

3.3. Selection of Modelling Approach

It is clear from Figure 6 that different stakeholders have different interests in SuDS; therefore, it is necessary to understand their interests before beginning any modelling exercise. Stakeholders may have different tolerances with respect to the reliability of the modelling results. For instance, 80% accurate results from a lumped model could be enough for a local council or resident to draw conclusions about SuDS performance. On the other hand, a reliability in model outputs of less than 90% may not be accepted by designers or researchers whose work is used to write guidelines or to construct SuDS where over design would result in unnecessary extra costs and under design may pose a risk of failure.

From the ANP results, it can be seen that the more SuDS benefits are of interest to a certain stakeholder, the less role they have in the actual implementation of SuDS projects and therefore the less modelling complexity they need to make their decisions. This suggests that the simplicity associated with the use of basic models such as conceptual or lumped designs can be acceptable by stakeholders such as residents, local councils and policy makers, because their role in the realisation of SuDS projects is at a high level. For instance, a simple scoring system is sufficient for citizens to learn about SuDS, whereas for environmental regulators and water companies, a more granular models giving high resolution results are needed to determine the level of SuDS performance. Lastly, the most complex physically-based models are used in the technical design of a SuDS when the individual components of the system are designed separately and then integrated to form a constructible and operational design within the existing urban landscape.

Therefore, the decision about what level of complexity should be adopted in SuDS modelling may now be attributed to stakeholders based on their priorities and according to their position in the decision-making process (Table 3). This can help tailor the modelling exercise to better manage the available resources without compromising the influence of the modelling results on stakeholder decision making.

Table 3. Classification of stakeholders involved in SuDS and the desired level of modelling complexity.

Stakeholder	N2-Group	Key Priorities Base on ANP	Model Complexity
Water Utilities	Adoption and Maintenance	Technical and Economical	Complex
Local Councils	Institutional	Economic and Social	Moderate
Policy makers	Institutional	Technical, Economical, Environmental and Social	Moderate
Designers	Design and Build	Technical	Very Complex
Researchers	Design and Build	Technical	Very Complex
Residents	Source	Economic, Environmental and Social	Light
Environmental regulators	Institutional	Environmental and Technical	Complex

Note: Analytic Network Process (ANP).

4. Conclusions

SuDS projects provide a broad array of benefits, affecting diverse stakeholders. This paper addresses the issue of providing a methodological framework to determine the optimal SuDS hydrological modelling approach for the different stakeholders involved in SuDS projects.

The confidence levels of different modelling approaches and the extent to which they fulfil the stakeholders' goals are not well established, and a systematic investigation towards establishing these thresholds is identified as a gap in the literature. Conducting complex and time-consuming modelling exercises are rather more utilised when addressing any stakeholder needs.

Soft System Methodology evaluates the system holistically to characterise the decision-making process and identify potential interventions to ameliorate the system in a loopback approach. The stakeholders' interests in SuDS were determined through a survey questionnaire addressed to a population of the clustered stakeholders. The results of the survey were analysed using the Analytical Network Process to determine their priorities.

With the N2 method it was possible to cluster 14 stakeholders into 4 groups where the Institutional and Adoption and Management groups are managed together, the consultancies and contractors are identified as separate groups and the users as the source group. The ANP results showed that for the stakeholders identified as Institutional group, the key priorities are the economic, social and environmental aspects of the SuDS. Whereas for the Adoption and Maintenance group, such as water companies, the most important aspects were the technical and economic benefits of the SuDS. For the consultancies and researchers, their first worry is the technical aspects of SuDS projects, among others, depending on the nature of a project. Lastly, the residents who are the users of any SuDS intervention are mostly interested in environmental and social benefits, and the economic impact on them.

By presenting a robust systems framework, it is possible to see how analysing the system of SuDS stakeholders may enable future optimisation of implementation tools, starting with better management of efforts utilised in hydrologic modelling of SuDS. The more direct action a stakeholder has in the system the more complex the model he will need, compared to a less complex model needed for a stakeholder who has a minor impact on the decision-making system.

While this study establishes a methodological framework, it is important to apply such methods in a computational case where different models are quantitatively compared and evaluated based on the level of accuracy of the information they provide for a decision-maker in the SuDS system.

Supplementary Materials: The following are available online at http://www.mdpi.com/2073-4441/12/3/632/s1, Questionnaire Survey, Figure S1: Un-clustered initial matrix of stakeholders and the possible links between them, Table S1: SuDS benefits categorised by the selected criteria for stakeholder priorities assessment, Table S2: Stakeholders names, numbers and grouping for the N2 method.

Author Contributions: Conceptualization, M.H.E.H. and A.M.; methodology, M.H.E.H., G.T. and X.R.; software, G.T.; validation, M.H.E.H. and A.M.; formal analysis, G.T., X.R. and M.H.E.H.; investigation, G.T., X.R. and M.H.E.H.; resources, M.H.E.H. and A.M.; data curation, M.H.E.H.; writing—original draft preparation, M.H.E.H., G.T. and X.R.; writing—review and editing, A.M.; visualization, M.H.E.H., G.T. and X.R.; supervision, A.M.; project administration, A.M.; funding acquisition, A.M. All authors have read and agreed to the published version of the manuscript.

Funding: This research was funded by the Center for Doctoral Training (CDT) in Sustainable Civil Engineering at Imperial College London and Thames Water Utilities Limited through a grant from the Engineering and Physical Sciences Research Council (EPSRC), grant number EP/L016826/1.

Acknowledgments: The authors would like to thank the stakeholders who participated in the questionnaire survey and the London Borough of Hammersmith and Fulham for giving us the platform to discuss the goals of this study. Also, the authors would like to thank Zorica Todorovic, from Atkins member of the SNC-Lavalin Group for sharing her thoughts with the authors. Finally, the authors would like to thank the anonymous reviewers whose feedback and comments helped strengthen the paper and make it more robust.

Conflicts of Interest: The authors declare no conflict of interest. The funders had no role in the design of the study; in the collection, analyses or interpretation of data; in the writing of the manuscript or in the decision to publish the results.

References

1. United Nations. The Speed of Urbanization around the World. Population Facts. 2018. Available online: https://www.un.org/en/development/desa/population/publications/pdf/popfacts/PopFacts_2018-1.pdf (accessed on 11 January 2019).
2. Rauch, W.; Seggelke, K.; Brown, R.; Krebs, P. Integrated Approaches in Urban Storm Drainage: Where Do We Stand? *Environ. Manag.* **2005**, *35*, 396–409. [CrossRef]
3. Bamford, T.B.; Balmforth, D.J.; Lai, R.H.H.; Martin, N. Understanding the Complexities of Urban Flooding through Integrated Modelling. In Proceedings of the 11th International Conference on Urban Drainage, Edinburgh, Scotland, UK, 31 August–5 September 2008.
4. Chen, A.S.; Djordjević, S.; Leandro, J.; Savić, D.A. An Analysis of the Combined Consequences of Pluvial and Fluvial Flooding. *Water Sci. Technol.* **2010**, *62*, 1491–1498. [CrossRef]
5. Sto. Domingo, N.D.; Refsgaard, A.; Mark, O.; Paludan, B. Flood Analysis in Mixed-Urban Areas Reflecting Interactions with the Complete Water Cycle through Coupled Hydrologic-Hydraulic Modelling. *Water Sci. Technol.* **2010**, *62*, 1386–1392. [CrossRef]
6. Ward, S.; Farmani, R.; Atkinson, S.; Butler, D.; Hargreaves, A.; Cheng, V.; Denman, S.; Echenique, M. Towards an Integrated Modelling Framework for Sustainable Urban Development. In Proceedings of the 9th International Conference on Urban Drainage Modelling, Belgrade, Serbia, 4–7 September 2012.
7. Bach, P.M.; Rauch, W.; Mikkelsen, P.S.; Mccarthy, D.T.; Deletic, A. Environmental Modelling & Software A Critical Review of Integrated Urban Water Modelling - Urban Drainage and Beyond. *Environ. Model. Softw.* **2014**, *54*, 88–107. [CrossRef]
8. Kroll, S.; Fenu, A.; Wambecq, T.; Weemaes, M.; Van Impe, J.; Willems, P. Energy Optimization of the Urban Drainage System by Integrated Real-Time Control during Wet and Dry Weather Conditions. *Urb. Water J.* **2018**, *15*, 362–370. [CrossRef]
9. Woods-Ballard, B.; Kellagher, R.; Martin, P.; Jefferies, C.; Bray, R.; Shaffer, P. *The SUDS Manual*; CIRIA: London, UK, 2015.
10. Lai, E.; Lundie, S.; Ashbolt, N.J. Review of Multi-Criteria Decision Aid for Integrated Sustainability Assessment of Urban Water Systems. *Urb. Water J.* **2008**, *5*, 315–327. [CrossRef]
11. Thampapillai, D.J. *Environmental Economics*; Oxford University Press: Melbourne, Australia, 1991.

12. Pearce, D.; Atkinson, G.; Mourato, S. *Cost Benefit Analysis and the Environment: Recent Developments*; Organisation for Economic Co-operation and Development: Paris, French, 2006.

13. Arrow, K.J.; Cropper, M.L.; Eads, G.C.; Hahn, R.W.; Lave, L.B.; Noll, R.G.; Stavins, R.N. Is there a role for benefit-cost analysis in environmental, health, and safety regulation? *Environm. Dev. Econom.* **1997**, *2*, 195–221. [CrossRef]

14. Ossa-Moreno, J.; Smith, K.M.; Mijic, A. Economic Analysis of Wider Benefits to Facilitate SuDS Uptake in London, UK. *Sustain. Cities Soc.* **2017**, *28*, 411–419. [CrossRef]

15. Liu, Y.; Bralts, V.F.; Engel, B.A. Evaluating the Effectiveness of Management Practices on Hydrology and Water Quality at Watershed Scale with a Rainfall-Runoff Model. *Sci. Total Environ.* **2015**, *511*, 298–308. [CrossRef]

16. Resource Assessment Commission. *Multi-Criteria Analysis as a Resource Assessment Tool*; Australian Government Publishing Service: Canberra, Australia, 1992.

17. Belton, V.; Stewart, T.J. *Multiple Criteria Decision Analysis: An Integrated Approach*; Kluwer Academic Publishers: Boston, NY, USA, 2005.

18. Hajkowicz, S.; Collins, K. A review of multiple criteria analysis for water resource planning and management. *Water Resour. Manag.* **2007**, *21*, 1553–1566. [CrossRef]

19. Ellis, J.B.; Deutsch, J.C.; Mouchel, J.M.; Scholes, L.; Revitt, M.D. Multicriteria Decision Approaches to Support Sustainable Drainage Options for the Treatment of Highway and Urban Runoff. *Sci. Total Environ.* **2004**, *334–335*, 251–260. [CrossRef] [PubMed]

20. Elkington, J. *Cannibals with Forks: The triple Bottom Line of 21st Century Business*; New Society Publishers: Gabriola Island, BC, Canada, 1998.

21. Viavattene, C.; Ellis, J.B. The Management of Urban Surface Water Flood Risks: SUDS Performance in Flood Reduction from Extreme Events. *Water Sci. Technol.* **2013**, *67*, 99–108. [CrossRef] [PubMed]

22. Parker, P.; Letcher, R.; Jakeman, A.; Beck, M.B.; Harris, G.; Argent, R.M.; Sullivan, P. Progress in integrated assessment and modelling. *Environm. Model. Softw.* **2002**, *17*, 209–217. [CrossRef]

23. Brouwer, R.; Georgiou, S.; Turner, R.K. Integrated assessment and sustainable water and wetland management. A review of concepts and methods. *Integr. Assess.* **2003**, *4*, 172–184. [CrossRef]

24. Saaty, T.L. —Dependence and Feedback in Decision-Making with a Single Network. *J. Syst. Sci. Syst. Eng.* **2004**, *13*, 129–157. [CrossRef]

25. Lee, J.W.; Kim, S.H. Using Analytic Network Process and Goal Programming for Interdependent Information System Project Selection. *Comput. Oper. Res.* **2000**, *27*, 367–382. [CrossRef]

26. Boran, S.; Goztepe, K. Development of a Fuzzy Decision Support System for Commodity Acquisition Using Fuzzy Analytic Network Process. *Expert Syst. Appl.* **2010**, *37*, 1939–1945. [CrossRef]

27. Nawaz, S.; Ali, Y. Factors Affecting the Performance of Water Treatment Plants in Pakistan. *Water Conserv. Sci. Eng.* **2018**, *3*, 191–203. [CrossRef]

28. Molinos-Senante, M.; Gómez, T.; Caballero, R.; Hernández-Sancho, F.; Sala-Garrido, R. Assessment of Wastewater Treatment Alternatives for Small Communities: An Analytic Network Process Approach. *Sci. Total Environ.* **2015**, *532*, 676–687. [CrossRef]

29. Wey, W.M.; Wei, W.L. Urban Street Environment Design for Quality of Urban Life. *Soc. Indic. Res.* **2016**, *126*, 161–186. [CrossRef]

30. Zografidou, E.; Petridis, K.; Arabatzis, G.; Dey, P.K. Optimal Design of the Renewable Energy Map of Greece Using Weighted Goal-Programming and Data Envelopment Analysis. *Comput. Oper. Res.* **2016**, *66*, 313–326. [CrossRef]

31. TWUL. Why We Need the Counters Creek Storm Relief Sewer. 2013. Available online: https://www.thameswater.co.uk/sitecore/content/counterscreek/counterscreek/theproblem/why-we-need-the-storm-relief-sewer (accessed on 12 March 2018).

32. Hattab, M.E.; Vernon, D.; Mijic, A. Performance Evaluation of retrofitted low impact development practices in urban environments: A case study from London, UK. In Proceedings of the International Conference on Sustainable Infrastructure 2017: Technology, New York, NY, USA, 26–28 October 2017. [CrossRef]

33. Ashley, R.; Woods-Ballard, B.; Shaffer, P.; Wilson, S.; Illman, S.; Walker, L.; D'Arcy, B.; Chatfield, P. UK Sustainable Drainage Systems: Past, Present and Future. *Proc. Inst. Civ. Eng.-Civ. Eng.* **2015**, *168*, 125–130. [CrossRef]

34. Alves, A.; Gómez, J.P.; Vojinovic, Z.; Sánchez, A.; Weesakul, S. Combining Co-Benefits and Stakeholders Perceptions into Green Infrastructure Selection for Flood Risk Reduction. *Environments* **2018**, *5*, 29. [CrossRef]

35. Thorne, C.R.; Lawson, E.C.; Ozawa, C.; Hamlin, S.L.; Smith, L.A. Overcoming Uncertainty and Barriers to Adoption of Blue-Green Infrastructure for Urban Flood Risk Management. *J. Flood Risk Manag.* **2018**, *11*, S960–S972. [CrossRef]

36. O'Donnell, E.C.; Lamond, J.E.; Thorne, C.R. Recognising Barriers to Implementation of Blue-Green Infrastructure: A Newcastle Case Study. *Urban. Water J.* **2017**, *14*, 964–971. [CrossRef]

37. Gasson, S. *The Use of Soft Systems Methodology (SSM) as a Tool for Investigation*; Warwick Business School: Coventry, UK, 1994.

38. Burge, S. The Systems Engineering Tool Box. Available online: https://www.burgehugheswalsh.co.uk/Uploaded/1/Documents/Functional-Modelling-Tool-Draft.pdf (accessed on 8 June 2017).

39. Bustnay, T.; Ben-Asher, J.Z. How Many Systems Are There?—Using the N2method for Systems Partitioning. *Syst. Eng.* **2005**, *8*, 109–118. [CrossRef]

40. Bird, D.K. The Use of Questionnaires for Acquiring Information on Public Perception of Natural Hazards and Risk Mitigation—A Review of Current Knowledge and Practice. *Nat. Hazards Earth Syst. Sci.* **2009**, *9*, 1307–1325. [CrossRef]

41. Saaty, T.L.; Takizawa, M. Dependence and Independence: From Linear Hierarchies to Nonlinear Networks. *Eur. J. Oper. Res.* **1986**, *26*, 229–237. [CrossRef]

42. Grimaldi, M.; Pellecchia, V.; Fasolino, I. Urban Plan and Water Infrastructures Planning: A Methodology Based on Spatial ANP. *Sustainability* **2017**, *9*, 771. [CrossRef]

43. Creative Decisions Foundation. Super Decision CDF. Available online: https://superdecisions.com/ (accessed on 10 July 2017).

44. Meade, L.M., Sarkis, J. Analyzing Organizational Project Alternatives for Agile Manufacturing Processes: An Analytical Network Approach. *Int. J. Prod. Res.* **1999**, *37*, 241–261. [CrossRef]

45. Saaty, R.W. *Decision Making in Complex Environment: The Analytic Hierarchy Process (AHP) for Decision Making and the Analytic Network Process (ANP) for Decision Making with Dependence and Feedback*; Super Decisions: Pittsburgh, PA, USA, 2003.

Review

Uncertainty Quantification in Water Resource Systems Modeling: Case Studies from India

Shaik Rehana [1], Chandra Rupa Rajulapati [2], Subimal Ghosh [3], Subhankar Karmakar [4] and Pradeep Mujumdar [5,*]

1. Lab for Spatial Informatics, International Institute of Information Technology Hyderabad, Hyderabad 500032, India; rehana.s@iiit.ac.in
2. Centre for Hydrology, University of Saskatchewan, Saskatoon, SK S7N1K2, Canada; chandra.rajulapati@usask.ca
3. Department of Civil Engineering, Indian Institute of Technology Bombay, Mumbai 400076, India; subimal@civil.iitb.ac.in
4. Environmental Science and Engineering Department, Indian Institute of Technology Bombay, Mumbai 400076, India; skarmakar@iitb.ac.in
5. Department of Civil Engineering, Interdisciplinary Centre for Water Research, Indian Institute of Science Bangalore 520011, India
* Correspondence: pradeep@iisc.ac.in; Tel.: +91-80-2360-0290

Received: 30 April 2020; Accepted: 18 June 2020; Published: 23 June 2020

Abstract: Regional water resource modelling is important for evaluating system performance by analyzing the reliability, resilience and vulnerability criteria of the system. In water resource systems modelling, several uncertainties abound, including data inadequacy and errors, modeling inaccuracy, lack of knowledge, imprecision, inexactness, randomness of natural phenomena, and operational variability, in addition to challenges such as growing population, increasing water demands, diminishing water sources and climate change. Recent advances in modelling techniques along with high computational capabilities have facilitated rapid progress in this area. In India, several studies have been carried out to understand and quantify uncertainties in various basins, enumerate large temporal and regional mismatches between water availability and demands, and project likely changes due to warming. A comprehensive review of uncertainties in water resource modelling from an Indian perspective is yet to be done. In this work, we aim to appraise the quantification of uncertainties in systems modelling in India and discuss various water resource management and operation models. Basic formulation of models for probabilistic, fuzzy and grey/inexact simulation, optimization, and multi-objective analyses to water resource design, planning and operations are presented. We further discuss challenges in modelling uncertainties, missing links in integrated systems approach, along with directions for future.

Keywords: reservoir operation; stochastic dynamic programming; fuzzy optimization; reservoir-river system; water quantity-quality management; climate change

---

1. Introduction

Water resource management is about the integration of various disciplines of hydrology for the planning, management and optimum utilization of water resources following the competing needs and demands of society. Integrated water resource management consists of four dimensions: (i) natural element of water resources, considering the entire hydrological cycle and various components of it such as rainfall, water in rivers, etc.; (ii) water users and stakeholders, including socioeconomic interests; (iii) variability of water resources and users, such as spatial mismatch of water availability between upstream and downstream river plains; (iv) temporal variability of water availability and demands [1].

Globally regional hydrologic systems have long struggled for many decades with the planning and management of water resources under growing population, increasing demands and climate change [2]. River water resource systems are under great stress as a result of unsustainable consumption patterns and poor management practices [3]. The need for a regional water resource management model accounting for water availability and demands, water quantity and quality has become prominent in recent years under climate signals [4]. Based on several scientific studies, climate change is likely to affect various subsystems of regional water resource systems, such as water availability for consumer needs and food production, irrigation water demands, hydropower, water quality, etc., under an increase in temperatures and changes in precipitation patterns [5]. Climate change has been identified as one of the major driving forces in regional water resource systems management by several studies globally [6–8] and in India [3,9,10].

A regional water resource management model is an integration of a water quantity and quality estimation model, a water demand estimation model along with a decision making model [11]. For instance, a hydrological model is used to estimate the water availabilities in terms of inflows; demand estimation models estimate factors such as drinking, irrigation and hydropower; water quantity and reservoir operation models are used to estimate the optimal release policy and water allocations of reservoir users; and water quality management models are used to estimate the optimal treatment policies (Figure 1). Figure 1 shows a single reservoir–river system accounting for the upstream catchment flows, evapotranspiration, overland flows, infiltration and in the downstream side water withdrawals and return flows from irrigation. An integrated operation of reservoirs of a complete river system, starting from the furthest upstream reservoir to the furthest downstream reservoir, should include inflows to each reservoir, evaporation losses, power draft, releases, withdrawals and overflows in the reservoir operation. While simulating and integrating the reservoir operation of major river systems, the inflow to any particular reservoir should account for uncontrolled intermediate catchment flows, irrigation return flows and controlled flows from the upstream reservoir [12].

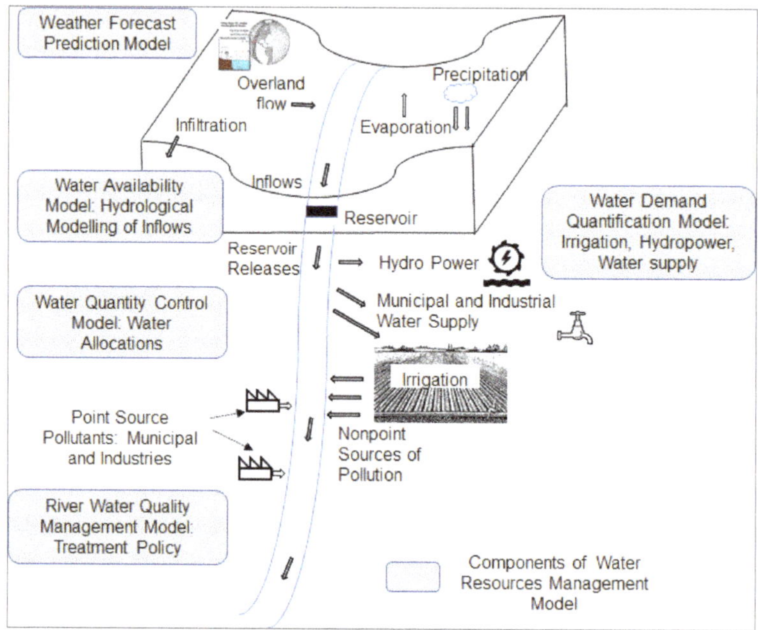

Figure 1. An integrated regional water resource system management model.

Integrated regional water resource management models have evolved to secure water resource systems at the basin scale in terms of water quantity and quality, accounting for water availability and the demands of various users [1]. In this context, water resource systems models have been evolved in the past four decades in several aspects of single and multi-purpose reservoirs, optimization models, knowledge-based decisions, real time operations, imprecision and uncertainty quantifications and climate change [13]. Several review papers have articulated the evolution of water resource management systems modeling, focusing on several key aspects in terms of optimization models, and concluded the research gaps between the developed models [11,14,15]. One of the pioneering review papers on reservoir systems analysis was by Simonovic [16], in which the gaps between research studies and application of systems approach in practice were discussed and an optimization model for reservoir sizing and the inclusion of knowledge-based technologies in single-multipurpose reservoir analysis was recommended. Furthermore, most of the earlier review papers articulated on the evolution of water resource management modeling at basin scale [17] integrated water resource optimization models [18,19]. Very recent review studies focused on the application of evolutionary algorithms and metaheuristic optimizations for optimal strategies of the planning and management of water resource systems [20–24]. In this context, Mohammad-Azari et al. [23] have reviewed the application of Genetic Programming to solve water resource systems analysis and stressed on the capability and superiority of evolutionary algorithms in solving reservoir operation problems. Few earlier review papers focused on reservoir operation challenges related to inflows [25], simulation and optimization techniques [26].

The integrated regional water resource management models are associated with various forms of uncertainties accumulating from various stages of decision making [27]. Uncertainties arise at each stage of the modelling and decision-making process due to random nature of input variables, various parameters and models, imprecise goals of the users, priorities and social importance in decision making by various stakeholders. Addressing these uncertainties is very important for precise decision making and to avoid the failure of water resource system management [16]. The inclusion of uncertainties of reservoir inflows in the water resource systems models was one of the basic studies and have implemented by several researchers by considering inflow as stochastic variable [28]. The next prevailing uncertainty in reservoir operation is imprecise goals of the users, which has been conventionally addressed using fuzzy set theory [29]. Identifying and addressing various sources of uncertainties is one of the crucial tasks in water resource modelling to have better operating policies with more dependability and flexibility in decision making. Review papers which can articulate various studies of water resource management and associated uncertainties are limited in the literature. Ahmad [15] reviewed reservoir operation models with fuzzy optimization along with other optimization methods such as Artificial Neural Network (ANN), Genetic Algorithm (GA), artificial bee colony and Gravitational Search Algorithm (GSA). A comprehensive review which can include the uncertainty quantification in water resource systems modeling, various approaches so far applied, research gaps and challenges is lacking in the literature. In this article, we review water resource management systems models to address various sources of uncertainties by highlighting key findings and identify important future research directions which can improve the understanding of water resource planning and management.

India has large regional mismatches between water availability and demands, with increasing withdrawals from surface and subsurface sources rising to unsustainable conditions [30]. India is an agriculture-dominated country and about 70% of the population's employment and economy depends on agriculture sector. The timely supply of irrigation water with sufficient quantity is challenging given the spatial and temporal mismatches of river water availabilities, increasing drinking and industrial water demands under population growth and pressure to increase crop yields. The determination of optimal water allocations for various sectors to fulfill various demands is of primary interest for most of the reservoirs of India. Tremendous population growth, rapid urbanization, alterations in agricultural patterns, unplanned growth of industries and failure of maintaining the environmental standards are the major causes for poor river water quality systems in India [31]. The present research article explores some of the Indian case studies carried out in the field of water resource management and uncertainty quantification. Sources and approaches to address uncertainties in the context of water resource management and modelling are discussed with a focus on Indian case studies. Furthermore, missing links in modelling, challenges remaining and future directions are noted.

2. Reservoir Operation and Associated Uncertainties

Reservoir operation has gained attention in water resource engineering for more than four decades [32]. The reservoir operation systems models vary according to various components of consideration such as drinking water supply, irrigation, hydropower, low-flow augmentation, aquaculture, navigation along with flood control and management. Fundamentally, a regional water resource systems management model is an integration of a reservoir operation model to define the possible releases following the storage continuity equation and an optimization model to define the optimal water allocation policies, accounting for the conflicting goals of the reservoir users and possible demands (Figure 2).

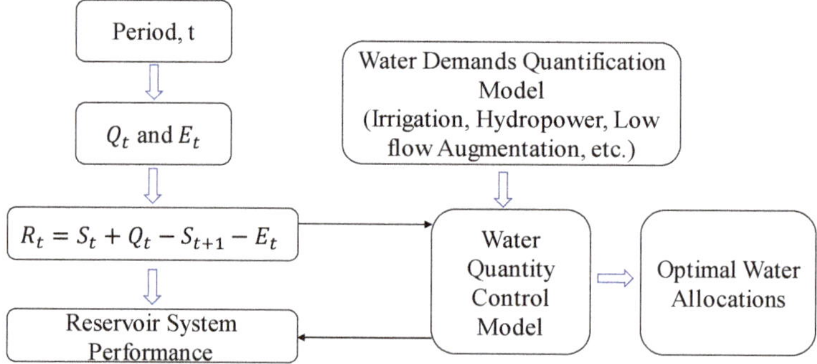

Figure 2. Water quantity control model of a reservoir.

For a given period of time, t, the inflows to the reservoir (Q_t) and evaporation loss from the reservoir during the period, t (E_t), storage at the beginning of the period (S_t), storage at the end of the period (S_{t+1}), the continuity equation forms the basis for the determination of the possible releases (Figure 2). The release during the period, t, R_t, is the decision variable, with storage at the beginning of the period, t, S_t, as the state variable in the reservoir optimization model with objective as to maximize the total net benefit, $B_t(S_t, R_t)$, during a year T [33]:

$$\text{Maximize} \sum_{t=1}^{T} B_t(S_t, R_t) \tag{1}$$

$$0 \leq R_t \leq S_t + Q_t \tag{2}$$

$$S_t + Q_t - R_t \leq K \tag{3}$$

Equations (2) and (3) represent constraints over the possible release, R_T, restricting it to the total water available in storage in period t (Equation (2)), and the end of period storage (S_{t+1}) is restricted to the live storage capacity, (K) (Equation (3)). In general, the optimization model has to be solved recursively until it yields a steady state policy within a few annual cycles [3]. In this single objective reservoir operation model, the most influential variable for optimal release is the reservoir inflows (water available for release) and it is highly uncertain, due to the upstream catchment rainfall uncertainty and other basin characteristics. In addition, other hydrological variables such as evapotranspiration, soil moisture, ground water flows, etc. which define crop water demands in the downstream command area are also burdened with uncertainty due to randomness which can cause stochastic or aleatory uncertainty in the reservoir operation [28].

In this context, various studies considered the input variables of a reservoir operation model as having a random nature and explicitly included in the optimization model through their probability distributions [34]. The hydrologic variable uncertainty due to randomness has been addressed by various authors by considering the reservoir inflow to follow a one-step Markov process through transition probabilities over Indian case studies [35–37]. Conventionally, the uncertainty due to the randomness of inflows in reservoir operation is addressed by applying stochastic dynamic programming (SDP) [32]. In one of the pioneering works by Vedula and Mujumdar [38], a reservoir operation model based on SDP was developed to find the optimal water allocations for irrigation under multiple crops scenarios, where reservoir storage, inflows, and soil moisture are treated as state variables in the decision-making process for Malaprabha reservoir, Krishna basin, Karnataka state, India. Ravikumar and Venugopal [39] developed an optimal operation model using simulation and SDP combination, where both demand and inflow are considered as stochastic and both are assumed to follow first order Markov chain model, which is demonstrated with the Periyar Vaigai irrigation system as one of the typical south Indian irrigation systems of India. Mujumdar and Kumar [12] developed an integrated reservoir operation tool for providing the operation of the eight major reservoirs of Narmanda river basin, India. The study developed a simulation model with eight major reservoirs, viz., Matiyari, Bargi, Barna, Tawa, Indira Sagar, Omkareshwar, Maheshwar and Sardar Sarvovar, as shown in Figure 3. A computer simulation model was developed starting from the furthest upstream reservoir (Matiyari) to the furthest downstream reservoir (Sardar Sarovar) by accounting for inflows to the reservoir (including uncontrolled intermediate catchment flows, irrigation return flows and controlled flows from upstream reservoirs), evaporation losses, power draft, releases, withdrawals and overflows during every period until the end of simulation.

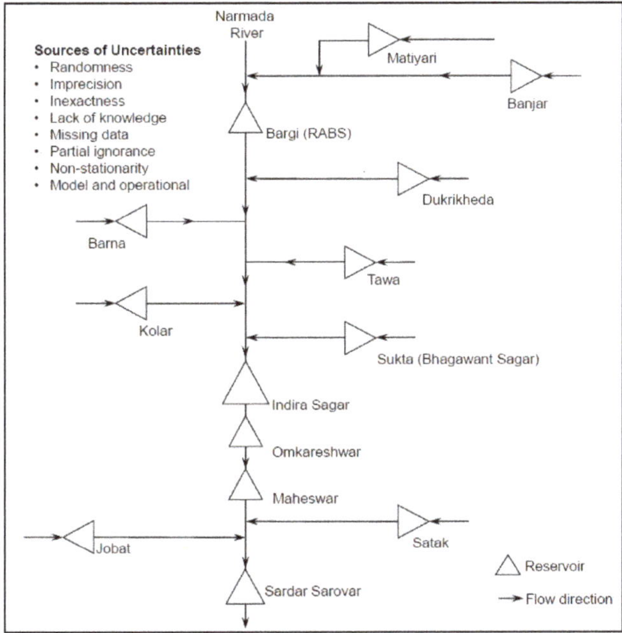

Figure 3. Integrated reservoir operation with major and medium reservoirs in Narmada river basin, India, modified from Mujumdar and Nagesh Kumar [12].

Reservoir operation based on SDP has emerged as a promising tool to address the uncertainty of reservoir input variables for various case studies of India, namely Hirakud reservoir [3,40], Malabrapha reservoir [38], Bhadra reservoir [4,37], Ukai reservoir [41], Kodaiyar Basin [42]. In recent years, various forms of SDP, such as folded dynamic programming [34], two-phase stochastic dynamic programming [43], deterministic dynamic programming (DDP) [44], etc., have become popular for reservoir operation and management in India based on the abilities to improve the net benefit and to overcome the dimensionality issues of SDP.

The reservoir operation stakeholders are often uncomfortable with sophisticated optimization techniques, and need flexibility in specifying the goals and decision making, which causes uncertainty due to imprecision in water resource systems models [45]. Fuzzy logic was identified as an appropriate tool to address the uncertainty due to imprecision in defining the goals of the stakeholders [29]. In this context, fuzzy water allocation models to address the uncertainty due to imprecision in defining the goals of the reservoir users has been widely used all over the world [46] and in India [47–49]. A typical fuzzy optimization model for reservoir operation works on specifying the goals of the users as fuzzy membership functions and the mathematical formulation of a typical reservoir operation as a water quantity control model, following Rehana and Mujumdar [4], which can be expressed as follows:

$$\text{Maximize } \lambda \tag{4}$$

Subject to

$$f(q_\alpha) \geq \lambda \tag{5}$$

$$f\left(q_\beta\right) \geq \lambda \tag{6}$$

$$f(q_\chi) \geq \lambda \tag{7}$$

$$q_\alpha^{Min} \leq q_\alpha \leq q_\alpha^{D} \tag{8}$$

$$q_\beta^{Min} \leq q_\beta \leq q_\beta^{D} \tag{9}$$

$$q_\chi^{Min} \leq q_\chi \leq q_\chi^{D} \tag{10}$$

$$q_\alpha + q_\beta + q_\chi \leq W_A \tag{11}$$

$$0 \leq \lambda \leq 1 \tag{12}$$

where W_A is the amount of water available for allocation, which is the reservoir release, Rt, for a given time period, t, from the reservoir operation model (Equations (1)–(3)) (Figure 2). The solution of the resulting optimization problem will be q^* and λ^* where $q^* = \{q_\alpha^*, q_\beta^*, q_\chi^*\}$ corresponds to the optimum water allocation among the water users; viz., irrigation (α), water quality (β) and hydropower (χ), and λ^* is the maximized minimum satisfaction level in the system. The imprecise goals of reservoir users will be represented using membership functions such as $f(q_\alpha), f\left(q_\beta\right)$ and $f(q_\chi)$ for irrigation, water quality and hydropower, respectively. For each reservoir user, the minimum (q_α^D, q_β^D and q_χ^D) and desirable (q_α^D, q_β^D and q_χ^D) limits will be specified. The fuzzy optimization model works with the objective function so as to maximize the minimum satisfaction level (Equation (4)) including the imprecise goals of the reservoir users as fuzzy membership functions (Equations (5)–(7)), constraints over water allocations to be within minimum and maximum limits (Equations (8)–(10)), constraints over the total water available for allocation (Equation (11)) and constraints over the satisfaction level ranging between 0 and 1 (Equation (12)).

In this context, fuzzy rule-based reservoir operation models have gained interest to address the uncertainty due to impression in specifying the goals of various reservoir users [50]. Many researchers adopted fuzzy optimization models for optimum water quantity allocations in reservoir operation [45,48,49,51]. A fuzzy rule-based model was developed by Panigrahi and Mujumdar [45] for Malaprabha irrigation reservoir in Karnataka, a single purpose reservoir, where the fuzzy membership functions have been constructed for inflow, storage, demand and release. The inclusion of fuzzy membership in reservoir operation can address the uncertainty due to imprecision but not the uncertainty due to the randomness of input variables. Therefore, the integration of SDP with fuzzy optimization for optimal reservoir operation has become a promising tool for the development of long-term operating policies in recent years [4,52]. These models are advantageous to address the uncertainty due to the randomness of reservoir inflows by applying SDP and due to the imprecision in specifying the goals of the stakeholders by applying fuzzy optimizations. In this context, a water quantity modelling method was developed by integrating SDP and fuzzy optimization model by Rehana and Mujumdar [4]. This model addresses uncertainty due to randomness and fuzziness combinedly in developing long-term operating policies which has been implemented on Bhadra reservoir, India (Figure 2). Furthermore, such a SDP-fuzzy model was extended by Kumari and Mujumdar [52] for Bhadra reservoir by considering the state variables of reservoir storage and soil moisture as fuzzy variables and reservoir inflow as a random variable in modelling reservoir operation using SDP. By considering the state variables as fuzzy variables in the formulation of SDP, uncertainty due to imprecision originating due to consideration of single representative value of the state variable can be addressed. One of the improvements in the developed model can be the consideration of rainfall and potential evapotranspiration also as stochastic variables along with reservoir inflows, but not as deterministic as considered in the study of Kumari and Mujumdar [52]. In another study by Kumari and Mujumdar [53], a fuzzy set-based performance measure for irrigation reservoir system in terms of fuzzy reliability, fuzzy resilience and fuzzy vulnerability to study the failure/success state of a reservoir system was developed by relating evapotranspiration deficit of the crops and applied on Bhadra reservoir system, Karnataka, India. To this end, the fuzzy-SDP reservoir operation models have advanced in several means in addressing uncertainties of probabilistic and imprecision combinedly. In this context, a few attempts have also been made by adopting fuzzy Markov chain-based SDP models to address the probabilistic and fuzzy uncertainty at the same time by introducing the concept of distribution with fuzzy probability to develop a fuzzy-Markov-chain-based SDP (e.g., [54,55]).

3. River Water Quality Management under Uncertainties

A water quality management model is essentially an integration of water quality simulation model and an optimization model to manage the quality of river systems without violating the standards specified by the pollution control agencies. A river water quality control model is necessarily a decision-making process to maintain the ecological stability of the riverine environment involving the pollution control boards (PCBs) and effluent dischargers. In this context, Waste Load Allocation (WLA) models have been evolved for determining the required treatment levels or fractional removal levels for various point and non-point sources of pollutants accounting for the water quality standards specified by PCBs in an economically efficient manner. Majorly, WLA models run with the integration of a river water quality simulation model and an optimization model dealing with the goals of dischargers and pollution control boards [56]. In this context, a surface water quality model is a tool for the better understanding of the mechanisms and interactions between anthropogenic residual inputs and resulting water quality [57]. Water quality simulation models run by accounting for river hydrology and hydraulic variables (streamflow, longitudinal slope, Manning's coefficient, etc.), river water quality parameters (dissolved oxygen (DO), biochemical oxygen demand (BOD), nitrates, temperature, etc.), climate data (air temperature, wind speed, etc.), effluent discharge characteristics (pollutant DO, BOD, temperature, etc.) to simulate the river water quality indicators along the river stretch under consideration [57,58]. Meanwhile, an optimization model considers the resulting water quality for a given pollutant loading along with the goals of the PCBs and industries releasing the effluents [59].

The input variables, such as streamflow, temperature, etc., of water quality simulation models are random variables and therefore are associated with uncertainty due to their randomness [60]. Conventionally, the uncertainty due to randomness in the river water quality variables has been addressed using probabilistic mathematical programming techniques [61]. Another major source of uncertainty is associated with the imprecise goals of the dischargers and PCBs, which is usually addressed by fuzzy membership functions to represent the satisfaction levels of the users by most of the Indian authors [56,59,62,63].

The pioneering work in the fuzzy river water quality management models was by Sasikumar and Mujumdar [56]. The study developed a Fuzzy Waste Load Allocation Model (FWLAM), addressing the uncertainty due to imprecision in specifying the goals of the dischargers and PCBs. In this context, FWLAMs were evolved to address the uncertainty due to imprecision in specifying the goals of the stakeholders [56] and fuzzy risk minimization waste load allocation model to address uncertainty due to the combined randomness of input variables and fuzziness of decision makers' requirements [64]. The mathematical formulation of a typical river water quality management model can be expressed following Sasikumar and Mujumdar [56] as follows:

$$\text{Maximize } \lambda \tag{13}$$

Subject to

$$f(C_l) \geq \lambda \tag{14}$$

$$f(x_m) \geq \lambda \tag{15}$$

$$C_l^L \leq C_l \leq C_l^D \tag{16}$$

$$x_m^L \leq x_m \leq x_m^D \tag{17}$$

$$0 \leq \lambda \leq 1 \tag{18}$$

where C_l is the concentration level of water quality parameter at check point, l; x_m is the fraction removal level for discharger, m; C_l^L and C_l^D are the minimum and maximum permissible levels set by PCBs, respectively; x_m^L and x_m^D are the minimum and maximum possible treatment levels specified by the dischargers, respectively; λ as the satisfaction level of PCBs and dischargers. $f(C_l)$ and $f(x_m)$ represent the membership functions of PCBs and dischargers, respectively. The solution of the resulting optimization problem will be $x_m{}^*$ and λ^* where $x_m{}^*$ corresponds to optimum fraction removal level for each discharger and λ^* is the maximized minimum satisfaction level in the system.

Some improvements in water quality management models were made by Singh et al. [65] by developing an interactive fuzzy multi-objective linear programming model to evaluate optimal treatment efficiencies for various drains located along Yamuna across New Delhi, India. The study allotted weights for DO deficits at each grid point to address the uncertainty in specifying the goals of the decision makers with continuous interaction with decision makers.

Many studies considered risk of low water quality (LWQ) as one of the criteria to represent the goal of the PCBs [59,63,64]. By considering this risk in the river water quality management models, uncertainty due to a combination of randomness in the water quality concentrations along with imprecision in defining the standards was addressed. The risk of LWQ is defined as the probability of a fuzzy event of LWQ [64]. The conventional definition of LWQ is any concentration less than a specified value, say, c_l^{min}, the minimum permissible level at check point, l. The crisp definition of risk of LWQ, with a water quality indicator as DO, is given as:

$$r_l = P(c_l < c_l^{min}) \tag{19}$$

where r_l is the risk of LWQ at check point, l; c_l is the DO level at check point, l; c_l^{min} is the minimum permissible level of DO at check point, l; $P(c_l < c_l^{min})$ is the probability associated with the occurrence of the LWQ event. The fuzzy risk of LWQ is defined as the probability of occurrence of the fuzzy event of LWQ. Fuzzy risk can be expressed as the expected degree of failure [64].

$$r_{il} = \int_0^\infty \mu_{wil}(c_{il}) \, f(c_{il}) \, dc_{il} \tag{20}$$

where $\mu_{wil}(c_{il})$ is the membership function of the fuzzy set, W_{il} of LWQ and $f(c_{il})$ is the probability density function (PDF) of the concentration level, c_{il}, for water quality indicator, i, at the checkpoint, l in the river system. Based on the PDF, $f(c_{il})$ of the LWQ indicator, i, and the membership function $\mu_{wil}(c_{il})$ of the fuzzy set, W_{il}, of LWQ, direct or numerical integration may be performed to evaluate the fuzzy risk, r_{il}.

Sasikumar and Mujumdar [64] developed a fuzzy risk approach to address both uncertainty due to randomness and uncertainty due to imprecision of the goals by considering the probability of risk of LWQ as fuzzy event. The study was implemented over the Tunga-Bhadra river stretch, India to estimate optimal fractional removal levels of the dischargers. A fuzzy risk minimization model was solved by Ghosh and Mujumdar [63] to minimize the risk of LWQ using a non-linear optimization model of Probabilistic Global Search Laussane applied to the Tunga-Bhadra river system, India.

In a conventional fuzzy optimization model, the membership parameters are assumed to be fixed and values are assigned based on experience and judgement and are thus highly subjective; for instance, the lower bound of DO is assigned as 5 mg/L and the upper bound is 8 mg/L. In general, such membership parameters are defined based on the minimum and maximum permissible levels of water quality standards, which may vary for each criterion such as public water supply, agricultural and industrial water supplies, etc. [62]. This results in uncertainty in the membership parameters, which can be considered as the next level of fuzziness in the fuzzy optimization models [62]. To address the uncertainty in the membership parameters, Karmakar and Mujumdar [62] developed a grey fuzzy waste load allocation model by considering the membership parameters as interval grey numbers to represent as imprecise membership function. The study was implemented over Tunga-Bhadra river system, India, by considering the imprecise fuzzy membership functions, which provided the optimal treatment policy and satisfaction levels, both in the form of interval numbers, allowing the decision-maker to select various alternatives required in a particular situation. A conventional approach to solve grey optimization models is the two-step sub model method [62,66,67], which bifurcates the parent uncertain model into two daughter models, one for the least favorable case and another for the most favorable case. However, Rosenberg [68] and Yadav et al. [69] found issues such as infeasibility, non-optimality and fat solutions in the two-step method. Any derived problem of an interval/grey model by fixing a deterministic value of available interval numbers is known as the subproblem of the parent model [70] or a deterministic equivalent of the interval/grey model. If the extreme optimum solutions of all such subproblems have significant differences with the solutions obtained from a given technique (two-step method in this case), then the solutions are known as fat solutions [71], which necessarily implies a set of very uncertain outputs.

Huang and Cao [72] further developed a three-step method to resolve the infeasibility of the solution in the two-step method, but made the issue of non-optimality more severe [73]. Yadav et al. [71] proposed an interval-valued integer programming model based on interval analysis to overcome the issues of two-step and three-step methods. Algorithms based on interval analysis are computationally more rigorous than grey analysis, but pave the way for an effective and powerful methodology to quantify the inexact or grey uncertainty. Therefore, interval analysis-based scalable algorithms have the potential to make conventional uncertainty quantification techniques such as probabilistic or fuzzy redundant. In this context, 'Imprecision' is a representation of disjunctive information, which is characterized by a set of possible values for which the actual values are known to exist [74]. This characteristic of ordered disjunctive information has been incorporated in the Fuzzy Set Theory. As per the literature of fuzzy mathematics, the 'imprecision' is analogous to 'vagueness,' [75], which is a linguistic uncertainty and is often represented with fuzzy membership functions. On the other hand, 'Inexactness' is another representation of uncertainty when the exact value is unknown; however, the range within this value exists is known [69]. The concept of inexact uncertainty is relatively new and is extensively used in grey/interval systems. Inexactness may be represented with interval grey numbers, where lower and upper bounds are known, but the distribution information is unknown.

Another source of uncertainty is partial ignorance resulting from missing or inadequate data in a time series of hydrological or water quality variables, which forms the input variables for a water quality simulation model. Rehana and Mujumdar [59] developed an Imprecise Fuzzy Waste Load Allocation Model (IFWLAM) to address the uncertainties not only due to randomness and fuzziness but also due to missing or inadequate data by considering the input variables as interval grey numbers. The developed model was implemented on Tunga-Bhadra river, India. A grey fuzzy risk of LWQ was introduced in the WLAM, which is capable of evaluating grey fuzzy risk with corresponding bounds of DO, rather than specifying a single value of risk. The consideration of fuzzy risk as an interval grey number results in a range of fractional removal levels for the dischargers, which enhances flexibility in decision making (Table 1).

Table 1. Results from the IFWLAM optimization models of upper and lower limit of fractional removal levels for various dischargers along Tunga-Bhadra River, India.

Discharger	Risk Minimization Model (Ghosh and Mujumdar [63])	Fractional Removal Levels	
		Lower Limit	Upper Limit
1	0.77	0.69	0.69
2	0.77	0.68	0.69
3	0.65	0.35	0.68
4	0.77	0.52	0.69
5	0.75	0.35	0.69
6	0.77	0.36	0.69
7	0.77	0.35	0.69
8	0.77	0.35	0.69

In recent years, the development of water quality index has become popular among government and related agencies for a quantitative measure of water quality status and for the evaluation of river systems as a river water quality management problem [76]. In a typical water quality index, various important water quality indicators will be integrated into a single water quality index, which can be easily communicated among the stakeholders [77]. However, such indexing methods with respect to water quality evaluation system are burdened with uncertainties originating from errors in measurement, imprecision in characterization, classification and weighting system [31]. In this context, few studies have considered fuzzy-based classifications in the evaluation of water quality indices to address the uncertainty in the quality evaluation [78]. Singh et al. [31] considered the attributes of the water quality parameters as linguistic variable and water quality index of a given location was estimated by aggregating the attributes based on degree of importance to develop fuzzy comprehensive water quality index. The study was implemented in various locations on the Yamuna river, India and tried to address the uncertainty due to natural stream flows originating from rainfall uncertainty and corresponding uncertainty in the prediction of water quality by considering the quality attributes as fuzzy variables. In another recent study by Chanapathi and Thatikonda [78], a fuzzy-based inference system was developed for defining the regional water quality index, the fuzzy-based regional water quality index (FRWQI), based on ten water quality parameters to address the uncertainty due to imprecision for the major rivers of India as: Wainganga, Bhima river, Subarnerekha river, Beas river, etc.

4. Water Resource Management under Climate Change Induced Uncertainties

Water resource systems management models have been advanced in recent years to consider climate change as a driving force to develop adaptive policies in the decision making [79]. In this context, climate change impact assessment in terms of reservoir operation and altered optimal policies has been widely developed by many researchers all over the world [10,80,81]. The most sophisticated and advanced techniques for the climate change impact assessment studies are statistical downscaling models using the most credible general circulation model (GCM) outputs to predict the projected scenarios of hydrological variables [82]. In this context, a few Indian case studies made efforts to integrate statistical downscaling models to predict the reservoir inflows under climate change and addressed the associated uncertainty due to various climate model projections. For example, Ghosh and Mujumdar [82] predicted monthly inflows to Hirakud dam, Mahanadi river basin, using fuzzy clustering and the relevance vector machine as a downscaling model. Raje and Mujumdar [83] used the conditional random field (CRF) downscaling model to predict the inflows of Hirakud Reservoir, Mahanadi basin, India. Rehana and Mujumdar [4] used canonical correlation analysis (CCA) to predict the monthly inflows of Bhadra reservoir, India. These studies predicted reservoir inflow projections by considering the influence of various climate variables using statistical downscaling models and GCM outputs. However, these models do not account for the uncertainties of rainfall, catchment characteristics, soil and land use changes in the reservoir inflow prediction. In this context, Shimola and Krishnaveni [84] studied the climate change impact on Periyar reservoir inflows, Vaippar river, by considering a combination of change of precipitation and temperature and regional climate change scenarios by integrating hydrological model, Soil Water Assessment Tool (SWAT) [84]. However, this model does not integrate the modeled reservoir inflow projections along with the reservoir operation model. Such integration can address the uncertainty originating from uncertain climate change projections of reservoir inflows and resulting operating policies. In another study, Adeloye et al. [10] evaluated the hedging-integrated reservoir rule curves on the current and climate-change-perturbed future performance for Pong reservoir, Beas river in Himachal Pradesh, using sequent peak algorithm and genetic algorithm as optimization model.

Climate change impact assessment on river water quality management has also gained much attention in recent years [85,86]. Rehana and Mujumdar [87] employed CCA as a statistical downscaling model with a threshold-based risk of LWQ model based on multiple logistic regression to develop adaptive treatment policies for the projected scenarios under climate change with the Tunga-Bhadra river system as a case study. The model considered uncertainty due to randomness and imprecision in terms of imprecise fuzzy risk with an integration of climate change projection model. The projected decrease in streamflows and increase in water temperatures tend to decrease DO levels and increase the risk of LWQ events along the Tunga-Bhadra river system. The extreme risk of LWQ was predicted to increase by 50.6% for the period of 2020–2040 compared with the current risk levels of 4.5% for the Tunga-Bhadra river system under climate signals [88]. The fractional removal policy may reach up to its maximum limits of 90% during the period 2070–2100, even though the effluents are at safe permissible levels, indicating revised current standards for better river water quality management for future scenarios under climate change uncertainty.

An integrated water resource management model under climate change, as shown in Figure 1, is subjected to a range of uncertainties, including uncertainty due to hydrological models [89], climate model and scenario uncertainty [90] and uncertainty due to downscaling models [91]. Such climate model and scenario uncertainty in the water resource systems can originate due to inadequate information of underlying geophysical processes, the variability of internal parameterization and boundary conditions [92]. Climate change impact assessment studies of water resource management are associated with various uncertainties originating from variation of climate change projections resulting from various climate models, leading to GCM and scenario uncertainty [93]. A few other sources of uncertainties are associated with climate model initial conditions, statistical downscaling models, hydrological models and parameters [94]. In this context, few studies have attempted to address the climate model uncertainties into water resource management [95].

Raje and Mujumdar [96] developed an uncertainty modeling framework for Mahanadi River at Hirakud Reservoir in Orissa, India, to address GCM scenario uncertainties along with uncertainty in the nature of the downscaling relationship with the Dempster–Shafer theory of evidence combination. The results suggest that by linking regional impacts to natural regime frequencies, uncertainty in regional predictions can be realistically quantified. Raje and Mujumdar [3] derived reservoir operating policy for Hirakud reservoir, Mahanadi Basin, India by considering the reliability of hydropower generation for the current scenario, with consideration of conflicts between hydropower, irrigation and flood control with the standard operating policy (SOP). The projected monsoon streamflows for current and future scenarios for a range of GCM-scenario combinations were used with an integration of a conditional random field (CRF)-downscaling model as the statistical downscaling model to address the uncertainty in the climate model projections. The results of the study found a decrease in hydropower and increase in vulnerability for the future, with a significant impact in terms of a decrease in reliability and increase in vulnerability. The study also suggested revising the reservoir rules under climate change with a projected decrease in inflow to the Hirakud reservoir.

Rehana and Mujumdar [4] developed an integrated regional water resource management model addressing various sources of uncertainties in the prediction of a hydro-climatic variable projection model, an irrigation demand quantification model, and a water quantity quality management model using SDP and fuzzy optimization for Bhadra Reservoir-River system in Karnataka, India. A SDP model is used to derive the optimal monthly steady state operating policy considering irrigation, hydropower and downstream river water quality as reservoir users. The uncertainty due to the randomness of reservoir inflows was addressed using SDP, and the imprecise goals of each reservoir user were addressed by considering fuzzy memberships. A fuzzy water allocation model was developed for obtaining the optimal allocations among various users of the reservoir under climate change.

Another prominent source of uncertainty is the variability of climate projections resulting from different climate models and scenarios, which has been identified as climate model uncertainty in water resource management modelling [93]. Certainly, a range of climate change projections resulting from various models will provide flexibility in decision making [97]. However, combining projections resulting from various GCMs and scenarios to have a single representative projection by deriving a multimodal weighted mean has been widely applied to address climate model uncertainty in water resource management [4] (Figure 4). In this context, Mujumdar and Ghosh [93] proposed a possibilistic approach to address climate model uncertainty, with Hirakud dam inflow climate change projections, located on Mahanadi river, Orissa, India, as a case study. The study developed the possibilistic mean cumulative distribution function (CDF) by assigning weights to GCMs and scenarios based on their performance in the recent years as well as for the future scenarios. The results of the study reveal that the amount of uncertainty for a given inflow projection will increase with time, due to different climate sensitivity among the models. Instead of using a single climate projection resulting from one GCM and scenario, the use of such multimodal ensembles may be promising in water resource management models under climate change.

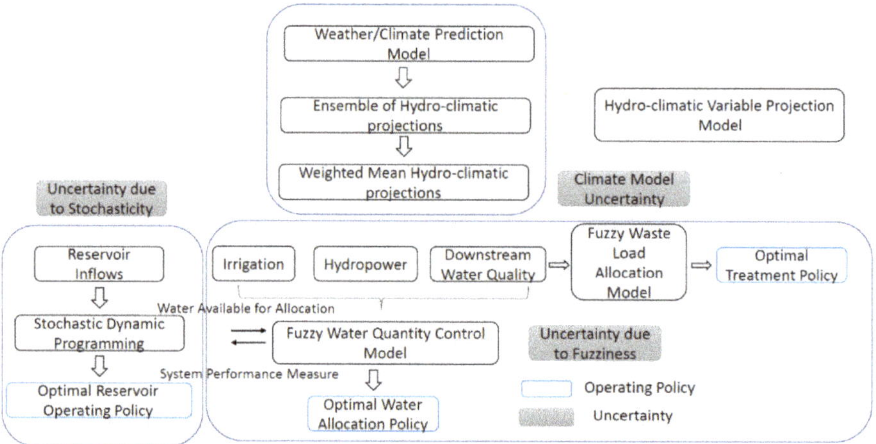

Figure 4. Basin-scale water resource systems modeling and associated uncertainties.

In another study by Rehana and Mujumdar [98], entropy weights to each GCM and scenario projections were assigned based on the performance of the GCM and scenario in reproducing the present climatology and deviation of each projections from the projected ensemble average. Entropy weights were assigned to each hydro-meteorological variable defining water availability (reservoir inflows) and demands (e.g., irrigation demands: rainfall and other meteorological variables affecting evapotranspiration, etc.) in the reservoir operation. The multimodal weighted mean (MWM) projections of various hydro-meteorological variables addressing the climate model uncertainty have been used in the water resource management model developed for Bhadra reservoir, India as case study (Figure 4).

Uncertainties are expected to occur at every stage of the water resource management models and their propagation at regional and local scales can lead to large uncertainty ranges and increasing the complexity in decision making [27]. The climate model uncertainty originating from the mismatch between various GCMs and scenarios can be considered as the first level of uncertainty, which can be modeled by using the weighted mean hydro-meteorological projections in reservoir inflow modelling (Figure 5a), and the estimation of projected demands (Figure 5b) in the reservoir operation. The second level of uncertainty originates due to the imprecision and conflicting goals of the reservoir users leading to uncertainty due to imprecision, which can be modeled by using fuzzy set theory. The third level of uncertainty can arise from the inherent variability of the reservoir inflow leading to uncertainty due to randomness, which can be modeled by considering the reservoir inflow as stochastic variable in SDP and consequent uncertainties in resulting operating policies (Figure 5c). Since uncertainties accumulate from various levels, their propagation up to the regional or local level leads to large uncertainty ranges at such scales [27].

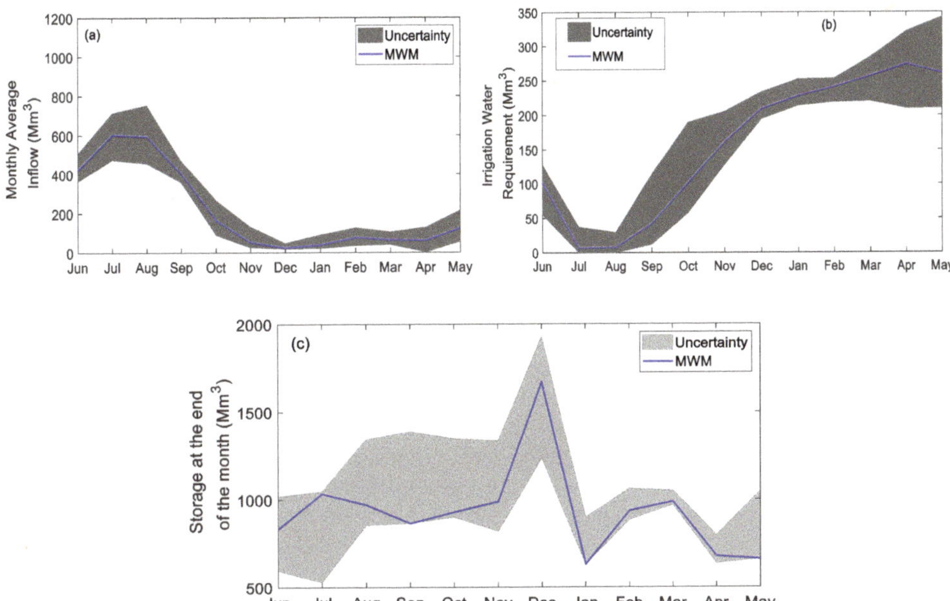

Figure 5. Regional water resource management model-associated uncertainties at each stage: (**a**) water availability model in terms of reservoir inflows; (**b**) demand estimation model such as irrigation water requirements; (**c**) operating policies in terms of storages. The results are for Bhadra river basin, India, showing maximum and minimum values along with multi weighted mean (MWM) hydro-climatology.

Overall, while most of the studies tried to address climate model uncertainties in reservoir operation models, the scope of improvements may be towards addressing hydrological model uncertainties, different downscaling models and various reanalysis data sources of uncertainties, along with the stochastic and fuzziness uncertainties. The assessment of uncertainties in decision making at each stage of reservoir operation has to be understood for the possible risk of failure of water resource systems. This necessitates the development of holistic approaches to include various sources of uncertainties at each stage of the water resource management model, from climate or weather predictions to operating policies.

5. Challenges, Missing Links and Directions for Future

Water resource systems management models have emerged as promising tools for the effective management of resources in economically efficient manner in recent years addressing various forms of uncertainties. However, constraints and challenges still remain in terms of the inadequacy of water to meet demands, rapidly growing population, urbanization, increased social and economic development and uncertain future climate [99].

Most of the studies of water resource systems in India have focused on a single reservoir with single or multi objective functions. Integrated water resource management can be developed, considering various reservoirs of a river system, accounting for inflows, uncontrolled intermediate catchment flows, irrigation return flows and controlled flows from an upstream reservoir [12]. The development of an integrated river basin management which can include various reservoirs inter-connected in a river basin and considering various forms of uncertainties can be potential area of research. Such integrated water resource management studies may have challenges in implementation in terms of ungauged sub basins of upstream catchments, a lack of detailed data on downstream releases, lack of information about the inauthentic water abstractions, etc.

A water resource systems model has to integrate various sources of information, such as hydrological, meteorological, pollutant, agricultural, demographic and socioeconomic. One of the major challenges in water resource management models is the integration of various sources of uncertain information which are burdened with spatial and temporal mismatches among scales. Synthesizing various sources of information needs careful attention in terms of validation with field observations not only at the individual scale but also at the integrated scale. Such holistic approaches have the capability to capture the association between various subsystems of regional water resources. In addition, integrated water resource management can improve agreement and cooperation between various stakeholders for sustainable water management.

One of the major challenges which arises in the implementation of holistic approaches at various spatial and temporal scales is the expected increase in climate extremes such as floods and droughts, as the developed water resource systems models are based on the observed and historical data and therefore bounded with the experiences faced in the past and do not consider the possible anthropogenic and climate extremes. The sophisticated climate change impact assessment models developed in recent years can provide a basis to understand projected changes in terms of hydrologic variability and possible adaptive policies. However, such climate change impact assessment studies are developed based on past historical observations under nonstationary assumptions with uncertain information. Future advancements have to be made towards the development of universal water resource management models under hydroclimate extremes along with operation and management by addressing various sources of uncertainties.

The existing water resource management models so far developed are able to address various forms of uncertainties such as randomness, imprecision, fuzziness, inexactness, lack of knowledge and missing data, climate model uncertainties, models and parameters, etc. Most of these uncertainties have been addressed at the individual scale but not in an integrated manner. There is a necessity to integrate various sources of uncertainties to study resulting combined uncertainty and impact on operating policies. Such uncertainty accumulation studies can be promising in the development of approaches representing uncertainties originating at each stage of decision making.

Another area for water resource management model evaluation is towards the development of decision support systems (DSS) for real time operating policies which can act as a bridge to link the model-generated decisions with practical water utility. Such models should work as real time holistic approaches with an integration of weather forecasting models, hydrological models, reservoir operation models and operating policies. The current real time water resource models are dedicated to a single purpose, majorly as a flood controlling devices, with quantity control as a priority. The land use and land cover changes of a natural landscape can intensify the sediments, nutrients and other organic pollutants entering into inland water bodies such as reservoirs, lakes etc. Furthermore, increasing pressure of crop yields has increased the use of fertilizers, which again has increased the number of non-point sources of pollution to the rivers. Under these consequences, real-time water management operating policies should work with an integration of quantity as well as quality as priorities. The development of a general approach which can integrate quantity and quality aspects integrating reservoir-river systems with a DSS in a web-based environment can be a promising tool for the development of real-time operating policies. Formulating such real-time holistic approaches necessitates close coordination and cooperation between various stakeholders, researchers, government bodies and policy makers. It is important to identify all the beneficial and adverse ecological, economic, environmental, and social effects in the context of long-term effects associated in water resource system planning and management.

6. Conclusions

Water resource systems models have advanced in several directions, starting with modelling approaches, uncertainty quantification, ease in decision making of stakeholders, along with climate change impact adaptation. Uncertainty quantification in water resource systems models has been identified as an active research topic in the research community. The major source of uncertainty identified in reservoir operation is the random nature of streamflows and this has been addressed using various forms of stochastic dynamic programming. Another major source of uncertainty considered in the research community is the imprecision and vagueness in defining the goals of stakeholders. Such uncertainty was defined as uncertainty due to fuzziness, which has been addressed by considering the goals as fuzzy membership functions and associated satisfaction levels. Fuzzy optimization has been used as a revolutionary algorithm in water resource management models that deals with the uncertainty arising due to fuzzy goals of decision-makers. Fuzzy optimization models in water resource systems have progressed further to address the next level of uncertainty associated with defining the membership parameters by considering them as interval grey numbers. Uncertainty due to a lack of knowledge and missing data has also been tackled by considering the grey fuzzy optimization models. The consideration of hydrological variables as interval grey numbers has resulted in a range of operating policies and provided flexibility to the stakeholders. Climate change-induced uncertainty has emerged as a major source of uncertainty in water resource management models in recent years under changes of hydrological extremes.

To summarize, we reviewed water resource management models and associated uncertainties originating in modelling and decision making. Water resource management models such as reservoir operation, water quantity allocation, waste load allocation, quantity-quality integrated models and water quality index models were discussed. The recent developments in water resource management under climate change were articulated. Several methods that deal with different sources of uncertainties originating in the water resources modelling and decision making were critically evaluated with a focus on Indian case studies. The research gaps, challenges, missing links and future directions in water resource management models under uncertainties were discussed. This review suggests that water resource management models are powerful computational tools that ought to be upgraded by synthesizing various sources of uncertainties for real-time operation and sustainable policy making.

Author Contributions: S.R., C.R.R., S.K., S.G., and P.M. conceptualized and designed the paper; S.R. and C.R.R. wrote the paper; P.M., S.K., and S.G. supervised the paper. All authors have read and agreed to the published version of the manuscript.

Funding: This research received no external funding.

Acknowledgments: We sincerely thank the editor and the three anonymous reviewers for reviewing the manuscript and offering critical comments to improve the manuscript. C.R.R. is funded by the Pacific Institute for the Mathematical Sciences and the Global Water Futures, Center for Hydrology, University of Saskatchewan, Canada.

Conflicts of Interest: The authors declare no conflict of interest.

References

1. Savenije, H.H.G.; Van der Zaag, P. Integrated water resources management: Concepts and issues. *Phys. Chem. Earth* **2008**, *33*, 290–297. [CrossRef]
2. Brown, C.M.; Lund, J.R.; Cai, X.; Reed, P.M.; Zagona, E.A.; Ostfeld, A.; Hall, J.; Characklis, G.W.; Yu, W.; Brekke, L. The future of water resources systems analysis: Toward a scientific framework for sustainable water management. *Water Resour. Res.* **2015**, *51*, 6110–6124. [CrossRef]
3. Raje, D.; Mujumdar, P.P. Reservoir performance under uncertainty in hydrologic impacts of climate change. *Adv. Water Resour.* **2010**, *33*, 312–326. [CrossRef]
4. Rehana, S.; Mujumdar, P.P. Basin Scale Water Resources Systems Modeling Under Cascading Uncertainties. *Water Resour. Manag.* **2014**, *28*, 3127–3142. [CrossRef]
5. Cunderlik, J.M.; Simonovic, S.P. Inverse flood risk modelling under changing climatic conditions. *Hydrol. Process.* **2007**, *21*, 563–577. [CrossRef]
6. IPCC. Climate change 2007: Impacts, Adaptation, and Vulnerability. In *Contribution of Working Group II to the Third Assessment Report of the Intergovernmental Panel on Climate Change*; Parry, M.L., Canziani, O., Palutikof, J., Van der Linden, P., Hanson, C., Eds.; Cambridge Univ. Press: Cambridge, UK, 2007.
7. Eum, H.-I.; Simonovic, S.P. Integrated Reservoir Management System for Adaptation to Climate Change: The Nakdong River Basin in Korea. *Water Resour. Manag.* **2010**, *24*, 3397–3417. [CrossRef]
8. Mandal, S.; Arunkumar, R.; Breach, P.A.; Simonovic, S.P. Reservoir Operations under Changing Climate Conditions: Hydropower-Production Perspective. *J. Water Resour. Plan. Manag.* **2019**, *145*, 04019016. [CrossRef]
9. Asokan, S.M.; Dutta, D. Analysis of water resources in the Mahanadi River Basin, India under projected climate conditions. *Hydrol. Process.* **2008**, *22*, 3589–3603. [CrossRef]
10. Adeloye, A.J.; Soundharajan, B.-S.; Ojha, C.S.P.; Remesan, R. Effect of Hedging-Integrated Rule Curves on the Performance of the Pong Reservoir (India) During Scenario-Neutral Climate Change Perturbations. *Water Resour. Manag.* **2016**, *30*, 445–470. [CrossRef]
11. Loucks, D.P. A comment on optimization methods for branching multistage water resource systems. *Water Resour. Res.* **1968**, *4*, 447–450. [CrossRef]
12. Mujumdar, P.P.; Nagesh Kumar, D. *Integrated Reservoir Operation Studies for Reservoirs in the Narmada Basin, Report Submitted to Narmada Control Authority*; Narmada Control Authority: Indore, India, 2006.
13. Loucks, D.P. Managing Water as a Critical Component of a Changing World. *Water Resour. Manag.* **2017**, *31*, 2905–2916. [CrossRef]
14. Yeh, C.H.; Labadie, J.W. Multiobjective Watershed-Level Planning of Storm Water Detention Systems. *J. Water Resour. Plan. Manag.* **1997**, *123*, 336–343. [CrossRef]
15. Ahmad, A.; El-Shafie, A.; Razali, S.; Mohamad, Z. Reservoir Optimization in Water Resources: A Review. *Water Resour. Manag.* **2014**, *28*, 3391–3405. [CrossRef]
16. Simonovic, S.P. Reservoir Systems Analysis: Closing Gap between Theory and Practice. *J. Water Resour. Plan. Manag.* **1992**, *118*, 262–280. [CrossRef]
17. McKinney, D.C.; Cai, X.; Rosegrant, M.W.; Ringler, C.; Scott, C.A. *Modeling Water Resources Management at the Basin Level: Review and Future Directions*; International Water Management Institute: Colombo, Sri Lanka, 1999.
18. Mayer, A.; Muñoz-Hernandez, A. Integrated Water Resources Optimization Models: An Assessment of a Multidisciplinary Tool for Sustainable Water Resources Management Strategies. *Geogr. Compass* **2009**, *3*, 1176–1195. [CrossRef]

19. Rani, D.; Moreira, M.M.; Moreira, M. Simulation–Optimization Modeling: A Survey and Potential Application in Reservoir Systems Operation. *Water Resour. Manag.* **2009**, *24*, 1107–1138. [CrossRef]

20. Hossain, S.; El-Shafie, A. Intelligent Systems in Optimizing Reservoir Operation Policy: A Review. *Water Resour. Manag.* **2013**, *27*, 3387–3407. [CrossRef]

21. Afshar, A.; Massoumi, F.; Afshar, A.; Mariño, M.A. State of the Art Review of Ant Colony Optimization Applications in Water Resource Management. *Water Resour. Manag.* **2015**, *29*, 3891–3904. [CrossRef]

22. Jahandideh-Tehrani, M.; Bozorg-Haddad, O.; Loáiciga, H.A. Application of non-animal–inspired evolutionary algorithms to reservoir operation: An overview. *Environ. Monit. Assess.* **2019**, *191*, 439. [CrossRef]

23. Mohammad-Azari, S.; Bozorg-Haddad, O.; Loáiciga, H.A. State-of-art of genetic programming applications in water-resources systems analysis. *Environ. Monit. Assess.* **2020**, *192*, 1–17. [CrossRef]

24. Rani, D.; Pant, M.; Jain, S.K. Dynamic programming integrated particle swarm optimization algorithm for reservoir operation. *Int. J. Syst. Assur. Eng. Manag.* **2020**, *11*, 515–529. [CrossRef]

25. Choong, S.-M.; El-Shafie, A. State-of-the-Art for Modelling Reservoir Inflows and Management Optimization. *Water Resour. Manag.* **2014**, *29*, 1267–1282. [CrossRef]

26. Fayaed, S.S.; El-Shafie, A.; Jaafar, O. Reservoir-system simulation and optimization techniques. *Stoch. Environ. Res. Risk Assess.* **2013**, *27*, 1751–1772. [CrossRef]

27. Vicuña, S.; Dracup, J.A.; Lund, J.R.; Dale, L.; Maurer, E. Basin-scale water system operations with uncertain future climate conditions: Methodology and case studies. *Water Resour. Res.* **2010**, *46*. [CrossRef]

28. Mujumdar, P.P.; Vedula, S. Performance evaluation of an irrigation system under some optimal operating policies. *Hydrol. Sci. J.* **1992**, *37*, 13–26. [CrossRef]

29. Russell, S.O.; Campbell, P.F. Reservoir Operating Rules with Fuzzy Programming. *J. Water Resour. Plan. Manag.* **1996**, *122*, 165–170. [CrossRef]

30. Jain, S.K. Water Resources Management in India–Challenges and the Way Forward. *Curr. Sci.* **2019**, *117*, 569. [CrossRef]

31. Singh, A.P.; Dhadse, K.; Ahalawat, J. Managing water quality of a river using an integrated geographically weighted regression technique with fuzzy decision-making model. *Environ. Monit. Assess.* **2019**, *191*, 378. [CrossRef]

32. Loucks, D.P.; Van Beek, E. *Water Resource Systems Planning and Management—An Introduction to Methods, Models, and Applications*; Springer: Dordrecht, The Netherlands, 2017; Available online: https://www.springer.com/gp/book/9783319442327 (accessed on 18 April 2020).

33. Vedula, S.; Mujumdar, P.P. *Water Resources Systems Modelling Techniques and Analysis*; Tata-McGraw Hill: New Delhi, India, 2005; Available online: https://www.scirp.org/(S(lz5mqp453edsnp55rrgjct55))/reference/ReferencesPapers.aspx?ReferenceID=40407 (accessed on 19 April 2020).

34. Kumar, D.N.; Baliarsingh, F.; Raju, K.S. Optimal Reservoir Operation for Flood Control Using Folded Dynamic Programming. *Water Resour. Manag.* **2009**, *24*, 1045–1064. [CrossRef]

35. Vedula, S.; Kumar, D.N. An Integrated Model for Optimal Reservoir Operation for Irrigation of Multiple Crops. *Water Resour. Res.* **1996**, *32*, 1101–1108. [CrossRef]

36. Mujumdar, P.P.; Nirmala, B. A Bayesian Stochastic Optimization Model for a Multi-Reservoir Hydropower System. *Water Resour. Manag.* **2006**, *21*, 1465–1485. [CrossRef]

37. Kumari, S.; Mujumdar, P.P. Fuzzy State Real-Time Reservoir Operation Model for Irrigation with Gridded Rainfall Forecasts. *J. Irrig. Drain. Eng.* **2016**, *142*, 04015042. [CrossRef]

38. Vedula, S.; Mujumdar, P.P. Optimal reservoir operation for irrigation of multiple crops. *Water Resour. Res.* **1992**, *28*, 1–9. [CrossRef]

39. Ravikumar, V.; Venugopal, K. Optimal Operation of South Indian Irrigation Systems. *J. Water Resour. Plan. Manag.* **1998**, *124*, 264–271. [CrossRef]

40. Chandramouli, S.; Nanduri, U. Comparison of stochastic and fuzzy dynamic programming models for the operation of a multipurpose reservoir. *Water Environ. J.* **2011**, *25*, 547–554. [CrossRef]

41. Sharma, P.J.; Patel, P.; Jothiprakash, V. Efficient discretization of state variables in stochastic dynamic programming model of Ukai reservoir, India. *ISH J. Hydraul. Eng.* **2016**, *22*, 1–12. [CrossRef]

42. Jothiprakash, V.; Shanthi, G. Comparison of Policies Derived from Stochastic Dynamic Programming and Genetic Algorithm Models. *Water Resour. Manag.* **2008**, *23*, 1563–1580. [CrossRef]

43. Umamahesh, N.V.; Sreenivasulu, P. Technical Communication: Two-Phase Stochastic Dynamic Programming Model for Optimal Operation of Irrigation Reservoir. *Water Resour. Manag.* **1997**, *11*, 395–406. [CrossRef]

44. Prasad, A.S.; Umamahesh, N.V.; Viswanath, G.K.; Nanduri, U. Optimal Irrigation Planning under Water Scarcity. *J. Irrig. Drain. Eng.* **2006**, *132*, 228–237. [CrossRef]

45. Panigrahi, D.P.; Mujumdar, P.P. Reservoir Operation Modelling with Fuzzy Logic. *Water Resour. Manag.* **2000**, *14*, 89–109. [CrossRef]

46. Simonovic, S.P. A spatial multi-objective decision-making under uncertainty for water resources management. *J. Hydroinform.* **2005**, *7*, 117–133. [CrossRef]

47. Regulwar, D.G.; Raj, P.A. Development of 3-D Optimal Surface for Operation Policies of a Multireservoir in Fuzzy Environment Using Genetic Algorithm for River Basin Development and Management. *Water Resour. Manag.* **2007**, *22*, 595–610. [CrossRef]

48. Teegavarapu, R.S.V.; Ferreira, A.R.; Simonovic, S.P. Fuzzy multiobjective models for optimal operation of a hydropower system. *Water Resour. Res.* **2013**, *49*, 3180–3193. [CrossRef]

49. Arunkumar, R.; Jothiprakash, V. A multiobjective fuzzy linear programming model for sustainable integrated operation of a multireservoir system. *Lakes Reserv. Res. Manag.* **2016**, *21*, 171–187. [CrossRef]

50. Shrestha, B.P.; Duckstein, L.; Stakhiv, E.Z. Fuzzy Rule-Based Modeling of Reservoir Operation. *J. Water Resour. Plan. Manag.* **1996**, *122*, 262–269. [CrossRef]

51. Kamodkar, R.U.; Regulwar, D.G. Multipurpose Reservoir Operating Policies: A Fully Fuzzy Linear Programming Approach. *J. Agric. Sci. Technol.* **2013**, *15*, 1261–1274.

52. Kumari, S.; Mujumdar, P.P. Reservoir Operation with Fuzzy State Variables for Irrigation of Multiple Crops. *J. Irrig. Drain. Eng.* **2015**, *141*, 04015015. [CrossRef]

53. Kumari, S.; Mujumdar, P.P. Fuzzy Set–Based System Performance Evaluation of an Irrigation Reservoir System. *J. Irrig. Drain. Eng.* **2017**, *143*, 04017002. [CrossRef]

54. Fu, D.Z.; Li, Y.P.; Huang, G. A fuzzy-Markov-chain-based analysis method for reservoir operation. *Stoch. Environ. Res. Risk Assess.* **2011**, *26*, 375–391. [CrossRef]

55. Mousavi, S.J.; Mahdizadeh, K.; Afshar, A. A stochastic dynamic programming model with fuzzy storage states for reservoir operations. *Adv. Water Resour.* **2004**, *27*, 1105–1110. [CrossRef]

56. Sasikumar, K.; Mujumdar, P.P. Fuzzy Optimization Model for Water Quality Management of a River System. *J. Water Resour. Plan. Manag.* **1998**, *124*, 79–88. [CrossRef]

57. Chapra, S.C. *Surface Water Quality Modeling*; McGraw-Hill Publisher: New York, NY, USA, 1997; Available online: https://www.scirp.org/(S(351jmbntvnsjt1aadkposzje))/reference/ReferencesPapers.aspx?ReferenceID=1651544 (accessed on 19 April 2020).

58. Chapra, S.C.; Pelletier, G.J. *QUAL2K: A Modeling Frame Work for Simulating River and Stream Water Quality: Documentation and Users Manual*; Civil and Environmental Department, Tufts University: Medford, MA, USA, 2003.

59. Rehana, S.; Mujumdar, P. An imprecise fuzzy risk approach for water quality management of a river system. *J. Environ. Manag.* **2009**, *90*, 3653–3664. [CrossRef] [PubMed]

60. Loucks, D.P.; Lynn, W.R. Probabilistic models for predicting stream quality. *Water Resour. Res.* **1966**, *2*, 593–605. [CrossRef]

61. Takyi, A.K.; Lence, B.J. Markov Chain Model for Seasonal-Water Quality Management. *J. Water Resour. Plan. Manag.* **1995**, *121*, 144–157. [CrossRef]

62. Karmakar, S.; Mujumdar, P. Grey fuzzy optimization model for water quality management of a river system. *Adv. Water Resour.* **2006**, *29*, 1088–1105. [CrossRef]

63. Ghosh, S.; Mujumdar, P. Risk minimization in water quality control problems of a river system. *Adv. Water Resour.* **2006**, *29*, 458–470. [CrossRef]

64. Mujumdar, P.P.; Sasikumar, K. A fuzzy risk approach for seasonal water quality management of a river system. *Water Resour. Res.* **2002**, *38*, 5-1–5-9. [CrossRef]

65. Singh, A.P.; Ghosh, S.K.; Sharma, P. Water quality management of a stretch of river Yamuna: An interactive fuzzy multi-objective approach. *Water Resour. Manag.* **2006**, *21*, 515–532. [CrossRef]

66. Huang, G.; Baetz, B.W.; Patry, G.G. Grey integer programming: An application to waste management planning under uncertainty. *Eur. J. Oper. Res.* **1995**, *83*, 594–620. [CrossRef]

67. Mujumdar, P.P.; Karmakar, S. Grey Fuzzy Multi-Objective Optimization. In *Fuzzy Multi-Criteria Decision Making: Theory and Applications with Recent Developments*; Springer: Berlin, Germany, 2008; Volume 16, pp. 453–482.

68. Rosenberg, D.E. Shades of grey: A critical review of grey-number optimization. *Eng. Optim.* **2009**, *41*, 573–592. [CrossRef]

69. Yadav, V.; Bhurjee, A.; Karmakar, S.; Dikshit, A. A facility location model for municipal solid waste management system under uncertain environment. *Sci. Total Environ.* **2017**, *603*, 760–771. [CrossRef] [PubMed]

70. Liu, C.; Lee, C.; Chen, H.; Mehrotra, S. Stochastic Robust Mathematical Programming Model for Power System Optimization. *IEEE Trans. Power Syst.* **2015**, *31*, 1–2. [CrossRef]

71. Yadav, V.; Karmakar, S.; Dikshit, A.; Bhurjee, A. Interval-valued facility location model: An appraisal of municipal solid waste management system. *J. Clean. Prod.* **2018**, *171*, 250–263. [CrossRef]

72. Huang, G. Analysis of Solution Methods for Interval Linear Programming. *J. Environ. Inform.* **2011**, *17*, 54–64. [CrossRef]

73. Chang, N.-B.; Pires, A. *Sustainable Solid Waste Management: A Systems Engineering Approach*; John Wiley & Sons: Hoboken, NJ, USA, 2015.

74. Dubois, J.-E.; Gershon, N. *Modeling Complex Data for Creating Information*, 1st ed.; Springer Publishing Company, Incorporated: Berlin, Germany, 2012; ISBN 978-3-642-80201-0.

75. Ross, T.J. *Fuzzy Logic with Engineering Applications*; John Wiley & Sons: Hoboken, NJ, USA, 2010.

76. Fathi, E.; Zamani-Ahmadmahmoodi, R.; Zare-Bidaki, R. Water quality evaluation using water quality index and multivariate methods, Beheshtabad River, Iran. *Appl. Water Sci.* **2018**, *8*, 210. [CrossRef]

77. Avvannavar, S.M.; Shrihari, S. Evaluation of water quality index for drinking purposes for river Netravathi, Mangalore, South India. *Environ. Monit. Assess.* **2007**, *143*, 279–290. [CrossRef]

78. Chanapathi, T.; Thatikonda, S. Fuzzy-Based Regional Water Quality Index for Surface Water Quality Assessment. *J. Hazard. Toxic Radioact. Waste* **2019**, *23*, 04019010. [CrossRef]

79. Stojkovic, M.; Simonovic, S.P. System Dynamics Approach for Assessing the Behaviour of the Lim Reservoir System (Serbia) under Changing Climate Conditions. *Water* **2019**, *11*, 1620. [CrossRef]

80. Haro-Monteagudo, D.; Palazón, L.; Beguería, S. Long-term sustainability of large water resource systems under climate change: A cascade modeling approach. *J. Hydrol.* **2020**, *582*, 124546. [CrossRef]

81. Dang, T.D.; Chowdhury, A.F.M.K.; Galelli, S. On the representation of water reservoir storage and operations in large-scale hydrological models: Implications on model parameterization and climate change impact assessments. *Hydrol. Earth Syst. Sci.* **2020**, *24*, 397–416. [CrossRef]

82. Ghosh, S.; Mujumdar, P. Statistical downscaling of GCM simulations to streamflow using relevance vector machine. *Adv. Water Resour.* **2008**, *31*, 132–146. [CrossRef]

83. Raje, D.; Mujumdar, P.P. Constraining uncertainty in regional hydrologic impacts of climate change: Nonstationarity in downscaling. *Water Resour. Res.* **2010**, *46*. [CrossRef]

84. Shimola, K.; Krishnaveni, M. Sensitivity of SWAT simulated reservoir inflow to climate change in a semiarid basin. *Mausam* **2015**, *66*, 181–186.

85. Tung, T.M.; Yaseen, Z.M. A survey on river water quality modelling using artificial intelligence models: 2000–2020. *J. Hydrol.* **2020**, *585*, 124670. [CrossRef]

86. Shrestha, N.K.; Wang, J. Water Quality Management of a Cold Climate Region Watershed in Changing Climate. *J. Environ. Inform.* **2019**, *35*, 56–80. [CrossRef]

87. Rehana, S.; Mujumdar, P. Climate change induced risk in water quality control problems. *J. Hydrol.* **2012**, *444*, 63–77. [CrossRef]

88. Rehana, S.; Dhanya, C.T. Modeling of extreme risk in river water quality under climate change. *J. Water Clim. Chang.* **2018**, *9*, 512–524. [CrossRef]

89. Teutschbein, C.; Wetterhall, F.; Seibert, J. Evaluation of different downscaling techniques for hydrological climate-change impact studies at the catchment scale. *Clim. Dyn.* **2011**, *37*, 2087–2105. [CrossRef]

90. Minville, M.; Brissette, F.; Leconte, R. Uncertainty of the impact of climate change on the hydrology of a nordic watershed. *J. Hydrol.* **2008**, *358*, 70–83. [CrossRef]

91. Khan, M.S.; Coulibaly, P.; Dibike, Y. Uncertainty analysis of statistical downscaling methods. *J. Hydrol.* **2006**, *319*, 357–382. [CrossRef]

92. New, M.; Hulme, M. Representing uncertainty in climate change scenarios: A Monte-Carlo approach. *Integr. Assess.* **2000**, *1*, 203–213. [CrossRef]

93. Mujumdar, P.P.; Ghosh, S. Modeling GCM and scenario uncertainty using a possibilistic approach: Application to the Mahanadi River, India. *Water Resour. Res.* **2008**, *44*. [CrossRef]

94. Simonovic, S.; Davies, E.G.R. Are we modelling impacts of climatic change properly? *Hydrol. Process.* **2006**, *20*, 431–433. [CrossRef]
95. Majone, B.; Bovolo, C.I.; Bellin, A.; Blenkinsop, S.; Fowler, H.J. Modeling the impacts of future climate change on water resources for the Gállego river basin (Spain): Climate Change Impacts on Water Resources. *Water Resour. Res.* **2012**, *48*. [CrossRef]
96. Raje, D.; Mujumdar, P. Hydrologic drought prediction under climate change: Uncertainty modeling with Dempster–Shafer and Bayesian approaches. *Adv. Water Resour.* **2010**, *33*, 1176–1186. [CrossRef]
97. Giorgi, F.; Mearns, L.O. Calculation of Average, Uncertainty Range, and Reliability of Regional Climate Changes from AOGCM Simulations via the "Reliability Ensemble Averaging" (REA) Method. *J. Clim.* **2002**, *15*, 1141–1158. [CrossRef]
98. Rehana, S. Regional Hydrologic Impacts of Climate Change. Ph.D. Thesis, Department of Civil Engineering, Indian Institute of Science, Bangalore, India, 2012.
99. Cosgrove, W.J.; Loucks, D.P. Water management: Current and future challenges and research directions. *Water Resour. Res.* **2015**, *51*, 4823–4839. [CrossRef]

Article

Risk and Resilience: A Case of Perception versus Reality in Flood Management

Nirupama Agrawal [1,*], Mark Elliott [2] and Slobodan P Simonovic [3]

[1]   Disaster & Emergency Management Program, York University, Toronto, ON M3J 1P3, Canada
[2]   Institute for Catastrophic Loss Reduction, Toronto, ON M5C 2R9, Canada; melliott@iclr.org
[3]   Department of Civil and Environmental Engineering, University of Western Ontario, London, ON N6A 5B9, Canada; simonovic@uwo.ca
*   Correspondence: nirupama@yorku.ca

Received: 28 March 2020; Accepted: 25 April 2020; Published: 28 April 2020

Abstract: Canada's vast regions are reacting to climate change in uncertain ways. Understanding of local disaster risks and knowledge of underlying causes for negative impacts of disasters are critical factors to working toward a resilient environment across the social, economic, and the built sectors. Historically, floods have caused more economical and social damage around the world than other types of natural hazards. Since the 1900s, the most frequent hazards in Canada have been floods, wildfire, drought, and extreme cold, in terms of economic damage. The recent flood events in the Canadian provinces of Ontario, New Brunswick, Quebec, Alberta, and Manitoba have raised compelling concerns. These include should communities be educated with useful knowledge on hazard risk and resilience so they would be interested in the discussion on the vital role they can play in building resilience in their communities. Increasing awareness that perceived risk can be very different from the real threat is the motivation behind this study. The main objectives of this study include identifying and quantifying the gap between people's perception of exposure and susceptibility to the risk and a lack of coping capacity and objective assessment of risk and resilience, as well as estimating an integrated measure of disaster resilience in a community. The proposed method has been applied to floods as an example, using actual data on the geomorphology of the study area, including terrain and low lying regions. It is hoped that the study will encourage a broader debate if a unified strategy for disaster resilience would be feasible and beneficial in Canada.

Keywords: disaster; risk; perception; community; resilience; Canada

---

1. Introduction

The impacts of geo-hydrologic hazards in the last two decades of the 20th century were felt by three-quarters of the population worldwide [1]. Since the 1900s, the most economically damaging disasters from natural causes in Canada were floods, wildfires, drought, and extreme cold. Recent research has linked specific flooding events, as well as a general rise in the intensity of wet weather in the northern hemisphere, to the effects of rising greenhouse gas levels and global climate change [2]. Past studies intended to focus on disaster response founded on a top-down approach, but the focus has now shifted to a community-based approach [3–9] that stresses resilience building. After the Hyogo Framework for Action (HFA), 2005–2015 [10] launched a movement for building the resilience of nations and communities to disasters. The Sendai Framework for Disaster Risk Reduction (SF-DRR) 2015–2030 [11] serves as a continuum to the commitment supported by the United Nations Office for Disaster Risk Reduction (UNDRR). The SF-DRR notes the need for improved understanding of disaster risk in all its dimensions of exposure, vulnerability, and hazard characteristics, as well as the strengthening of disaster risk governance [11]. Potential variations in the understanding of risk, resilience, susceptibility, and coping capacity allow for interpretations and applications in managing disasters. Scholars have analyzed and explained perceptions of risk, as a component of the social factors, which interacts with the geographic context to create vulnerability [12], as part of the cultural arrangements of society to cope with flood events [13], and as determinants of vulnerability [14,15].

The role of risk perceptions in improving the resilience of people and communities is widely recognized as an essential component [16]. While it is not possible for societies and citizens to directly influence the natural sources of hazards, much can be done to mitigate their risk by understanding and managing the consequences they could experience should a disaster occur—this is addressed by resilience [17]. The idea of disaster-resilient communities goes beyond estimating monetary losses alone but also accounting for multiple dimensions, including technical, organizational, social, and economic facets [18]. The world risk index (WRI) focuses on the understanding of risk, which is defined as the interaction of physical hazards and the vulnerability of exposed elements [19]. It also demonstrates through the assessed risk for 173 countries that the vulnerability of a society or a country is not the same as exposure to natural hazards [19].

The most comprehensive definition of resilience is developed by the UNDRR, "The ability of a system, community or society exposed to hazards to resist, absorb, accommodate, adapt to, transform and recover from the effects of a hazard in a timely and efficient manner, including through the preservation and restoration of its essential basic structures and functions through risk management." Communities or systems that can creatively reorganize themselves in the wake of disruptive events are considered resilient [20,21]. Studies have argued that community resilience is one of the main priority mechanisms for disaster risk reduction [22–27].

Coping capacity is deeply intertwined with resilience building and is a component of the broader term "capacity" that includes capacity assessment and capacity development [28]. It is the ability of people, organizations, and systems, using available skills and resources, to manage adverse conditions, risk, or disasters. The capacity to cope requires continuing awareness, resources, early warnings, and good management, both in normal times as well as during disasters or adverse conditions. Coping capacities relate to complex inter-linked factors, including coordination between institutions designed to provide support in disasters and consolidation of knowledge and methodologies to assess and deal with the identified risks at the local and national levels [11].

Disaster resilience has been the main focus in disaster-prone regions in recent times [29–36]. The latest emergency management (EM) strategy for Canada [37] also focuses on disaster resilience in the wake of intensifying natural disasters in the country [38,39] and recommends prioritizing resilience building, disaster prevention, and mitigation activities. According to this program, the concept of resilience is defined as "the capacity of a system, community or society to adapt to disturbances resulting from hazards by persevering, recuperating or changing to reach and maintain an acceptable level of functioning." The EM strategy [37] identified five priority areas and activities approved by federal/provincial/territorial (FPT) governments in Canada to strengthen overall resilience:

i.   Enhance whole-of-society collaboration and governance to strengthen resilience.
ii. Improve understanding of disaster risks in all sectors of society.
iii. Increase focus on whole-of-society disaster prevention and mitigation activities.
iv. Enhance disaster response capacity and coordination and foster the development of new capabilities.
v.   Strengthen recovery efforts by building back better to minimize the impacts of future disasters.

The EM strategy supports the FPT governments' vision to strengthen Canada's EM capabilities to prevent/mitigate, prepare for, respond to, and recover from disasters, to reduce disaster risk and increase the resiliency of all individuals and communities in Canada. Frequent flooding is a serious concern in Canada, as was evident from the recurring spring floods in Ontario, Quebec, and Toronto Islands during 2017–2019. Previous recent significant floods include the 2013 southern Alberta floods, the 2013 Toronto urban flood, the 2014 Saskatchewan and Manitoba floods, the 2011 Manitoba flood, and the 2017 British Columbia flood. A potential flooding concern for Toronto Islands is in place in 2020 as Lake Ontario levels may rise in spring.

Although the importance of community perception of risk in the decision-making process has been extensively discussed in the literature [4,8,40–44], the concept has not been applied in practice in a significant way across disciplines. Some of the reasons for that include difficulty in measuring the perception of different actors in the system, assessing the impact of the inclusion of opinion in risk assessment and disaster risk reduction (DRR) policy, and, most of all, designing and collecting relevant data. Perceived resilience is key to comprehend and estimate as it relates to how people perceive risk (exposure to hazards), vulnerability (susceptibility), and their capacity to cope including the institutional support. In survey-based qualitative research, it has been established that post-disaster experiences of affected communities show their preference for coping actions align with their individual personalities [3,45–48]. It is clear that pre-disaster preparedness and capacity building very much depends on how people "feel" about the probability of another disaster to occur in their community [3,49]. A recent study has used a questionnaire survey to evaluate risk perception for risk awareness and to increase resilience in schools [50]. It is worth noting here that people's participation in community-based disaster management must not merely be an illusion of inclusion [51,52].

There is considerable interest in disaster resilience as a mechanism for mitigating the impacts on local communities, yet the identification of metrics and standards for measuring resilience remains a challenge [53]. By measuring baseline characteristics of communities, changes in disaster resilience over time can be monitored by estimating the individual drivers of the disaster resilience (or lack thereof)—social, economic, institutional, infrastructure, and community capacities [53]. Conventionally, quantitative methods for disaster risk assessment are better understood and also preferred as they tend to be based on numbers and indices [19,54–57]. However, interactions between the makeup of the communities, their priorities, and general vulnerabilities, and the geomorphology of the region are natural and must be integrated holistically in risk and resilience assessment [12,58–64]. On the other hand, qualitative methods are generally based on surveys, focus groups, questionnaires, and interviews. The data collected from these methods is then examined, thematized, and analyzed for the understanding and interpretation of a variety of phenomena [65–67]. Qualitative techniques are useful for measuring the impact of policy, needs-assessment of communities, and understanding of people's behaviour and preferences during and post-disaster [68,69]. While the most common approach in research is to apply one of the two methods, quantitative or qualitative, this study integrates both techniques with the intent to capture people's perception as well as the reality on the ground. Therefore, the focus of this study is to explore how to evaluate and incorporate people's perceptions of risk, exposure, susceptibility, and coping capacity to realize gaps between perceived disaster resilience and objectively assessed disaster resilience. Once these gaps are identified, mitigation measures, coping capacity building efforts, and adaptivity initiatives can be developed and implemented with greater success. Based on these principles, a national strategy must be considered and prepared for a more comprehensive application for disaster risk management. The following sections outline the data used in the study, the methodology developed and used on a Canadian city, discussion of the results, and concluding remarks.

2. Materials and Methods

The method proposed in this study is an adapted and extended version of the world risk index (WRI) method [19]. While the WRI measures disaster resilience only objectively, the technique proposed here evaluates people's perception of disaster resilience (function of exposure, susceptibility, and coping capacity) and incorporates their opinion into the assessment of perceived resilience. Previous studies have shown that a questionnaire survey is a useful tool to capture risk awareness and perception of the population to help plan future risk management efforts and encourage a resilience culture in the community [47,50]. In this study, objective and perceived resilience are being evaluated and compared, gaps between perceived and actual resilience are identified, and an integrated resilience is calculated, as well. Community participation is essential in effective and successful disaster management in communities, and this point has been validated by various studies that are based on simulation-based planning tools, basically to engage stakeholders in a user-friendly manner [55,70].

Perceived measurements—In order to incorporate people's perceptions, we have used questionnaire surveys to collect data from four different locations in the City of Brampton in the Greater Toronto Area (GTA) of Canada (Figure 1). The selection of the survey sites was made based on a good representation of the community. Specifically, a sports and community centre (survey location 1) that also houses the public library and a swimming pool, a multicultural community centre (survey location 2), a church (survey location 3), and a restaurant (survey location 4), are places regularly used by the community. We received 100 responses to the questionnaire consisting of 29 questions designed to reflect people's perceptions of threats from natural hazards, how they would cope in emergencies, their background, and how they engage in their local environment. We explained the nature and intent of the study to each participant, including what is meant by risk, how it is a function of different elements, including exposure to hazards. However, it should be noted that at the time of the survey, the exact buffer zones to estimate exposure had not been determined and, therefore, were not communicated to the survey participants. The precise nature of the questions and how they relate to model parameters is given later in this section. People's perceptions are treated as representative of the entire city for demonstrating the methodology leading to the perceived assessment of community resilience (Table 1).

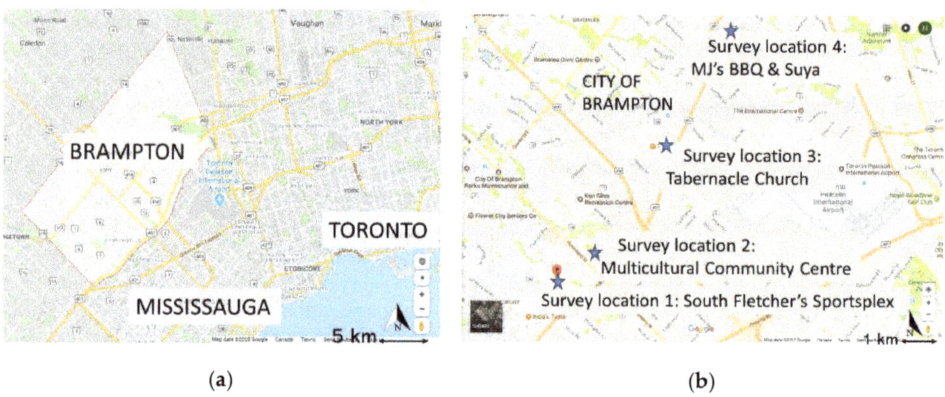

(a)　　　　　　　　　　　　　　　　(b)

Figure 1. (a) Part of the Greater Toronto Area (GTA) for perspective purposes showing the City of Mississauga and the City of Toronto in the south and southeast of Brampton; (b) the four survey locations in Brampton: survey location 1: South Fletcher's Sportsplex community centre, survey location 2: Brampton multicultural community centre; survey location 3: New Birth Tabernacle, a non-denominational faith gathering center; and survey location 4: a local restaurant.

We have used the Thiessen polygons method to aggregate the surveyed areas for the entire city by dissemination areas (DAs)—areas of equal density of population, for perceived assessment of the parameters used in the methodology (Figure 2). The Thiessen polygon technique appropriately assigns areal significance to each survey site by constructing perpendicular bisectors to the lines joining each site with those immediately surrounding it. These bisectors form a series of polygons, each polygon containing one site. The data collected at a site gets assigned to the whole area covered by the enclosing polygon.

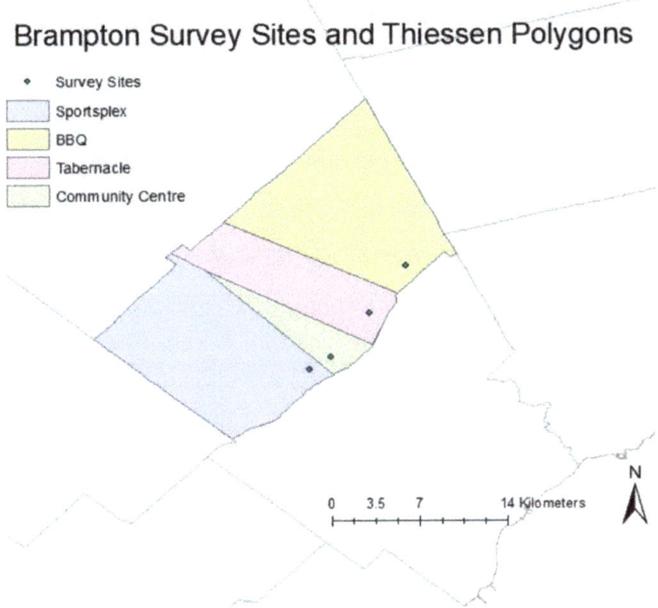

Figure 2. The City of Brampton survey sites and Thiessen polygons (TP) for perceived parameters' assessment. The TP technique appropriately assigns areal significance to each survey site.

It is worth noting here that two survey locations, BBQ and Tabernacle church, shown in Figure 1b, were lumped together as one entity, "BBQ and Tabernacle" for calculations in GIS. The reason being, the total number of responses collected from the two locations combined was rather small (14) compared to the other sites.

As demonstrated in Table 1, three parameters, namely exposure, susceptibility, and lack of coping capacity, are appropriately extracted from the responses to the survey questions by assigning binary values to them. Each parameter is comprised of several variables, representative of the parameter [19,71–73]. Although we have presented a select set of variables to represent each of the three parameters to demonstrate the method, these should not be taken as exhaustive. Disaster types, geomorphology, landuse of the region, and demographics of the population must be taken into account in the determination of these variables. The three parameters are allocated weights according to their assumed influence in this particular case study [55]. The weights can vary depending on the impact of each parameter on community resilience. For example, a known dangerous environment (high exposure), a prosperous and educated neighbourhood (low susceptibility), and an accessible network of emergency services (high coping capacity) must guide how the parameters can be weighted. For example, the residential development in the Barker reservoir in Houston, Texas, that got flooded during Hurricane Harvey [74], will have a high exposure. Table 1 summarizes the proposed process, including a guide to assigning binary values to individual variables as part of the methodology developed to account for people's input in the process of resilience assessment. For example, the parameter exposure is based on dangerous locations such as:

- River (flood risk)
- The transportation network (risk from derailment, explosion, oil spill)
- Chemical plants and hazardous industries
- Transmission lines (elevated health risk from high voltage corridors)

- Oil and gas pipelines (toxic spills)
- Garbage dumpsites (a potential failing of large piles, health, and environmental hazard)

The parameter, susceptibility, comprises the following seven variables:

- Ownership/renting status of the residence
- Crowdedness factor based on the number of members in the household
- Language ability—English or French
- Employment status
- Job satisfaction
- Age—assuming that the very young and the very old would be more susceptible to the exposure to the disaster risk
- Persons with disability

Similarly, the parameter, lack of coping capacity comprises the following variables:

- Level of education
- Family income
- Means of transportation (personal or public)
- Social network support
- Disaster experience by family or friends
- Importance of disaster preparedness
- Engagement in local politics as a proxy of participation

Objective measurements—We have used the 2011 census of Canada for demographic information, Municipal Property Assessment Corporation (MPAC) average property values, slopes and terrain of the region, and landuse. The landuse data is valuable for determining the location of critical infrastructure and critical facilities for the assessment of objective parameters. The GIS software, ArcGIS, is used for data processing and analysis according to the dissemination area (DA) map of the study area. Appendix A lists all the data sources. Figure 3 shows the municipal boundaries and dissemination areas in the GTA. Figure 4 presents datasets, cropped for the City of Brampton, used for parameters estimation proposed in this method.

Table 1. Information extracted from the four surveys toward perceived resilience along with assigned weights and binary values, where the total of all assigned weights $\sum W_i = 1$.

Parameter	Guidance for Assigning Binary Values 0 or 1 Based on Response to the Survey			
Exposure Weight W_1	Question	Binary value = 0	Binary value = 1	Remarks
	I live near a hazardous situation such as a river, chemical plant, etc.	If false	If true	Add all binary numbers to get total exposure
Susceptibility Weight W_2	Question/variable	Binary value = 0 if response is as below	Binary value = 1 if response is as below	Remarks
	My home is	Owned	Rented or other	Add all binary numbers to get total susceptibility
	# members in household (crowdedness)	<4	>4	
	Language (English)	very well/good	moderate/poor/blank	
	Employment	Full time/self-employed	Part time/retired	
	Job satisfaction	Very satisfied/satisfied	Somewhat satisfied/not satisfied/blank	
	Age	>20 and <65	<20 and >65	
	Disability	No	Yes	
Lack of Coping Capacity Weight W_3	Question/variable	Binary value = 0 if response is as below	Binary value = 1 if response is as below	Remarks
	Education	College and higher	Less than college	Add all binary values to get total lack of coping capacity
	Income	>50 K	<50 K	
	Transportation	Personal vehicle	Public/rideshare	
	Social network	Very important	Important/somewhat/other	
	Disaster experience	Yes	No/blank	
	Disaster preparedness option	Family/friend	Public shelter/blank	
	Voted in the past election	Yes	No/blank/n/a	

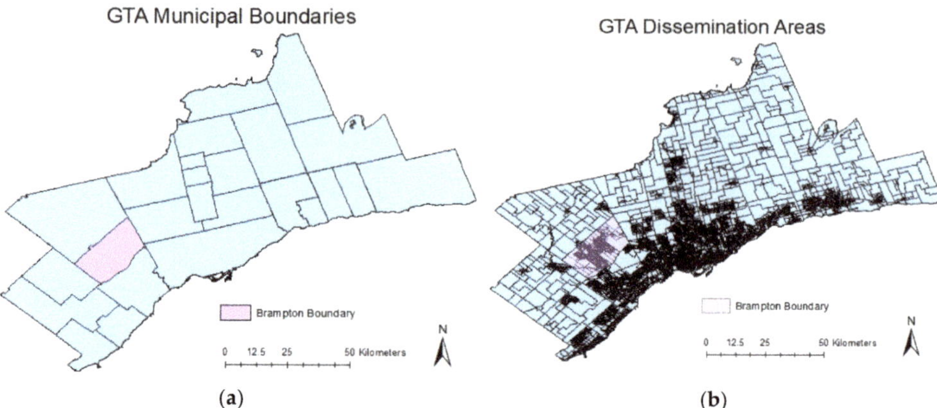

Figure 3. (**a**) GTA municipal boundaries, and (**b**) dissemination areas. City of Brampton is shown in purple.

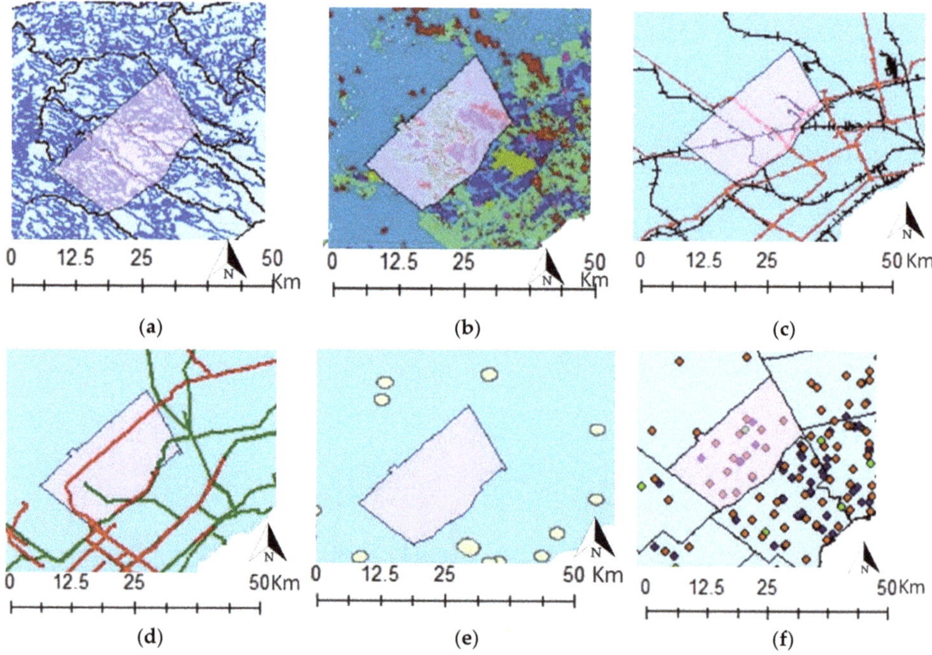

Figure 4. Objective datasets for the City of Brampton (purple boundary): (**a**) watercourses, (**b**) landuse, (**c**) pipelines in red and transmission lines in green, (**d**) the highways in red and the railways in black, (**e**) dumpsites in yellow circles, and (**f**) emergency services.

The landuse data in Figure 4b includes residential (green), commercial (dark purple), government and institutional infrastructure (green), industrial (dark blue), parks and recreational (red), and open area (blue). Similarly, in Figure 4f, emergency services include fire stations (red), police stations (dark blue), and hospitals in green. It should be noted that none of the dumpsite buffers fall within the boundary of the city and, therefore, do not contribute toward exposure in this case study.

Table 2 summarizes the process of quantifying individual variables within each of the three parameters as part of the methodology. Real datasets explained and illustrated in Figure 4 earlier have been used in the measurement of the model parameters: exposure, susceptibility, and lack of coping capacity. Some of the processed data are shown in Figure 5. For example, exposure is determined based on whether or not a resident is located in proximity of (within a kilometer of) a highway, railway, river, industrial site, pipeline, transmission line, oil and gas facility, or dumpsite. Similarly, objective assessment of susceptibility is derived from census data, including the residence type, the age of its construction, the property value, language skill, employment status, and disability rate. The lack of coping capacity is derived from census and GIS data; it includes, income, education, a distance of more than one kilometer from emergency services, a fire station, police service, or ambulance service. See Figure 5 for the visuals of the buffer zones.

Table 2. List of parameters and variables for the objective assessment of community resilience. The total of all assigned weights $\sum W_i = 1$.

Parameter	Criteria to Assign Binary Values to Variables			
	Variable	Binary value = 0	Binary value = 0	Remarks
Exposure Weight W_1	Highways, Railway tracks, River and creeks, Industries, Oil and gas pipelines, Dumpsites–stockpiles, Low lying areas (terrain/slope)	If no exposure to a potential hazard	If within 1 km buffer zone of any of the potentially hazardous situations	Add all binary assigned values to get total exposure
	Variable	Binary value = 0	Binary value = 1	Remarks
Susceptibility Weight W_2	Residence type	Detached/semi	Rented apartment	Add all binary assigned values to get total susceptibility
	Age of property construction	Post-1980	Pre 1980	
	Language	very well/good	moderate/poor/blank	
	Employment	Full time/self-employed	Part time/retired	
	Age	>20 and <65	< 20 and >65	
	Disability	No	Yes	
	Property value	>400 K	≤400 K	
	Variable	Assign value = 0	Assign value = 1	Remarks
Lack of Coping Capacity Weight W_3	Education	College and higher	Less than college	Add all binary assigned values to get total lack of coping capacity
	Income	>50 K	≤50 K	
	Disaster preparedness: Hospital	≤1 km	>1 km away	
	Disaster preparedness: ambulance service	≤1 km	>1 km away	
	Disaster preparedness: Health emergency services	≤1 km	>1 km away	
	Disaster preparedness: Police station	≤1 km	>1 km away	

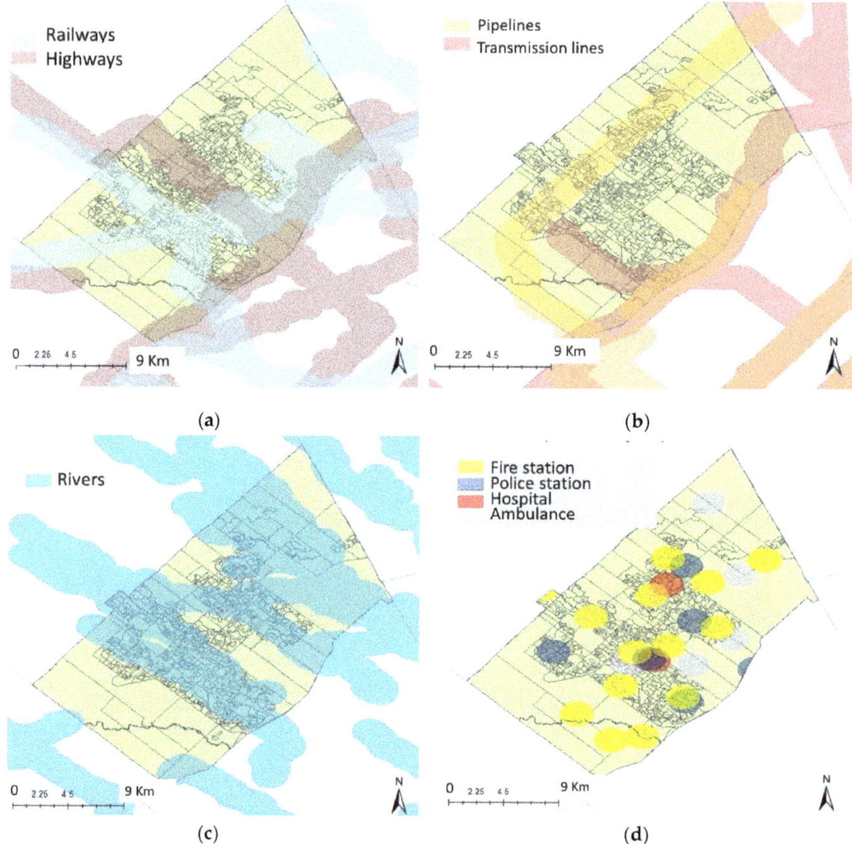

Figure 5. The City of Brampton: dissemination areas superimposed (**a**) with buffer zones of 1 km around the transportation network, railways in blue and highways in pink; (**b**) transmission lines in pink and gas pipelines in blue; (**c**) rivers in blue to demarcate exposure zones; and (**d**) buffer zones of 1km around emergency services to assess lack of coping capacity including fire stations in yellow, police stations in blue, hospitals in red, and ambulance services in a light purple.

The overall resilience is calculated using Equations (1) to (6), derived from the WRI method [19]. Precisely, Equation (1) calculates the Lack of Resilience using objective data for the three parameters that are individually normalized. The parameters can be weighted using weights thought to be appropriate for this case study for demonstration purposes. The weights (W_1, W_2, W_3) are open to adjustment in individual cases depending on the potential influence of the parameters, as well as the objective of the study. For example, a vulnerable community is perceived as more susceptible in comparison to an affluent neighbourhood. Residential development well outside of flood zones indicates a lower level of exposure even though the population may be regarded as vulnerable. If the community is at a substantial distance from a healthcare facility, it may reflect a lack of coping capacity in emergencies. Therefore, in this scenario, it will make sense to allocate a lower percentage of weight to "Exposure" and higher percentage to "Lack of Coping Capacity" and "Susceptibility." For the application and demonstration of the methodology, we have assigned weights to the parameters as given in Equation (2) for both objective and perceived calculations.

Equation (3) gives the estimate of Resilience using objective data, Equation (4) estimates Resilience using perceived data, and Equation (5) determines the combined Resilience by summing up the perceived and objective Resilience estimates. We propose here that the combined Resilience is a way to account for perceived and real measures of community resilience.

$$\begin{aligned}\text{Lack of Resilience}_{\text{Objective}} \\ = W_1 \times \text{Exposure}_{\text{Objective}} + W_2 \times \text{Susceptibility}_{\text{Objective}} \\ + W_3 \times \text{Lack of Coping Capacity}_{\text{Objective}},\end{aligned} \qquad (1)$$

where $\sum W_i = 1$

$$\begin{aligned}\text{Lack of Resilience}_{\text{Objective}} \\ = 0.2 \times \text{Exposure}_{\text{Objective}} + 0.4 \times \text{Susceptibility}_{\text{Objective}} \\ + 0.4 \times \text{Lack of Coping Capacity}_{\text{Objective}}\end{aligned} \qquad (2)$$

$$\text{Resilience}_{\text{Objective}} = 1 - (\text{Lack of Resilience})_{\text{Objective}} \qquad (3)$$

$$\text{Resilience}_{\text{Perceived}} = 1 - (\text{Lack of Resilience})_{\text{Perceived}} \qquad (4)$$

$$\text{Resilience}_{\text{combined}} = \text{Resilience}_{\text{Perceived}} + \text{Resilience}_{\text{Objective}} \qquad (5)$$

The estimated total resilience has been calculated by normalizing the combined resilience estimates using Equation (6):

$$\text{Resilience} = \frac{\text{Resilience}_{\text{combined}} - \text{MIN}\,(\text{Resilience}_{\text{combined}})}{(\text{MAX}\,(\text{Resilience}_{\text{combined}}) \quad \text{MIN}\,(\text{Resilience}_{\text{combined}}))} \qquad (6)$$

It is noteworthy that Equations (3) and (4) were calculated using normalized and unweighted individual parameters in Equation (2), leading to an adjusted maximum total value of perceived (and objective) Resilience. The objective and perceived resilience calculations are done using the same formula.

3. Results

The findings of the research are illustrated through Figures 6–8. Objective and perceived measures were made for the survey sites. The actual datasets on the geomorphology of the study area, including rivers and low lying regions, indicate that the method has been applied to a flooding scenario as an example. As mentioned earlier, two survey locations, namely, BBQ and Tabernacle church, were lumped together as one entity, "BBQ and Tabernacle" for calculations in GIS due to the small number of responses from the two locations. Each of the three parameters was calculated separately to achieve perceived and objective estimates of them for each survey location. Figure 6 shows objective measures of the three model parameters, namely, Exposure, Susceptibility, and the Lack of Coping Capacity calculated for survey location 2, the Community Centre. Similar maps were obtained for perceived measures of the three parameters. Also, similar maps were developed for other survey locations. In total, 12 maps were obtained for objective and perceived estimates of the three parameters for the three survey locations. Although it may be desirable for the readers to visualize the step-by-step development of these maps, to avoid overcrowding of illustrations, selected graphs are presented here.

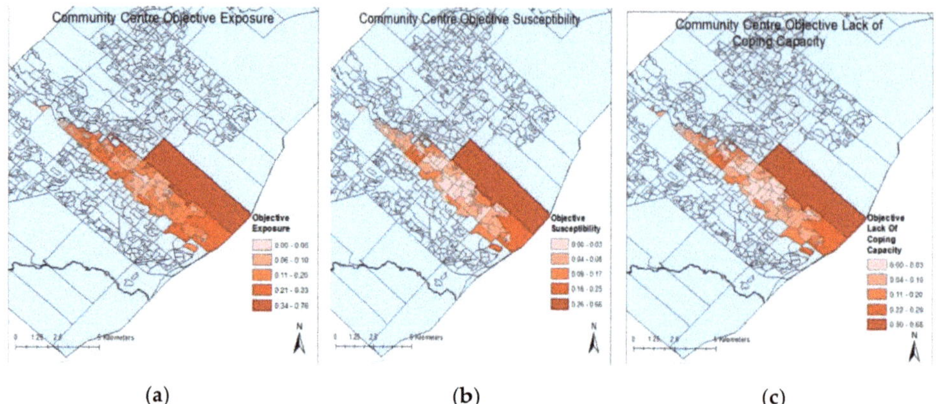

(a)   (b)   (c)

Figure 6. Objective measures of the three parameters for one of the three survey areas: (**a**) exposure, (**b**) susceptibility, and (**c**) lack of coping capacity. Similar maps were obtained for the other two survey areas but are not shown here.

Figure 7 shows the integrated (perceived plus objective) measure of Resilience for the three survey locations, separately. Figure 8 is a picture of the perceived and objective Resilience measures for the entire City of Brampton, and Figure 9 represents the integrated resilience map for the city. Each of the three parameters was normalized before being weighted in Equation (1). Therefore, as can be seen in Figure 8, the two resilience maps show estimates of up to 63 for perceived Resilience and up to 89 for the objective Resilience. It is noteworthy that the method suggests the ideal resilience value to be 100. Therefore, objective resilience gives a better level of resilience in the study area as compared to perceived measure of resilience. In essence, the maximum value of objective resilience is higher than perceived resilience, and the higher the value of resilience, the better in each case scenario.

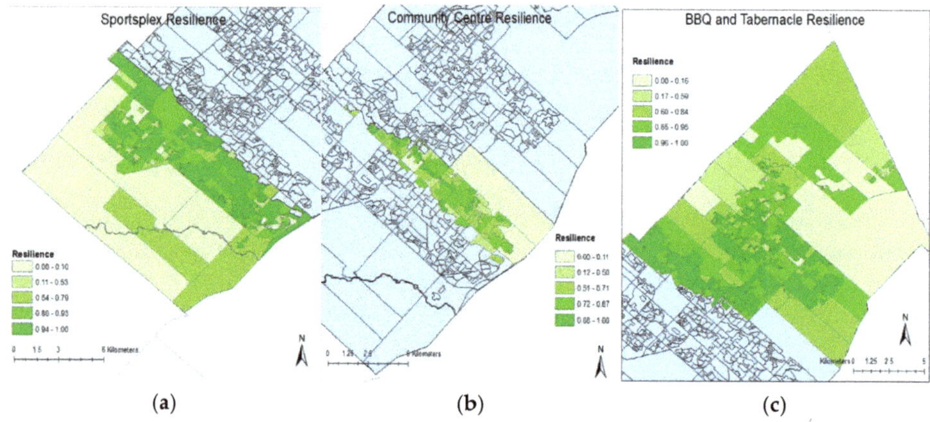

(a)   (b)   (c)

Figure 7. The measure of total resilience (perceived plus objective) for the three survey sites: (**a**) Location 1, (**b**), Location 2, and (**c**) Location 3.

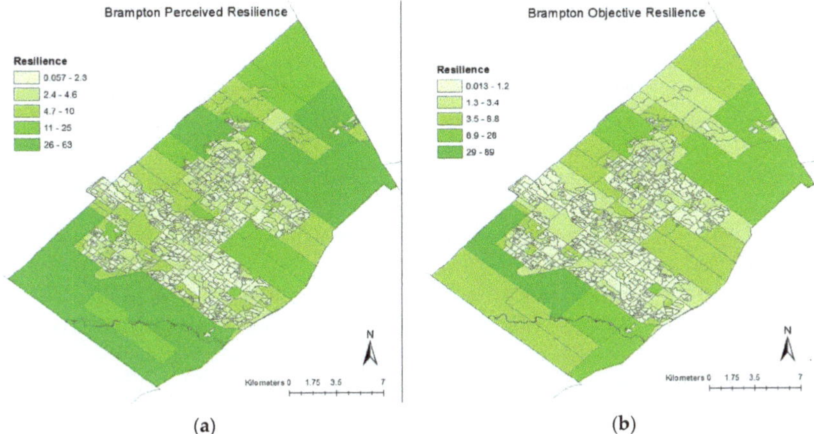

(**a**)   (**b**)

Figure 8. Resilience maps aggregated for the City of Brampton (in percentage): (**a**) perceived, and (**b**) objective.

Although Figure 8 is the key outcome of this study, Figure 9 has been arrived at by combining the two resilience measures, showing the regions of variable resilience. It is useful to see how the two measures of resilience may have aligned with each other, positively or negatively. In future research, it would be helpful to subtract the two resilience measures from each other and visualize where the gaps are between the public perception and the reality on the ground. The shades of green indicate levels of resilience of the community, the darker, the better. The shades of yellow and parts of red are in regions where rivers and creeks flow. The areas shown in red indicate low resilience due to a variety of reasons such as industrial areas, flood zones, low lying areas, and a dense network of watercourses.

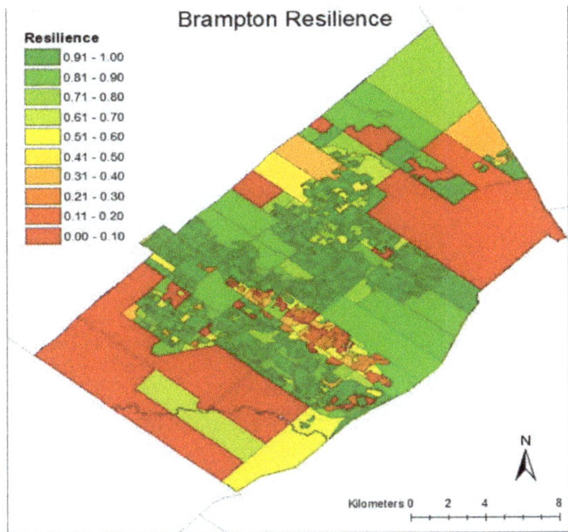

Figure 9. Integrated resilience map for the City of Brampton (normalized); higher resilience value indicates a higher level of resilience.

4. Discussion and Conclusions

The core objective of this study was to develop a methodology to capture community perceptions of flood hazard risk based on their exposure, susceptibility, and lack of coping capacity, and incorporate those perceptions into estimating community resilience. Inspired by the world risk index (WRI), we have successfully demonstrated that the WRI determinants can be used to explain the difference between how risk is perceived and the factual picture presented by the objective data. We estimate objective resilience using actual data related to exposure (dangerous locations), susceptibility (derived from demographic data), and lack of coping capacity (derived from a combination of census and landuse data). The two resilience measures, perceived and objective, were combined by adding them and normalizing the resulting combined resilience. It should be noted that the perceived and objective parameters were normalized at the unweighted stage that justifies the similar treatment of the two (perceived and objective). The study should be used as a decision-making tool to enhance the resilience of communities based on their circumstances.

We used a questionnaire to engage people, but it should be noted that some of the nuances of the methodology were under development at the time of the survey. The randomly selected respondents were residents of the City of Brampton, as reflected by their use of the community centre, sports centre, the church and the restaurant used as survey locations. We acknowledge the small sample size of 100 participants that may have led to a rather underwhelmingly visual difference in the perceived and objective resilience measures. The reason for that to happen could be the fact that a small sample from each survey site was used to represent an entire region determined by the Thiessen polygon method. Furthermore, with a limited representation of the community in the study area, the perceived data estimates are bound to have uncertainties that can be addressed with extensive survey data. The determination of the buffer zones for different exposure variables was made by treating the variables such as highways and rivers in a similar manner. However, this aspect needs to be improved by allowing different buffer zones for various variables.

In summary, this research makes a reasonable preliminary attempt toward achieving a better representative measure of the resilience of a community in the context of disasters and emergencies. Future research is recommended for examining current disaster mitigation policies in Ontario, engaging representative communities to capture their perceptions, and looking at how specific changes can be made in those policies to improve their relevance and outcome for a diverse population. Solutions to disaster risk reduction and preparedness strategies lie in meaningful consultation with stakeholders, no matter how insignificant some may seem. We recommend future research to refine the methodology along with its various aspects presented here. The approach should also be user-friendly for a broader application. In Canada, the most frequent hazard, flooding, and other disasters are managed by the local authorities first; the province assists if local capacities prove to be insufficient; and eventually, the federal government helps if the consequences are too severe. In many cases, the federal help arrives when it is too late due to unclear guidance, inconsistent messaging, political and ideological differences among actors, and conflicting priorities. There is a need, now more than ever before, to develop a national strategy for resilience to all disasters to enable an environment of swift impact assessment of events and allocation of resources across the nation.

Author Contributions: Conceptualization, methodology, and data curation by N.A.; data processing and analysis, M.E. and N.A.; visualization, M.E.; supervision, N.A. and S.P.S.; funding acquisition, S.P.S. and N.A. All authors have read and agreed to the published version of the manuscript.

Funding: Funding from NSERC-CREATE-ADERSIM and York University Minor Research Grant.

Acknowledgments: Funding from NSERC-CREATE-ADERSIM and York University Minor Research Grant made the study possible. It offered high-quality training opportunities for York University students, Mark Elliott from Environmental Studies for the GIS work, and Judith Jubril and Sarah-Maude Guindon from Master of Disaster and Emergency Management (MDEM) toward the surveys. Thanks to Alain Normand, Manager, Emergency Management, the City of Brampton, for helping with logistics in conducting community surveys. Sabrina Daddar, MDEM, also contributed in liaising with community partners in the initial stages.

Conflicts of Interest: The authors declare no conflict of interest.

Appendix A List of Data Sources

Data	Source
Exposure	
Highways	DMTI Spatial, 2015 http://geo.scholarsportal.info/#r/details/_uri@=2347499980
Rail Lines	Ontario Ministry of Natural Resources, 2012
Industrial Sites	Canada, Federal Government Open Data Program, created by request, 2016
GTA Pipes, Transmission Lines	DMTI Spatial, 2014
Slope and DEM	DMTI Spatial, 2015 (retired)
Major Rivers	Provided by Toronto and Region Conservation Authority
Susceptibility	
Home Ownership	Statscan, 2011 census data via http://dc1.chass.utoronto.ca/census/
Age of Construction	Statscan, 2011 census data via http://dc1.chass.utoronto.ca/census/
Language Skills	Statscan, 2011 census data via http://dc1.chass.utoronto.ca/census/
Employment	Statscan, 2011 census data via http://dc1.chass.utoronto.ca/census/
Age	Statscan, 2011 census data via http://dc1.chass.utoronto.ca/census/
Disability	Statistics Canada, by request, 2017
Property Value	online real estate listings (ReMax), geocoded by address in ArcGIS Online
Coping Capacity	
GTA Fires Stations	http://geo.scholarsportal.info/#r/details/_uri@=3739967620
GTA Police Stations	http://geo.scholarsportal.info/#r/details/_uri@=3739967620
GTA Hospitals	http://geo.scholarsportal.info/#r/details/_uri@=3570906326
Ambulance Stations	Addresses gathered from publicly available information at municipal websites, Wikipedia, and Google; geocoded using ArcGIS Online
Income	Statscan, 2011 census data via http://dc1.chass.utoronto.ca/census/
Education	Statscan, 2011 census data via http://dc1.chass.utoronto.ca/census/
Miscellaneous	
Watercourses	Ontario Ministry of Natural Resources, 2011
Watersheds	Ontario Ministry of Natural Resources, 2011
Dissemination Areas	University of Toronto Census Analyzer http://dc1.chass.utoronto.ca/census/
Land Use	DMTI Spatial, 2014 http://geo.scholarsportal.info/#r/details/_uri@=2785150059$DMTI_2014_CanMapRL_Topo_LUR_ALL_PROV
Census Tracts	University of Toronto Census Analyzer http://dc1.chass.utoronto.ca/cgi-bin/census/2011nhs/displayCensus.cgi?year=2011&geo=ct
GTA Municipalities	DMTI Spatial, 2014 http://geo.scholarsportal.info/#r/details/_uri@=4044335176$DMTI_2014_CanMapRL_Streets_MUN_ALL_PROV
Education	DMTI Spatial, 2015 http://geo.scholarsportal.info/#r/details/_uri@=4062179246

References

1. Pazzi, V.; Morelli, S.; Pratesi, F.; Sodi, T.; Valori, L. Assessing the safety of schools affected by geo-hydrologic hazards: The geohazard safety classification (GSC). *Int. J. Disaster Risk Reduct.* **2016**, *15*, 80–93. [CrossRef]
2. Schiermeier, Q. Increased flood risk linked to global warming. *Nature* **2011**, *470*, 316. [CrossRef]
3. Bempah, S.A.; Oyhus, A.O. The role of social perception in disaster risk reduction: Beliefs, perception, and attitudes regarding flood disasters in communities along the Volta River, Ghana. *Int. J. Disaster Risk Reduct.* **2017**, *23*, 104–108. [CrossRef]
4. Ranjan, E.S.; Abenayake, C.C. A study on community's perception on disaster resilience concept. *Procedia Econom. Financ.* **2014**, *18*, 88–94. [CrossRef]
5. Cutter, S. Preface. In *Disaster Resilience: A National Imperative, Committee on Increasing National Resilience to Hazards and Disasters*; The National Academies Press: Washington, DC, USA, 2011; p. 244.
6. Cutter, S.L.; Barnes, L.; Berry, M.; Burton, C.; Evans, E.; Tate, E. A place-based model for understanding community resilience to natural disasters. *Glob. Environ. Chang.* **2008**, *18*, 598–606. [CrossRef]

7. Birkholz, S.; Muro, M.; Jeffrey, P.; Smith, H.M. Rethinking the relationship between flood risk perception and flood management. *Sci. Total Environ.* **2014**, *478*, 12–20. [CrossRef] [PubMed]

8. Santoro, S.; Pluchinotta, I.; Pagano, A.; Pengal, P.; Cokan, B.; Giordano, R. Assessing stakeholders' risk perception to promote Nature Based Solutions as flood protection strategies: The case of the Glinščica River (Slovenia). *Sci. Total Environ.* **2019**, *655*, 188–201. [CrossRef] [PubMed]

9. Lei, Y.; Wang, J.; Yue, Y.; Zhou, H.; Yin, W. Rethinking the relationships of vulnerability, resilience, and adaptation from a disaster risk perspective. *Nat. Hazards* **2014**, *70*, 609–627. [CrossRef]

10. Hyogo Framework for Action (HFA) 2005–2015 Building the Resilience of Nations and Communities to Disasters. Available online: www.unisdr.org (accessed on 4 April 2020).

11. Sendai Framework for Disaster Risk Reduction (SF-DRR) 2015–2030; The United Nations Office for Disaster Risk Reduction: Geneva, Switzerland. Available online: www.unisdr.org (accessed on 4 April 2020).

12. Cutter, S.L.; Boruff, B.J.; Shirley, W.L. Social vulnerability to environmental hazards. *Soc. Sci. Q.* **2003**, *84*, 242–261. [CrossRef]

13. Few, R. Flooding, vulnerability and coping strategies: Local responses to a global threat. *Prog. Dev. Stud.* **2003**, *3*, 43–58. [CrossRef]

14. Messner, F.; Meyer, V. Flood damage, vulnerability and risk perception—Challenges for flood damage research. In *Flood Risk Management: Hazards, Vulnerability and Mitigation Measures*; NATO Science Series; Schanze, J., Zeman, E., Marsalek, J., Eds.; Springer: Dordrecht, The Netherlands, 2006; Volume 67.

15. Kuhlick, C.; Scolobig, A.; Tapsell, S.; Steinführer, A.; Marchi, B.D. *Contextualizing Social Vulnerability: Findings from Case Studies across Europe*; Springer Science & Business Media B.V.: New York, NY, USA, 2011.

16. Burns, J.W.; Slovic, P. Risk perception and behaviors: Anticipating and responding to crises. *Risk Anal.* **2012**, *32*, 579–582. [CrossRef] [PubMed]

17. Paton, D.; Johnston, D. *Disaster Resilience: An Integrated Approach*, 2nd ed.; Charles, C., Ed.; Thomas Publisher: Springfield, IL, USA, 2017; p. 438.

18. Chang, S.E.; Shinozuka, M. Measuring improvements in the disaster resilience of communities. *Earthq. Spectra* **2004**, *20*, 739–755. [CrossRef]

19. Welle, T.; Birkmann, J. The World Risk Index—An approach to assess risk and vulnerability on a global scale. *J. Extreme Events* **2015**, *2*, 1550003. [CrossRef]

20. Folke, C.; Carpenter, S.R.; Walker, B.; Scheffer, M.; Chapin, T.; Rockström, J. Resilience thinking: Integrating resilience, adaptability and transformability. *Ecol. Soc.* **2010**, *15*, 20–29. [CrossRef]

21. Folke, C. Resilience: The emergency of a perspective for social-ecological systems analyses. *Glob. Environ. Chang.* **2006**, *16*, 253–267. [CrossRef]

22. Kuhlicke, C.; Steinführer, A. Preface: Building social capacities for natural hazards: An emerging field for research and practice in Europe. *Nat. Hazards Earth Syst. Sci.* **2015**, *15*, 2359–2367. [CrossRef]

23. McEwen, L.J.; Krause, F.; Jones, O.; Garde-Hansen, J. Sustainable flood memories, informal knowledge and the development of community resilience to future flood risk. *WIT Trans. Ecol. Environ.* **2012**, *159*, 253–264.

24. Ashley, R.M.; Blanskby, J.; Newman, R.; Gersonius, B.; Poole, A.; Lindley, G.; Smith, S.; Ogden, S.; Nowell, R. Learning and action alliances to build capacity for flood resilience. *J. Flood Risk Manag.* **2012**, *5*, 14–22. [CrossRef]

25. Schelfaut, K.; Pannemans, B.; van der Craats, I.; Krywkow, J.; Mysiak, J.; Cools, J. Bringing flood resilience into practice: The FREEMAN project. *Environ. Sci. Policy* **2011**, *14*, 825–833. [CrossRef]

26. Hung, H.C.; Yang, C.Y.; Chien, C.Y.; Liu, Y.C. Building resilience: Mainstream community participation into integrated assessment of resilience to climatic hazards in metropolitan land use management. *Land Use Policy* **2016**, *50*, 48–58. [CrossRef]

27. WMO. *Building Resilience through Community Participation—A Report on the Pilot Project on Community Flood Management in Bangladesh, India and Nepal*; WMO: Geneva, Switzerland, 2017.

28. UN-DRR. United Nations Office for Disaster Risk Reduction. Available online: https://www.undrr.org/terminology/capacity (accessed on 4 April 2020).

29. Birkmann, J. (Ed.) Basic principles, and theoretical concepts. In *Measuring Vulnerability to Natural Hazards: Towards Disaster Resilient Societies*; United Nations University Press: Tokyo, Japan, 2006; p. 524.

30. Birkmann, J.; Cardona, O.D.; Carreno, M.L.; Barbat, A.H.; Pelling, M.; Scheiderbauer, S.; Kienberger, S.; Keiler, M.; Alexandar, D.; Zeil, P.; et al. Framing vulnerability, risk and societal responses: The MOVE framework. *Nat. Hazards* **2003**, *67*, 193–211. [CrossRef]

31. Boin, A.; Comfort, L.K.; Demchak, C.C. The rise of resilience. In *Designing Resilience: Preparing for Extreme Events*; Comfort, L.K., Boin, A., Demchak, C.C., Eds.; University of Pittsburgh Press: Pittsburgh, PA, USA, 2010.

32. Britton, N.R.; Clark, G.J. From response to resilience: Emergency management reform in New Zealand. *Nat. Hazards Rev.* **2000**, *1*, 145–150. [CrossRef]

33. Brooks, N. Vulnerability, risk, and adaptation: A conceptual framework. *Tyndall Cent. Clim. Chang. Res. Work. Pap.* **2003**, *38*, 1–16.

34. Simonovic, S.P.; Peck, A. Dynamic resilience to climate change caused natural disasters in coastal megacities quantification framework. *Br. J. Environ. Clim. Chang.* **2013**, *3*, 378–401. [CrossRef]

35. Timmerman, P. Vulnerability, resilience and the collapse of society: A review of models and possible climatic applications. In *Environmental Monograph No. 1*; Institute for Environmental Studies, University of Toronto: Toronto, ON, Canada, 1981.

36. Turner, B.L. Vulnerability and resilience: Coalescing or paralleling approaches for sustainability science? *Glob. Environ. Chang.* **2010**, *20*, 570–576. [CrossRef]

37. Government of Canada. *Emergency Management Strategy for Canada: Toward a Resilient 2030*; Public Safety Canada Report; Public Safety Canada: Ottawa, ON, Canada, 2019; ISBN 978-0-660-29248-9.

38. EM-DAT. The International Disaster Database, Centre for Research on the Epidemiology of Disasters 2019. Available online: http://www.emdat.be/disaster_trends/index.html (accessed on 16 March 2020).

39. Canadian Disaster Database. Public Safety Canada 2019. Available online: https://www.publicsafety.gc.ca/cnt/mrgnc-mngmnt/ntrl-hzrds/index-en.aspx (accessed on 16 March 2020).

40. Nirupama, N.; Maula, A. Engaging public for building resilient communities to reduce disaster impact. *Nat. Hazards* **2013**, *66*, 51–59. [CrossRef]

41. Nirupama, N.; Etkin, D. Institutional perception and support in emergency management in Ontario, Canada. *Disaster Prev. Manag.* **2012**, *21*, 599–607. [CrossRef]

42. Nirupama, N.; Etkin, D. Emergency managers in Ontario: An exploratory study of their perspective. *J. Homel. Secur. Emer. Manag.* **2009**, *6*. [CrossRef]

43. Nirupama, N. Risk, and vulnerability assessment—A comprehensive approach. *Int. J. Disaster Resil. Built Environ.* **2012**, *3*, 2. [CrossRef]

44. Bruneau, M.; Chang, S.E.; Eguchi, R.T.; Lee, G.C.; O'Rourke, T.D.; Reinhorn, A.M.; Shinozuka, M.; Tierney, K.; Wallace, W.A.; von Winterfeldt, D. A framework to quantitatively assess and enhance the seismic resilience of communities. *Earthq. Spectra* **2003**, *19*, 733–752. [CrossRef]

45. Building Resilient Communities. *Fort McMurray Wildfire: Learning from Canada's Costliest Disaster*; Institute for Catastrophic Loss Reduction: Zurich, Switzerland, 2019.

46. McGee, T.K. Public engagement in neighbourhood level wildfire mitigation and preparedness: Case studies from Canada, the US and Australia. *J. Environ. Manag.* **2011**, *92*, 2524–2532. [CrossRef]

47. Koksal, K.; McLennan, J.; Bearman, C. Living with bushfires on the urban-bush interface. *Aust. J Emerg. Manag.* **2020**, *35*, 21–28.

48. Miller, F.; Osbahr, H.; Boyd, E.; Thomalla, F.; Bharwani, S.; Ziervogel, G.; Walker, B.; Birkmann, J.; Vander Leeuw, S.; Rockstro"m, J.; et al. Resilience and vulnerability: Complementary or conflicting concepts? *Ecol. Soc.* **2010**, *15*, 11. [CrossRef]

49. Slovic, P. *The Feeling of Risk: New Perspectives on Risk Perception*; Routledge: London, UK, 2010; p. 425.

50. Bandecchi, A.E.; Pazzi, V.; Morelli, S.; Valori, L.; Casagli, N. Geo-hydrological and seismic risk awareness at school: Emergency preparedness and risk perception evaluation. *Int. J. Disaster Risk Reduct.* **2019**, *40*, 101–280. [CrossRef]

51. Arnstein, S.R. A ladder of citizen participation. *J. Am. Inst. Plan.* **1969**, *35*, 216–224. [CrossRef]

52. Few, R.; Brown, K.; Tompkins, E. Public participation and climate change adaptation: Avoiding the illusion of inclusion. *Clim. Policy* **2007**, *7*, 46–59. [CrossRef]

53. Cutter, S.L.; Burton, C.G.; Emrich, C.T. Disaster Resilience indicators for benchmarking baseline conditions. *J. Homel. Secur. Emerg. Manag.* **2010**, *7*, 51. [CrossRef]

54. Armenakis, C.; Nirupama, N. Estimating spatial disaster risk in urban environments. *Geomat. Nat. Hazards Risk* **2013**, *4*, 289–298. [CrossRef]

55. Irwin, S.; Simonovic, S.P.; Nirupama, N. *Introduction to ResilSIM: A Decision Support Tool for Estimating Disaster Resilience to Hydro-Meteorological Events*; Water Resources Research Report No. 094; Facility for Intelligent Decision Support, Department of Civil and Environmental Engineering, Western University: London, ON, Canada, 2016; p. 66. ISBN (print) 978-0-7714-3115-9; (online) 978-0-7714-3116-6.
56. Kovacs, P.; Kunreuther, H. Managing Catastrophic Risk: Lessons from Canada 2001. Available online: http://www.iclr.org/images/Managing_Catastrophic_Risk.pdf (accessed on 4 April 2020).
57. Smith, K. (Ed.) *Environmental Hazards: Assessing Risk and Reducing Disaster*, 2nd ed.; Routledge: London, UK, 1996.
58. Holling, C.S. Resilience and stability of ecological systems. *Annu. Rev. Ecol. Syst.* **1973**, *4*, 1–23. [CrossRef]
59. Kasperson, J.X.; Kasperson, R.E. *The Social Contours of Risk*; Earthscan: London, UK, 2005.
60. Klein, R.J.T.; Nicholls, R.J.; Thomalla, F.T. Resilience to natural hazards: How useful is this concept? *Environ. Hazards* **2003**, *5*, 35–45. [CrossRef]
61. McLaughlin, P. Climate change, adaptation, and vulnerability: Reconceptualizing societal-environment interaction within a socially constructed adaptive landscape. *Organ. Environ.* **2011**, *24*, 269–291. [CrossRef]
62. Paton, G. Disaster resilience: Building capacity to co-exist with natural hazards and their consequences. In *Disaster Resilience*; Charles C. Thomas Publishers: Springfield, IL, USA, 2006; p. 321.
63. Walker, B.; Holling, C.S.; Carpenter, S.R.; Kinzig, A. Resilience, adaptability and transformability in social-ecological systems. *Ecol. Soc.* **2004**, *9*, 5. [CrossRef]
64. Wisner, B.; Blaikie, P.; Cannon, T.; Davis, I. *At Risk: Natural Hazards, People's Vulnerability, and Disasters*, 2nd ed.; Routledge: London, UK, 2004; p. 471.
65. Bazeley, P. *Qualitative Data Analysis: Practical Strategies*; Sage Publications: Thousand Oaks, CA, USA, 2013; p. 472.
66. Creswell, J.W.; Shope, R.; Plano Clark, V.L.; Green, D.O. How interpretive qualitative research extends mixed methods research. *Res. Sch.* **2006**, *13*, 1–11.
67. Ratner, C. Subjectivity and objectivity in qualitative methodology. *Forum Qual. Soc. Res.* **2002**, *3*, 16.
68. Mamuji, A.; Rozdilsky, J.L. Canada's 2016 Fort McMurray Wildfire Evacuation: Experiences of the Muslim Community. *Int. J. Emerg. Manag.* **2019**, *15*, 125–146. [CrossRef]
69. Siegrist, M.; Gutscher, H. Natural hazards and motivation for mitigation behavior: People cannot predict the affect evoked by a severe flood. *Risk Anal.* **2008**, *28*, 771–778. [CrossRef]
70. Simonovic, S.; Nirupama, N. A spatial multi-objective decision making under uncertainty for water resources management. *J. Hydroinf.* **2005**, *7*, 117–133. [CrossRef]
71. Minguez, C.; Fermina, G.; Perez, R., II. Doctorate in Geography Workshop. In Proceedings of the XXV Congress of the Association of Spanish Geographers, Madrid, Spain, 25–27 October 2017.
72. Malik, S.; Awan, H.; Khan, N. Mapping vulnerability to climate change and its repercussions on human health in Pakistan. *Glob. Health* **2012**, *8*, 31–41. [CrossRef]
73. Twigg, J. *Characteristics of a Disaster-Resilient Community: A Guidance Note*; DFID Disaster Risk Reduction Interagency Coordination Group, Hazard Research Centre, Benfield UCL: London, UK, 2007; p. 39.
74. Wallace, T.; Watkins, D.; Park, H.; Singhve, A.; Williams, J. *How One Houston Suburb Ended up in a Reservoir*; New York Times: New York, NY, USA, 2018.

Article

Developing a Risk-Based Consensus-Based Decision-Support System Model for Selection of the Desirable Urban Water Strategy: Kashafroud Watershed Study

Reza Javidi Sabbaghian [1] and A. Pouyan Nejadhashemi [2,*]

[1] Department of Civil Engineering, Hakim Sabzevari University (HSU), Sabzevar 96179-76487, Iran; r.javidi.s@hsu.ac.ir
[2] Department of Biosystems and Agricultural Engineering, Michigan State University (MSU), East Lansing, MI 48824, USA
* Correspondence: pouyan@msu.edu; Tel.: +1-517-432-7653

Received: 18 March 2020; Accepted: 1 May 2020; Published: 5 May 2020

Abstract: In recent years, complexities related to a variety of sustainable development criteria and several preferences of stakeholders have caused a serious challenge for selecting the more desirable urban water strategy within watershed. In addition, stakeholders might have several risk attitudes depending on the number of criteria satisfied by water strategies. Accordingly, a risk-based consensus-based group decision-support system model is proposed for choosing the more desirable water strategy, using the external modified ordered weighted averaging (EMOWA) and internal modified ordered weighted averaging (IMOWA) operators. The operators calculate the scores of strategies in several risk-taking attitudes of group decision-making, considering the sustainable development criteria. Additionally, the consensus-seeking phase is considered using a risk-based weighted Minkowski's method. This model is successfully implemented for the Kashafroud urban watershed in Iran, for selecting the more desirable urban water strategy in 2040. Accordingly, in the completely risk-averse viewpoint, the stakeholders select the combined supply-demand management strategy satisfying all of the criteria. In contrast, in the completely risk-prone standpoint, the stakeholders choose the demand management strategy satisfying at least one criterion. Developing the risk-based consensus-based group decision-support system model is suggested for integrated urban watershed management for selecting the more desirable strategy, satisfying the sustainable development criteria.

Keywords: integrated urban watershed management; group decision-support system; risk analysis; group consensus; Kashafroud watershed

---

1. Introduction

According to the recent reports published by United Nations, the population of the world has been estimated at 7.7 billion people in 2019 and projected to continue its increasing trend to around 8.5 billion in 2030, 9.7 billion in 2050, and 10.9 billion in 2100 [1]. Furthermore, it is predicted that most of the world population will live in urban regions rather than rural areas. The growth of the urban population faces some challenges, including unsuitable urban planning and management, insufficient public services, social and cultural anomalies, economic problems associated with urban poverty, environmental contamination, and supplying secure and sustainable water [1,2].

Amongst the aforementioned urban challenges, the water supply is one of the most serious. Concerning water supply, water withdrawal from renewable resources, water transfer, treatment of wastewater, water allocation for several demands, satisfying security and sustainability, and consensus-seeking among urban stakeholders with different preferences are the main issues [2,3]. Therefore, the water supply issues should be met by implementing desirable water strategies, which consider multiple sustainable development criteria [4]. Additionally, the several preferences of multiple stakeholders regarding the relevant water demands should be satisfied and group consensus could be achieved [5]. Accordingly, Parkinson et al. have presented the integrated urban watershed management (IUWM) approach, which has been developed for better management of water and wastewater strategies in an urban setting [6].

One of the most complicated challenges for implementing IUWM is increasing the variety of sustainable development objectives, including water resources sustainability, environmental sustainability, socio-economic sustainability, and the related criteria [4,7]. This has led to a serious problem for selecting the more desirable urban water strategy, by which the sustainable development objectives and the important relevant criteria should be satisfied, in addition to achieving the group consensus among the stakeholders. Accordingly, implementing IUWM requires evaluation of the water strategies for supplying the urban demands, while considering the sustainable development criteria and the final group agreement [8,9].

The other significant challenge for implementing IUWM is related to the variety of risk-taking attitudes of stakeholders' groups [10]. The risk-taking attitudes represent the number of criteria that should be satisfied by the urban water strategies [11,12]. The risk-taking cases, which are identified by the risk-taking degrees, are expressed by some linguistic phrases such as "selecting the more desirable strategy for satisfying all criteria" in the completely risk-averse viewpoint, "selecting the more desirable strategy for satisfying at least one criterion" in the completely risk-prone standpoint, and the other cases between these two limits [13,14].

In order to take on the aforementioned challenges, an appropriate model for IUWM should be developed to consider the sustainable development objectives, risk-taking attitudes of stakeholders, and a final group consensus in evaluation of urban water strategies. Simonovic and Bender analyzed the collaborative planning-support system (CPSS) model, as the subset of the decision-support system (DSS) model, which considers all relevant aspects of sustainable water resources planning and management, especially in the process of criteria selection [15]. In the group decision support system (GDSS) approach, the main three issues, such as selection of criteria, generation of alternatives, and evaluation of alternatives based on the criteria are considered based on the balancing and reinforcing aspects for better decision analysis [16–18]. Accordingly, developing a group decision-support system (GDSS) model within an urban watershed needs to analyze a multiple criteria decision-making (MCDM) process, in which the final criteria and water strategies are selected, the strategies are evaluated with respect to the criteria, and the water strategies are ranked for several risk-taking cases.

For analyzing the MCDM process in a GDSS model, a large variety of methodologies have been utilized, of which the most frequently used methods have been well demonstrated in the literature [19]. The most important methods are classified into four categories, including: scoring methods [20], distance-based methods [20–23], outranking methods [20,24–26], and pair-wise comparisons methods [27,28]. The differences between these methods are related to their strategies for solving MCDM problems.

For risk analysis in the MCDM process, some of the risk-based methodologies have been utilized. The most commonly used method is using the family of the ordered weighted averaging (OWA) operator, which considers the risk analysis in the decision-making process [4,8,9]. The OWA family includes a group of operators with several properties. In this family, the most frequently used operators are the OWA, induced ordered weighted averaging (IOWA), and hybrid weighted averaging (HWA). The OWA operator considers only the risk-taking attitudes and disregards the criteria weights. The IOWA operator considers the risk-taking attitudes and the importance orders of criteria, whereas it ignores the criteria weights. The HWA operator considers the risk-taking attitudes and criteria weights, while disregards the stakeholders' power weights [11,12].

In recent years, related to the context of urban water management, several studies have been done based on MCDM methodologies [29–37]. However, in this paper, a risk-based consensus-based GDSS model is developed for IUWM within the study area of an urban watershed. Accordingly, the following improvements are performed in this study that are distinctive comparing to similar works in IUWM:

- Developing a comprehensive GDSS model for IUWM based on group risk considerations and group consensus measuring.
- Improving the OWA operator properties, considering the risk-taking attitudes of decision-making, importance degrees of criteria, and the stakeholders' power weights simultaneously.
- Considering the two types of aggregation, including the external modified ordered weighted averaging (EMOWA) and the internal modified ordered weighted averaging (IMOWA), for selecting the more desirable urban water strategy in several risk-taking attitudes of stakeholders.
- Selecting final criteria for the MCDM process by use of a risk-based group consensus method.
- Seeking group consensus among stakeholders during the GDSS process by use of a risk-based weighted Minkowski's method.

Accordingly, in order for sustainable water resources management, this research makes the connection between the outputs of watershed modeling and the inputs of a GDSS model for analyzing the risk-based MCDM process. This paper can be used to select the most effective sustainable development criteria of watershed by the stakeholders of the watershed. Additionally, it can assist water scientists and analysts of water resources management to analyze the several impacts of implementing water strategies on the selected sustainable development criteria. Furthermore, it can help all stakeholders to identify the conditions of watershed including several demands, probable water supply resources, and the related impacts on the criteria, which result in better decision making for a sustainable watershed. Ultimately, this study leads to a collaborative group consensus among stakeholders and, consequently, facilitates integrated watershed management.

This paper is organized as follows: Section 2 proposes the flow diagram and explains the complete analysis of the risk-based consensus-based GDSS model for the urban watershed. In addition, this section introduces the study area of the urban watershed, the criteria, the urban water strategies, and the participating stakeholders in the decision-making process. The methodology is also applied for IUWM of the study area. In Section 3, the results, including the scores of urban water strategies, the group consensus measurements, and the final ranking of strategies, are obtained in several risk-taking cases. Section 4 discusses the results and effects of several risk-taking cases on the scores and ranking of the strategies. Finally, Section 5 concludes this paper and proposes future research.

2. Materials and Methods

2.1. Overview of the Methodology

In watershed planning and management, especially for IUWM, the stakeholders have several opinions about the importance degrees of sustainable development criteria. Accordingly, the final most effective criteria should be selected based on stakeholders' group consensus. Additionally, the several water strategies, which are classified in three categories of supply management, demand management, and combined supply-demand management, should be evaluated with respect to the selected criteria. The impacts of watershed modeling outputs related to each water strategy on each criterion is considered as the evaluation value of that corresponding strategy with regard to that corresponding criteria. Therefore, the evaluation values of water strategies should enter to the MCDM process of the GDSS model as its inputs for analyzing the model.

In order to analyze the GDSS model based on the risk-taking considerations of the group of stakeholders and group consensus, the risk-based consensus-based GDSS model is developed for urban watershed management in this study. In the risk analysis, a type of OWA operator is proposed to improve the properties of the OWA, IOWA, and HWA operators. Therefore, the stakeholders can evaluate the water strategies with respect to the selected criteria and rank the strategies in each risk-taking attitude of the group. Indeed, each stakeholder and the group of stakeholders can determine that, in each risk-taking case, which water strategy is more desirable and how many criteria are satisfied by that strategy. By using a risk-based weighted Minkowski's method, the stakeholders' group consensus is controlled, and the level of group consensus is determined in each risk-taking case. If the final group agreement is reached, the GDSS process is terminated; otherwise, the threshold level of agreement is reconsidered, or the iterated GDSS process continues based on evaluating other water strategies until the final agreement is achieved.

2.1.1. The Proposed Risk-Based Consensus-Based GDSS Process

The proposed flow diagram of the risk-based consensus-based GDSS process for the IUWM is represented in Figure 1, which includes the six phases of identification and selection, weighting, evaluation, aggregation and risk analysis, consensus-seeking, and ranking:

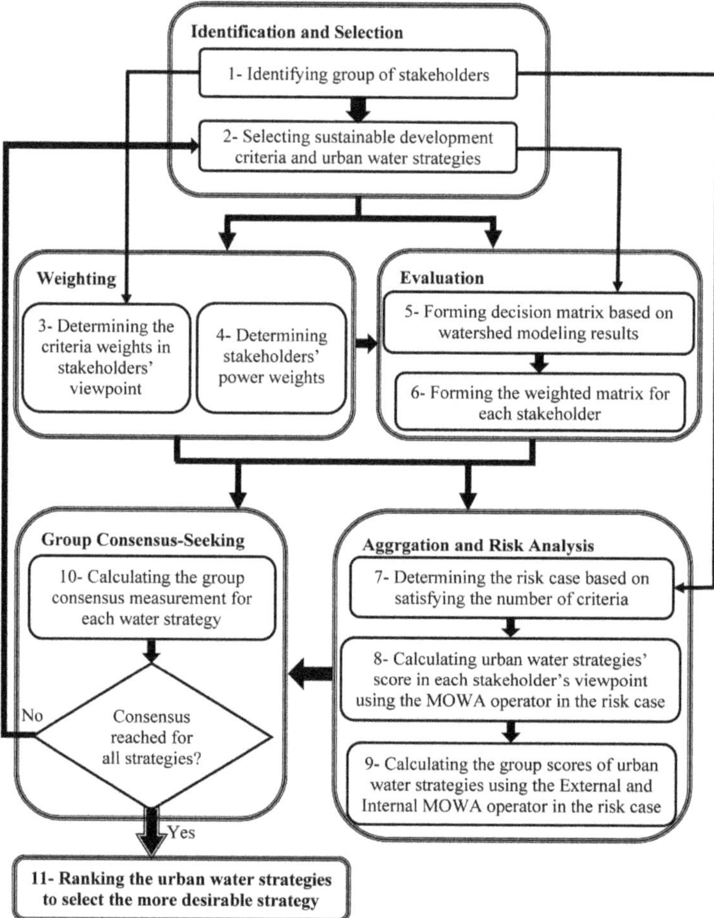

Figure 1. Proposed risk-based consensus-based group decision support system (GDSS) process for the integrated urban water management (IUWM). MOWA: modified ordered weighted averaging.

In the process, first, m' initial water strategies are scored by each of the p stakeholders. Additionally, n' initial criteria are weighted by each of the p stakeholders. Accordingly, $C' = \{C'_1, \ldots, C'_{i'}, \ldots, C'_{n'}\}$ is assumed as the set of initial criteria; $S' = \{S'_1, \ldots, S'_{j'}, \ldots, S'_{m'}\}$ is considered as the set of initial feasible strategies; and $Sth. = \{Sth._1, \ldots, Sth._k, \ldots, Sth._p\}$ is the set of stakeholders. After selection of final strategies and criteria, m final strategies are evaluated with regard to n final criteria by each of the p stakeholders. For convenience, $C = \{C_1, \ldots, C_i, \ldots, C_n\}$ is assumed as the set of final criteria, and $S = \{S_1, \ldots, S_j, \ldots, S_m\}$ is considered as the set of final feasible strategies. Additionally, $\lambda = (\lambda_1, \lambda_2, \ldots, \lambda_p)^T$ is the vector of the stakeholders' power weights, where $\lambda_k \geq 0$. In addition, $w^{(k)} = \left(w_1^{(k)}, w_2^{(k)}, \ldots, w_n^{(k)}\right)^T$ is the vector of criteria weights in the kth stakeholder's viewpoint ($w_i^{(k)} \geq 0$, $\sum_{i=1}^n w_i^{(k)} = 1$, $k = 1, 2, \ldots, p$).

2.1.2. Identification and Selection Phase

In Steps 1 and 2 of the proposed process (Figure 1), the stakeholders are identified to select final sustainable development criteria and choose final water strategies based on the stakeholders' group consensus.

In order to select the final appropriate criteria from a large number of criteria, first the Delphi methodology is used to extract the initial criteria from the large number of sustainable development criteria by obtaining the opinions of stakeholders through a survey process [15,38,39]. After that, considering the watershed facts comprises meteorological, hydrological, and hydrogeological characteristics of the watershed, priorities of the watershed, and concepts of sustainable development criteria, all stakeholders are asked about the relevant preferences of the initial criteria. The final sustainable development criteria are selected from the set of initial criteria based on the primitive consensus-based weighted Minkowski's method using Equation (1):

$$Consensus^{(G)}\left(C'_{i'}\right) = 1 - \left\{\sum_{k=1}^{p}\left\{\lambda_k \times \left|w'^{(G)}_{i'} - w'^{(k)}_{i'}\right|^2\right\}\right\}^{\frac{1}{2}}, i' = 1, 2, \ldots, n' \tag{1}$$

where λ_k is each stakeholder's final power weight. The power weight is determined primarily by the linguistic variable followed by defuzzifying the equivalent fuzzy number and obtaining each stakeholder's final power weight. $w'^{(k)}_{i'}$ and $w'^{(G)}_{i'}$ denote the preference values of the i'th initial sustainable criteria based on the kth stakeholder's viewpoint and group viewpoint, respectively, where $w'^{(k)}_{i'}$ is determined like the stakeholders' final power weights, and $w'^{(G)}_{i'} = \sum_{k=1}^{p} \lambda_k \times w'^{(k)}_{i'}$. In addition, $Consensus^{(G)}\left(C'_{i'}\right)$ is the consensus measurement for the i'th initial criteria. According to the group consensus-seeking literature, a threshold level of agreement (TLA) is determined by group of stakeholders to control the final agreement level between the individual stakeholders' viewpoints and the overall group opinion related to the initial criteria. The criteria that satisfy the condition of $Consensus^{(G)}\left(C'_{i'}\right) \geq TLA$ are selected as the final sustainable water criteria and considered as the inputs of the risk-based GDSS model.

Regarding the generate water strategies in the group decision-making process, the design theory has been widely accepted, as it is one of the most frequently used methodologies [16,40]. Accordingly, the C-K theory (concepts–knowledge) has been considered as a generative process that allows stakeholders to describe and analyze innovative design processes for generating strategies [41,42]. For operationalizing the C-K theory, the method of K-C-P (knowledge-concepts-proposals) has been proposed to manage the GDSS design process, in which multiple stakeholders could be included [43].

In this study, all details about watershed conditions, including meteorological, hydrological, and hydrogeological characteristics of watershed, water resources, water demands, and properties of sustainable development criteria, are provided for stakeholders within the questionnaire during the survey process [16]. Post-survey, all stakeholders are asked to comment about the initial water strategies. The final water strategies are chosen from the set of initial strategies according to the primitive consensus-based weighted Minkowski's method using Equation (2):

$$Consensus^{(G)}\left(S'_{j'}\right) = 1 - \left\{\sum_{k=1}^{p}\left\{\lambda_k \times \left|a_{j'}^{(G)} - a_{j'}^{(k)}\right|^2\right\}\right\}^{\frac{1}{2}}, j' = 1, 2, \ldots, m' \tag{2}$$

where $a_{j'}^{(k)}$ and $a_{j'}^{(G)}$ represent the preference values of the j'th initial water strategy based on the kth stakeholder's viewpoint and the group viewpoint, respectively, where $a_{j'}^{(k)}$ is determined like the stakeholders' final power weights, and $a_{j'}^{(G)} = \sum_{k=1}^{p}\lambda_k \times a_{j'}^{(k)}$. In addition, $Consensus^{(G)}\left(S'_{j'}\right)$ is the consensus measurement for the j'th initial water strategy. According to the group consensus-seeking literature, the strategies that satisfy the condition of $Consensus^{(G)}\left(S'_{j'}\right) \geq TLA$ are chosen as the final water strategies and considered as the inputs of the risk-based GDSS model. The other strategies are not chosen but have the chance to be reconsidered in the iterative process of the GDSS model. Additionally, these strategies could be analyzed in the K-C-P methodology for generating new strategies.

2.1.3. Weighting Phase

Regarding Step 3, the criteria weights are determined. In the MCDM problems, several methods have been applied for calculating criteria weights [38,44]. One of the most commonly used methodologies is the entropy method, which represents the dispersion of a criterion in evaluations of strategy [39,45]. In this study, the entropy method is utilized to calculate the entropy weight of each criterion by using Equation (3):

$$u_i^{(k)} = \frac{1 + K\sum_{j=1}^{m}\left\{\overline{a}_{ij}^{(k)} \times \log\left(\overline{a}_{ij}^{(k)}\right)\right\}}{\sum_{i=1}^{n}\left\{1 + K\sum_{j=1}^{m}\left\{\overline{a}_{ij}^{(k)} \times \log\left(\overline{a}_{ij}^{(k)}\right)\right\}\right\}} \tag{3}$$

where $K = 1/\log(n)$ is a constant value, n and m are the numbers of final criteria and final strategies, respectively, and $\overline{a}_{ij}^{(k)}$ is the normalized value of $a_{ij}^{(k)}$. $a_{ij}^{(k)}$ represents the evaluation value of the jth strategy with respect to the ith criterion based on the kth stakeholder's viewpoint. $u_i^{(k)}$ is the entropy weight of the ith criterion in the kth stakeholder's viewpoint.

In this paper, in addition to the entropy weight of each criterion as an objective weight, the linguistic importance degree of each criterion is also considered as a subjective weight, which represents the stakeholders' preferences related to that corresponding criterion.

In order to express the stakeholders' viewpoints, some of the methodologies have been proposed based on the fuzzy set theory and fuzzy logic [46]. In water resources management problems, the three types of response, such as crisp response, linguistic fuzzy response, and conditional fuzzy response, could be utilized for analyzing input values [47].

In this study, the linguistic fuzzy response is used for determining the importance degrees of criteria, which utilizes fuzzy membership functions and concludes accurate outputs [48]. Accordingly, each stakeholder determines the importance degree of each criterion by using one of the linguistic members from the set of S = (No importance, Very low importance, Low importance, Slightly low importance, Moderate importance, Slightly high importance, High importance, Very high importance, Perfect importance) [49,50]. The linguistic importance degrees of criteria are fuzzified by the trapezoidal-triangular fuzzy membership functions [51,52]. The trapezoidal-triangular fuzzy membership functions, which are used for importance degrees of criteria and the stakeholders' power weights, are represented in Figure 2:

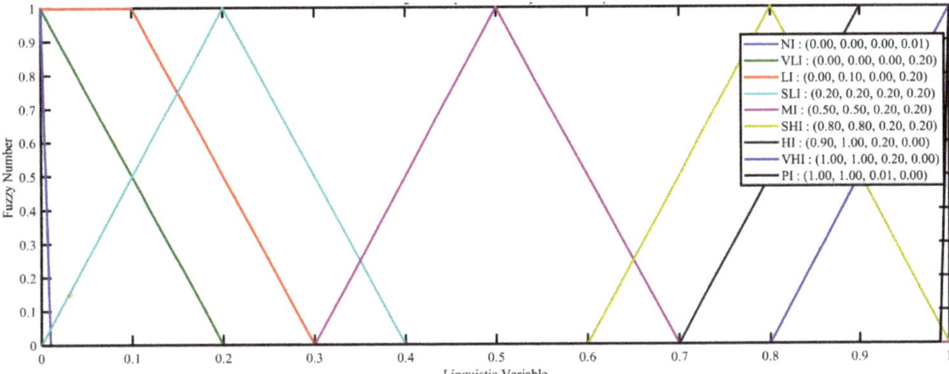

Figure 2. Trapezoidal-triangular fuzzy membership functions of linguistic variables. No importance (NI), very low importance (VLI), low importance (LI), slightly low importance (SLI), moderate importance (MI), slightly high importance (SHI), high importance (HI), very high importance (VHI), and perfect importance (PI).

The fuzzified variables are defuzzified by using the centroid method [51,53,54]. Accordingly, the defuzzified importance degree of each criterion (subjective weight) is determined using Equation (4):

$$\hat{w}_i^{(k)} = \frac{\int \mu\left(\hat{w}_i^{(k)}\right) \times \hat{w}_i^{(k)} \times d\left(\hat{w}_i^{(k)}\right)}{\int \mu\left(\hat{w}_i^{(k)}\right) \times d\left(\hat{w}_i^{(k)}\right)} , \ i = 1,\dots,n \ ; k = 1,\dots,p \tag{4}$$

where $\hat{w}_i^{(k)}$ is the defuzzified importance degree of the ith criterion in the kth stakeholder's viewpoint. $\mu\left(\hat{w}_i^{(k)}\right)$ is the trapezoidal-triangular fuzzy membership function of $\hat{w}_i^{(k)}$.

The linguistic variables and the defuzzified values are presented in Table 1.

Table 1. Linguistic variables and the equivalent fuzzy interval and defuzzified values [51].

Linguistic Variable	Fuzzy Numbers	Defuzzified Value
No importance (NI)	(0.00, 0.00, 0.00, 0.01)	0.001
Very low importance (VLI)	(0.00, 0.00, 0.00, 0.20)	0.063
Low importance (LI)	(0.00, 0.10, 0.00, 0.20)	0.106
Slightly low importance (SLI)	(0.20, 0.20, 0.20, 0.20)	0.200
Moderately importance (MI)	(0.50, 0.50, 0.20, 0.20)	0.500
Slightly high importance (SHI)	(0.80, 0.80, 0.20, 0.20)	0.800
High importance (HI)	(0.90, 1.00, 0.20, 0.00)	0.894
Very high importance (VHI)	(1.00, 1.00, 0.20, 0.00)	0.937
Perfect importance (PI)	(1.00, 1.00, 0.01, 0.00)	1.000

The final weight of the ith criterion in the kth stakeholder's viewpoint is determined by using Equation (5):

$$w_i^{(k)} = \frac{\hat{w}_i^{(k)} \times u_i^{(k)}}{\sum_{i=1}^{n} \hat{w}_i^{(k)} \times u_i^{(k)}} \tag{5}$$

Like the process illustrated for determination of the subjective weight for each criterion, the final power weight for each stakeholder (λ_k) is also determined using Equation (6):

$$\lambda_k = \frac{\int \mu(\lambda_k) \times \lambda_k \times d(\lambda_k)}{\int \mu(\lambda_k) \times d(\lambda_k)} , \; i = 1, \ldots, n \,; k = 1, \ldots, p \tag{6}$$

2.1.4. Evaluation Phase

Regarding Step 5, a decision matrix is formed for each stakeholder, in which m strategies are evaluated with regard to n criteria. Each element of decision matrix $(a_{ij}^{(k)})$ represents the evaluation value of the jth strategy with regard to the ith criterion based on the kth stakeholder's viewpoint.

The stakeholders' decision matrices are then normalized by using the first type of linear normalization method, which is applicable for both the positive and negative criteria based on Equations (7) and (8), respectively [55]. $\bar{a}_{ij}^{(k)}$ is the normalized evaluation value of $a_{ij}^{(k)}$:

$$\bar{a}_{ij}^{(k)} = \frac{a_{ij}^{(k)}}{a_i^{(k)*}} \text{ where } a_i^{(k)*} = max_j\left\{a_{ij}^{(k)}\right\} \tag{7}$$

$$\bar{a}_{ij}^{(k)} = \frac{a_i^{(k)\sim}}{a_{ij}^{(k)}} \text{ where } a_i^{(k)\sim} = min_j\left\{a_{ij}^{(k)}\right\} \tag{8}$$

2.1.5. Aggregation and Risk Analysis Phase

In GDSS for watershed management, a group of stakeholders have several risk-taking attitudes towards decision-making, which are expressed by linguistic phrases such as "selecting the more desirable water strategy based on satisfying all of criteria" in the completely risk-averse (completely conservative or completely pessimistic) viewpoint and "selecting the more desirable water strategy based on satisfying at least one criterion" in the completely risk-prone (completely nonconservative or completely optimistic) standpoint. In addition, the other risk-taking attitudes such as "most of, many of, half of, some of, and a few of" are applied between these two cases [13,56]. Accordingly, the risk-taking degree of θ has been assigned for each of the risk-taking cases [57,58]. Several risk-taking cases, equivalent linguistic phrases, and the relevant risk-taking degrees are presented in Table 2.

Table 2. Risk-taking cases, equivalent linguistic phrases, and relevant risk-taking degrees [58].

Risk-Taking Case	Equivalent Linguistic Phrase	Risk-Taking Degree (θ)
Completely risk-averse	Satisfies all of the criteria	0.001
Risk-averse	Satisfies most of the criteria	0.091
Fairly risk-averse	Satisfies many of the criteria	0.333
Neutral risk	Satisfies half of the criteria	0.500
Fairly risk-prone	Satisfies some of the criteria	0.667
Risk-prone	Satisfies few of the criteria	0.909
Completely risk-prone	Satisfies at least one criterion	0.999

Regarding Step 7, for each risk-taking case, a corresponding risk-based order weights vector of $v = (v_1, v_2, \ldots, v_n)^T$, $v_i \geq 0$, $\sum_{i=1}^n v_i = 1$ is determined. The order weights are determined for several risk cases and the relevant risk-taking degrees of θ based on the regular increasing monotone (RIM) fuzzy linguistic quantifier and using Equation (9) [13,56,58,59]:

$$v_i = \left(\frac{i}{n}\right)^{\left(\frac{1}{\theta}\right)-1} - \left(\frac{i-1}{n}\right)^{\left(\frac{1}{\theta}\right)-1} , \; i = 1, 2, \ldots, n \tag{9}$$

2.1.6. External Aggregation

In the external aggregation, the order weights vector for each risk-taking case is utilized to calculate the scores of water strategies in each stakeholder's opinion. In order to complete Step 8, an n-dimensional function of $F : I^n \rightarrow J$ is used for a weighted normalized matrix related to each stakeholder for aggregating its evaluation values within the first aggregation. In this function, I denotes the set of evaluation values of each strategy, and J represents the corresponding score.

Therefore, according to the external risk analysis through the EMOWA operator, the evaluation values of each strategy associated with each stakeholder are aggregated to calculate the score of that corresponding strategy in several risk-taking cases using Equation (10):

$$F_{EMOWA}^{(k)}\left(w_1^{(k)}\overline{a}_{1j}^{(k)}, \ldots, w_n^{(k)}\overline{a}_{nj}^{(k)}\right)\left(S_j\right) = \sum_{i=1}^{n}\left\{\left\{\left(\frac{i}{n}\right)^{\left(\frac{1}{\theta}\right)-1} - \left(\frac{i-1}{n}\right)^{\left(\frac{1}{\theta}\right)-1}\right\} \times b_i^{(k)}\right\} \tag{10}$$

where $v = (v_1, v_2, \ldots, v_n)^T$ is the risk-based order weights vector associated with n criteria, for which $v_i \geq 0$, $\sum_{i=1}^{n} v_i = 1$. Additionally, $b_i^{(k)}$ is the ith largest value of the $\left(w_1^{(k)}\overline{a}_{1j}^{(k)}, w_2^{(k)}\overline{a}_{2j}^{(k)}, \ldots, w_n^{(k)}\overline{a}_{nj}^{(k)}\right)$ vector related to each stakeholder's weighted normalized evaluation matrix. Finally, $F_{EMOWA}^{(k)}\left(S_j\right)$ is the score of the jth strategy from the kth stakeholder's viewpoint. In Equation (10), the scores of strategies from each stakeholder's viewpoint is calculated for several risk-taking cases.

Regarding Step 9, in the second aggregation, a p-dimensional function of $F^G : I^p \rightarrow J$ is applied to a group of stakeholders for aggregating their scorings related to each strategy. In this function, I denotes the set of stakeholders' scorings related to each strategy, and J represents the corresponding group score.

Therefore, the second aggregation step is accomplished, in which the stakeholders' scorings related to each strategy are aggregated to calculate the group score of that corresponding strategy in several risk-taking cases by using Equation (11):

$$F_{EMOWA}^G\left(S_j\right) = \sum_{k=1}^{p}\left\{\lambda_k \sum_{i=1}^{n}\left\{\left\{\left(\frac{i}{n}\right)^{\left(\frac{1}{\theta}\right)-1} - \left(\frac{i-1}{n}\right)^{\left(\frac{1}{\theta}\right)-1}\right\} \times b_i^{(k)}\right\}\right\} \tag{11}$$

where λ_k is the kth stakeholder's power weight, and $F_{EMOWA}^G\left(S_j\right)$ is the score of the jth strategy from the viewpoint of the group. In Equation (11), the scores of strategies from the group of stakeholders' viewpoints is calculated for several risk-taking cases.

2.1.7. Internal Aggregation

In the internal aggregation, the order weights vector for each risk-taking case is directly used to calculate the scores of water strategies in the group of stakeholders' viewpoints. Accordingly, in the one-step aggregation, an n-dimensional function of $F^G : I^p \rightarrow J$ is used for the group weighted normalized matrix related to the group of stakeholders for aggregating its evaluation values. In this function, I denotes the set of group evaluation values of each strategy, and J represents the corresponding group score.

Therefore, with respect to the internal risk analysis performed by the IMOWA operator, the evaluation values of each strategy associated with the group of stakeholders are aggregated to calculate the score of that corresponding strategy in several risk-taking cases using Equation (12):

$$F_{IMOWA}^{(G)}\left(w_1^{(G)}\overline{a}_{1j}^{(G)}, \ldots, w_n^{(G)}\overline{a}_{nj}^{(G)}\right)\left(S_j\right) = \sum_{i=1}^{n}\left\{\left\{\left(\frac{i}{n}\right)^{\left(\frac{1}{\theta}\right)-1} - \left(\frac{i-1}{n}\right)^{\left(\frac{1}{\theta}\right)-1}\right\} \times b_i^{(G)}\right\} \tag{12}$$

where $v = (v_1, v_2, \ldots, v_n)^T$ is the risk-based order weights vector related to n criteria, for which $v_i \geq 0$, $\sum_{i=1}^{n} v_i = 1$. Additionally, $b_i^{(G)}$ is the ith largest value of the $\left(w_1^{(G)}\bar{a}_{1j}^{(G)}, w_2^{(G)}\bar{a}_{2j}^{(G)}, \ldots, w_n^{(G)}\bar{a}_{nj}^{(G)}\right)$ vector related to the group weighted normalized evaluation matrix. Finally, $F_{IMOWA}^{(G)}(S_j)$ is calculated as the score of the jth strategy from the group of stakeholders' viewpoints for several risk-taking cases.

2.1.8. Group Consensus-Seeking Phase

Regarding Step 10 (Figure 1), group consensus should be controlled to confirm that a final agreement is reached among stakeholders about water strategies. Accordingly, the consensus measurement for each strategy is calculated in order to control the final agreement amongst stakeholders associated with all water strategies.

In recent years, various methodologies have been utilized for calculating consensus measurements. Most of the frequently used methodologies have been classified in the two general approaches [60–65]. The first approach has been developed based on the hard consensus, in which the consensus measurements are calculated concerning the similarity of individual preferences compared with the group opinion [5]. Next, this is compared with the threshold level of agreement (TLA) index. The second approach has been developed according to the soft consensus, in which the individuals change their opinions collaboratively, until a consensus is reached [66,67].

In this paper, a hard consensus approach is utilized for seeking consensus among stakeholders for the first implementation of the risk-based GDSS process. After the first implementation, a soft consensus approach is used in the iterative implementation of the risk-based GDSS process if a final agreement is not reached. First, the risk-based weighted Minkowski's method is applied to calculate the consensus measurements for water strategies. In this study, the Euclidean Minkowski's distance is used for calculating the consensus measurement of each strategy, which implies a simple squared weighting and the related parameter of q equals 2 ($q = 2$). Regarding the relationship between the Minkowski's parameter of q and the risk-taking degree of decision-making [68], the Euclidean Minkowski's method minimizes the distance between the individual viewpoints and the group opinion regarding water strategies leading to a consensus amongst the majority of stakeholders [69]. By using the Euclidean distance, the score of each water strategy ($F_{EMOWA}^{(k)}(S_j)$, $j = 1, 2, \ldots, m$), determined by individual stakeholders, is compared with the score of that corresponding strategy, determined by the group of stakeholders ($F_{EMOWA}^{(G)}(S_j)$ or $F_{IMOWA}^{(G)}(S_j)$, $j = 1, 2, \ldots, m$). The consensus measurement for each water strategy is calculated based on the EMOWA and IMOWA results, using Equations (13) and (14):

$$Consensus_{EMOWA}^{(G)}(S_j) = 1 - \left\{ \sum_{k=1}^{p} \lambda_k \times \left| F_{EMOWA}^{(G)}(S_j) - F_{EMOWA}^{(k)}(S_j) \right|^2 \right\}^{\frac{1}{2}}, \; j = 1, 2, \ldots, m \qquad (13)$$

$$Consensus_{IMOWA}^{(G)}(S_j) = 1 - \left\{ \sum_{k=1}^{p} \lambda_k \times \left| F_{IMOWA}^{(G)}(S_j) - F_{EMOWA}^{(k)}(S_j) \right|^2 \right\}^{\frac{1}{2}}, \; j = 1, 2, \ldots, m \qquad (14)$$

where $Consensus_{EMOWA}^{(G)}(S_j)$ and $Consensus_{IMOWA}^{(G)}(S_j)$ are the consensus measurements for the jth water strategy, where its score is calculated based on using EMOWA or IMOWA in several risk-taking cases, respectively.

According to Equations (13) and (14), it is considered that the lower distances between the individual stakeholders' viewpoints and the overall group opinion associated with each water strategy leads to higher consensus measurement for that strategy.

To control the hard consensus in this study, the TLA index is determined as the linguistic variable of "slightly high" and defuzzified to the corresponding crisp value of 0.800. The consensus measurements for strategies are compared with the selected TLA. Accordingly, the final agreement amongst stakeholders is achieved when $\forall j$, $Consensus_{EMOWA}^{(G)}(S_j) \geq TLA$ for external aggregation or

$\forall\, j, Consensus_{IMOWA}^{(G)}(S_j) \geq TLA$ for internal aggregation. Otherwise, the risk-based GDSS process is iterated, and the soft consensus approach is implemented. According to this issue, all stakeholders are asked about their preferences related to the generation of new strategies, considering the combination of rejected strategies. The generation of strategies' process could be modeled by the K-C-P methodology. The iterative risk-based GDSS process is then implemented based on the evaluation of newly generated strategies, the combined rejected strategies, and the previously agreed strategies with respect to the final selected criteria. This process is iterated until a sufficient level of agreement is achieved amongst all stakeholders.

Ultimately, after a final agreement among all stakeholders, the water strategies are ranked based on the group scores in the several risk-taking cases.

2.2. Study Area

The study of the risk-based GDSS model is performed on the Kashafroud urban watershed area, which is located in North-Eastern Iran with a longitude of 58°20′ up to 60°08′ and latitude of 35°40′ up to 36°03′ (Figure 3). The Kashafroud watershed is one of the largest and the most populated watersheds in Iran. The mean, minimum, and maximum watershed elevations above sea level are 1846 m, 390 m, and 3302 m, respectively. The watershed has a total area of 1,565,000 ha and a growing population that is estimated to reach 5,100,000 by 2040 [70]. The total urban water demand is predicted to reach 490 million cubic meters (MCM) by 2040. This watershed has a cold and arid climatic, and the mean annual precipitation is less than 250 mm [71].

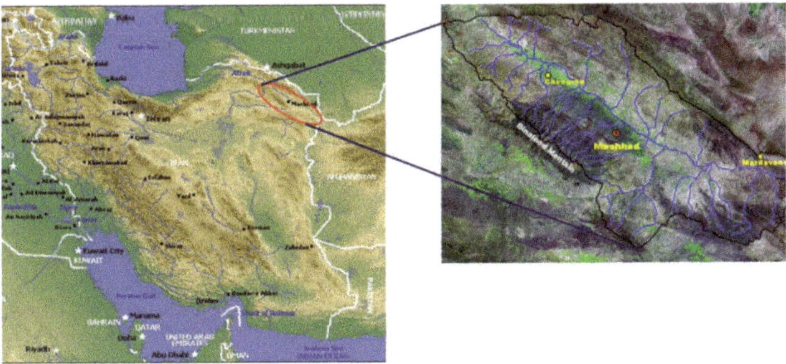

Figure 3. Location of the Kashafroud watershed in North-Eastern Iran.

In recent years, the Kashafroud urban watershed has encountered challenges, including an increase in the variety of water demands, quantitative and qualitative degradation of water resources, and relevant conflicts among stakeholders [72,73]. In efforts to resolve the challenges, the integrated water resources management (IWRM) approach for the Kashafroud watershed was proposed by the Iran Ministry of Energy in 2010. Since 2015, the IWRM project for this watershed has been analyzed based on the MODSIM modeling by common collaboration between the ToossAb Water Engineering Consultant Company and Iran Water Resources Management Company. The summary of the average 40-year long-term hydrological and hydrogeological budget entail results from comprehensive studies performed for this project, including the meteorological and climatic, hydrologic, hydrogeologic, and socio-economic issues [74–77] (see Appendix A, Table A1). Additionally, for several urban, agricultural, industrial, and environmental water demands of the Kashafroud watershed, the current water consumptions have been specified, and the water demands of 2040 have been predicted [71,78–80] (see Appendix A, Table A2).

According to the detailed data obtained from reports and several analysis on the watershed data, the most competitive water strategies have been modeled by the collaboration of the Iran

Water Resources Management Company and ToossAb consultant company based on the iterative calibration-validation process within the MODSIM modeling project [81].

However, a GDSS model should be developed for the Kashafroud watershed based on the IWRM project data and MODSIM modeling outputs considering the stakeholders' participation in a group MCDM process. This study proposes the risk-based GDSS model for evaluating the predefined water strategies with respect to the criteria while improving the properties of the risk-based operator for modeling GDSS, analyzing the effects of several risk-taking attitudes of stakeholders based on strategies ranking, and investigating the stakeholders' consensus.

2.2.1. Stakeholders

A thorough and extensive study was performed in efforts to analyze the risk-based consensus-based GDSS process for the Kashafroud watershed. The six most influential stakeholders in the urban watershed decision-making process, including governmental stakeholders and non-governmental organizations (NGOs), were selected based on the study. The governmental stakeholders' members include experts, deputies, and chief executive officers (CEOs). Details on the six identified stakeholders and the relevant members for Kashafroud urban watershed are presented in Table 3.

Table 3. List of stakeholders and the relevant members collaborated in the Kashafroud GDSS model. NGOs: non-governmental organizations.

Stakeholder's ID	Stakeholder's Name	Stakeholder's Role	Number of Members	Members' Roles
$Sth._1$	The state regional water company	Regional water management authority	9	1 CEO, 2 deputies, 6 experts
$Sth._2$	The state agricultural organization	Agricultural water user	2	1 deputy, 1 expert
$Sth._3$	The urban water and wastewater company	Potable urban water and wastewater user	4	1 CEO, 1 deputy, 2 experts
$Sth._4$	The state industrial township company	Industrial water user	1	1 CEO
$Sth._5$	The state environmental protection agency	Environmental water user	2	2 experts
$Sth._6$	The NGOs as the representative of people	Water and environmental resources defenders	3	2 faculty members, 1 farmers' representative

2.2.2. Initial Criteria

In order to qualify the four sustainable development objectives for the Kashafroud watershed, including water resources sustainability, environmental sustainability, economic sustainability, and social sustainability, a detailed survey was distributed in the urban watershed to collect viewpoints from the relevant stakeholder members. The survey was conducted through one-on-one interviews, collaborative workshop meetings in the presence of all members, and responses from the provided questionnaires. Fifty-three multiple criteria in the four categories of sustainable development objectives were reviewed by the stakeholders in the primitive screening process. Considering the watershed conditions and related priorities, 21 criteria were voted as the initial criteria and are represented in Table 4. These criteria are defined according to reports provided by the United Nations Educational, Scientific, and Cultural Organization (UNESCO), International Association of Hydrogeologists, and the national reports provided by Iran Water Resources Management Company in the IWRM project [82–86].

Table 4. The initial sustainable development criteria for the Kashafroud watershed [82–85].

Objective	Initial Criterion	Criterion ID
Water resources sustainability	Water stress	C'_1
	Groundwater dependency	C'_2
	Adjustable protentional of Surface Water Resources	C'_3
	Development of groundwater	C'_4
	Percentage of water supply for agricultural demand	C'_5
	Percentage of water supply for potable urban demand	C'_6
	Percentage of water supply for industrial demand	C'_7
	Percentage of water supply for environmental demand	C'_8
	Renewable water resources per capita	C'_9
	Potable water consumption per capita	C'_{10}
	Industrial water consumption per capita	C'_{11}
	Reliability for water supply	C'_{12}
	Balancing between using surface water and groundwater	C'_{13}
	Groundwater unsustainability	C'_{14}
	Surface water dependency on other watersheds	C'_{15}
Environmental sustainability	Purified sewerage ratio	C'_{16}
Economic sustainability	Potable water losses	C'_{17}
	Benefit per cost ratio	C'_{18}
Social sustainability	Conflict resolution amongst water stakeholders	C'_{19}
	Creating job opportunities	C'_{20}
	Social equity	C'_{21}

The initial criteria were weighted in the final screening process to select the final criteria based on a group consensus.

2.2.3. Water Strategies

After the investigation of several water strategies by the Ministry of Energy and the watershed stakeholders, the most competitive urban water strategies were selected by the stakeholders for the IWRM project to make a decision about choosing the more desirable strategy within the Kashafroud watershed [81]. Table 5 presents the five final water strategies for the Kashafroud urban watershed classified by supply management, demand management, and combined supply-demand management.

The main reasons for the selection of these five competitive strategies by the stakeholders include:

- Classification of the strategies within supply management, demand management, and combined supply-demand management.
- Investigating the effects of Doosti Dam on supplying the several demands, as well as the influences of substituting other water strategies instead of this project.
- Comparing the supply management and demand management approaches in regard to the several sustainable development criteria.
- Comparing the role of the two under-studied supply management projects, including the Idelik inter-basin water transfer and the utilization of purified wastewater on agricultural lands.

Table 5. The urban water strategies for the Kashafroud watershed by the 2040 vision.

Type of Strategy	Strategy ID	The Existing water Resources	The Under-Studying Supply Management Strategies		The Under-Studying Demand Management Strategies	Dependency on Water Transfer from Doosti Dam
		Withdrawal from Existing Dams and Groundwater Reservoir	Utilization of Purified Wastewater on Agricultural lands	Idelik Inter-Basin Water Transfer	Improving Water Network Efficiency and Modifying Cropping Pattern	
Supply management	S_1	Yes	No	No	No	Yes
	S_2	Yes	No	Yes	No	Yes
	S_3	Yes	Yes	No	No	Yes
Demand management	S_4	Yes	No	No	Yes	Yes
Combined supply-demand management	S_5	Yes	Yes	Yes	Yes	No

The existing water resources include the Ardak, Kardeh, Torogh, Dolatabad, Chalidarreh, and Esjil dams, as well as the groundwater reservoir. The under-studying supply management strategies include the utilization of purified wastewater on agricultural lands and the Idelik inter-basin water transfer. However, the Idelik project might cause conflicts between the stakeholders of the northern watershed and the Kashafroud watershed. The multi-criteria effects of this project and utilization of purified wastewater on agricultural lands are compared for the strategies S_2 and S_3.

The Doosti Dam is considered as a structural supply management that plays an active role in supplying water for the Kashafroud watershed. However, this project has high operation and maintenance costs, and its implementation could lead to dependence on the transboundary river basin. Therefore, the stakeholders' approach is to substitute more reliable water strategies instead of the dam for providing urban water.

The under-studying demand management strategies include improving water network efficiency and modifying cropping patterns, which are typical for strategies S_4 and S_5. The difference between these two strategies is that the strategy S_4 is considered as just a demand management approach with dependency on the Doosti Dam, while the strategy S_5 is considered as both a supply and demand management approach with no dependency on water transfer from the Doosti Dam.

The existing and under-studying strategies of the Kashafroud watershed are shown in Figure 4.

2.2.4. Data Collection

In order to analyze the risk-based GDSS model for selecting the more desirable water strategy, two types of data were collected. The first type of data is related to the criteria, including the selection of final criteria and weighting of the final criteria. The second type of data is associated with the strategies, including the evaluation of strategies with respect to the final criteria.

Accordingly, for collecting the first type of data, a survey questionnaire was prepared, and the 21 members of the six stakeholders were interviewed to capture their viewpoints about the importance degrees of the initial criteria and the final criteria, using the linguistic answers (no importance, very low importance, low importance, slightly low importance, moderately importance, slightly high importance, high importance, very high importance, and perfect importance) (see Appendix B, Table A3)

For collecting the second type of data, the results of the MODSIM modeling project and the data from the meteorology and climatology, hydrology, hydrogeology, and socio-economic reports [74–77,79], as well as information from the reports of urban, agricultural, industrial, and environmental water demands for the Kashafroud watershed, were utilized for evaluation of the water strategies with respect to the sustainable development criteria (see Appendix A, Tables A1 and A2; see Appendix C, Table A6).

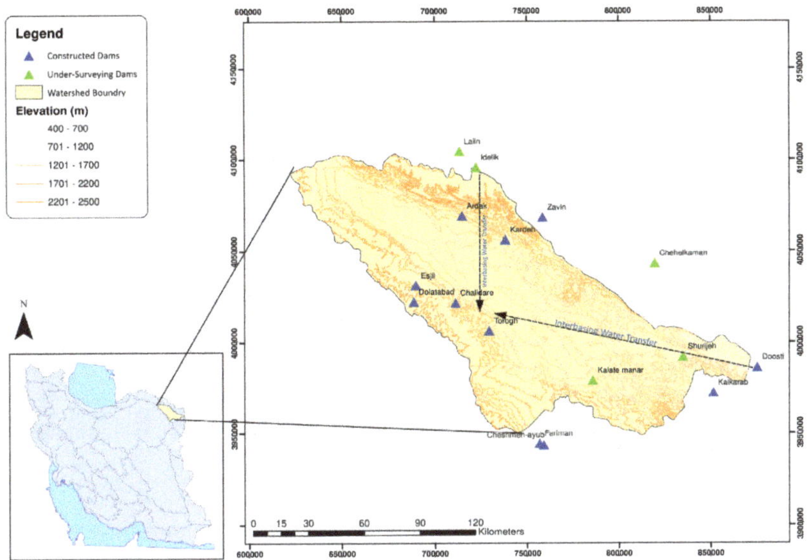

Figure 4. Location of Kashafroud in Iran and the existing and the under-studying water strategies.

2.2.5. Final Criteria Selection

Regarding Step 2 of the GDSS model (Figure 1), in order to consider the four sustainable development objectives for evaluating the five water strategies, the final criteria should be selected from the initial criteria. The first step in the survey process is an interview with the stakeholders, where the definitions of the initial criteria are explained. Next, the provided questionnaires are completed by the 21 members of the six stakeholders, in which the linguistic importance degrees of the initial criteria are assigned (see Appendix B, Table A4). In the end, the members' viewpoints of each of the six stakeholder's community are aggregated. The aggregated results related to the six stakeholders on the defuzzified weights of the initial criteria and the stakeholders' defuzzified normalized weights are presented in Figure 5.

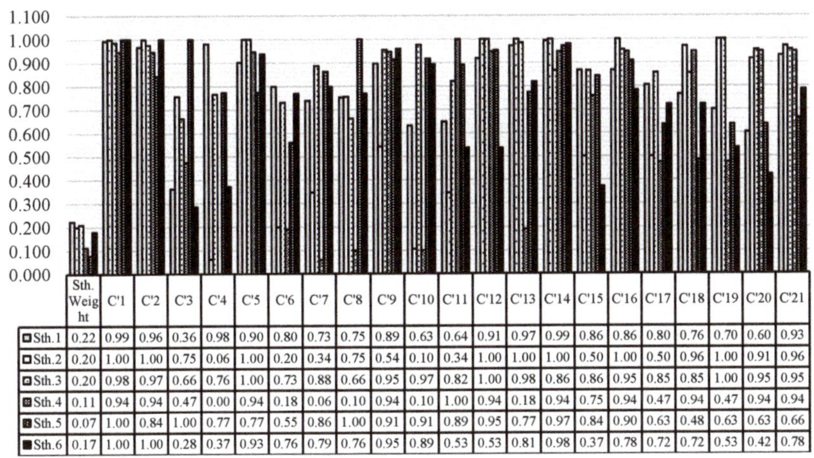

	Sth. Weight	C'1	C'2	C'3	C'4	C'5	C'6	C'7	C'8	C'9	C'10	C'11	C'12	C'13	C'14	C'15	C'16	C'17	C'18	C'19	C'20	C'21
Sth.1	0.22	0.99	0.96	0.36	0.98	0.90	0.80	0.73	0.75	0.89	0.63	0.64	0.91	0.97	0.99	0.86	0.86	0.80	0.76	0.70	0.60	0.93
Sth.2	0.20	1.00	1.00	0.75	0.06	1.00	0.20	0.34	0.75	0.54	0.10	0.34	1.00	1.00	1.00	0.50	1.00	0.50	0.96	1.00	0.91	0.96
Sth.3	0.20	0.98	0.97	0.66	0.76	1.00	0.73	0.88	0.66	0.95	0.97	0.82	1.00	0.98	0.86	0.86	0.95	0.85	0.85	1.00	0.95	0.95
Sth.4	0.11	0.94	0.94	0.47	0.00	0.94	0.18	0.06	0.10	0.94	0.10	1.00	0.94	0.18	0.94	0.75	0.94	0.47	0.94	0.47	0.94	0.94
Sth.5	0.07	1.00	0.84	1.00	0.77	0.77	0.55	0.86	1.00	0.91	0.91	0.89	0.95	0.77	0.97	0.84	0.90	0.63	0.48	0.63	0.63	0.66
Sth.6	0.17	1.00	1.00	0.28	0.37	0.93	0.76	0.79	0.76	0.95	0.89	0.53	0.53	0.81	0.98	0.37	0.78	0.72	0.72	0.53	0.42	0.78

Figure 5. Assigned weights for the initial criteria by the stakeholders of the Kashafroud watershed.

The initial criteria weights are utilized to calculate the relevant group consensus measurements using Equation (1). The group consensus measurement results of the initial criteria are presented in Figure 6.

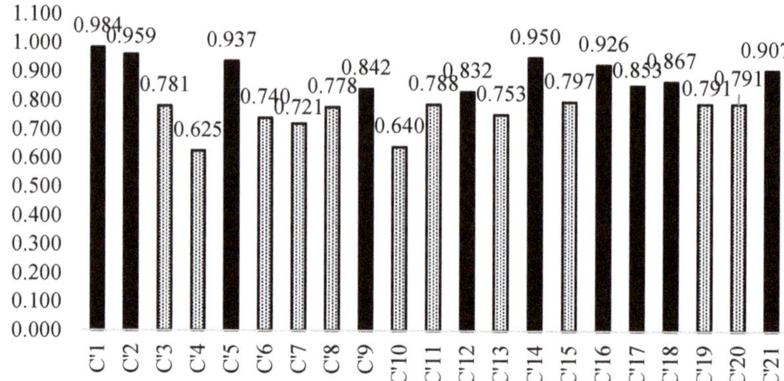

Figure 6. The group consensus measurements of the initial criteria for the Kashafroud watershed.

In order to select the final criteria from the 21 initial criteria, the group consensus measurements that are higher than the determined TLA ($TLA = 0.800$) are selected as the final criteria. According to Figure 6, the 10 black-filled criteria of $C\prime_1$, $C\prime_2$, $C\prime_5$, $C\prime_9$, $C\prime_{12}$, $C\prime_{14}$, $C\prime_{16}$, $C\prime_{17}$, $C\prime_{18}$, and $C\prime_{21}$ have been selected as the final sustainable development criteria, which are the most preferable criteria in the group viewpoints for the decision-making process within the watershed. The final selected criteria that are marked by C_1, C_2, C_3, C_4, C_5, C_6, C_7, C_8, C_9, and C_{10}, are defined in Table 6 [82–85].

Table 6. The final sustainable development criteria for the Kashafroud watershed by the 2040 vision. *TWW* is the total water withdrawal (includes urban, agricultural, and industrial and water withdrawal); *TWS* is the total water storage; *EGW* is the exploitation from groundwater; *TEWR* is the total exploitation of water resources (*TWW* + environmental water withdrawal); *SAWD* is the supplied agricultural water demand; *AWD* is the agricultural water demand; *SWR* is the surface water resources; *RGW* is the renewable groundwater resources; *P* is the population; *SPWD* is the supplied potable water demand; *PWD* is the potable water demand; *SIWD* is the supplied industrial water demand; *IWD* is the industrial water demand; *SEWD* is the supplied environmental water demand; *EWD* is the environmental water demand; *GWD* is the groundwater discharge; *GWR* is the groundwater recharge; *PS* is the purified sewerage; *US* and *IS* are the urban sewerage and industrial sewerage, respectively; and *B* and *C* are the amount of benefit and cost values of the implementation of the water strategies, respectively.

Criterion	ID	Criterion Definition	Data Resource
Water stress	C_1	TWW/TWS	[81,86]
Groundwater dependency	C_2	$EGW/TEWR$	[71,75,76,78–81]
Percentage of water supply for agricultural demand	C_3	$SAWD/AWD$	[78,81]
Renewable water resources per capita	C_4	$(SWR + RGWR)/P$	[70,81,86]
Reliability for water supply	C_5	$0.25 \times \left(\frac{SAWD}{AWD} + \frac{SPWD}{PWD} + \frac{SIWD}{IWD} + \frac{SEWD}{EWD} \right)$	[71,78–81]
Groundwater unsustainability	C_6	GWD/GWR	[81,86]
Purified sewerage ratio	C_7	$PS/(US + IS)$	[71,79,81]
Potable water losses	C_8	Percentage of water distribution losses	[71]
Benefit per cost ratio	C_9	B/C	[77]
Social equity	C_{10}	$0.333 \times \left(0.970 \times \frac{SAWD}{AWD} + 1 \times \frac{SPWD}{PWD} + 0.970 \times \frac{SIWD}{IWD} \right.$	[71,78–81]

2.2.6. Final Criteria Weights

Following Step 3, each stakeholder determines the importance degree for each selected criterion using linguistic variables (see Appendix B, Table A5). The linguistic variables are fuzzified and defuzzified. The defuzzified weights of the criteria in viewpoints of the stakeholders are presented in Figure 7.

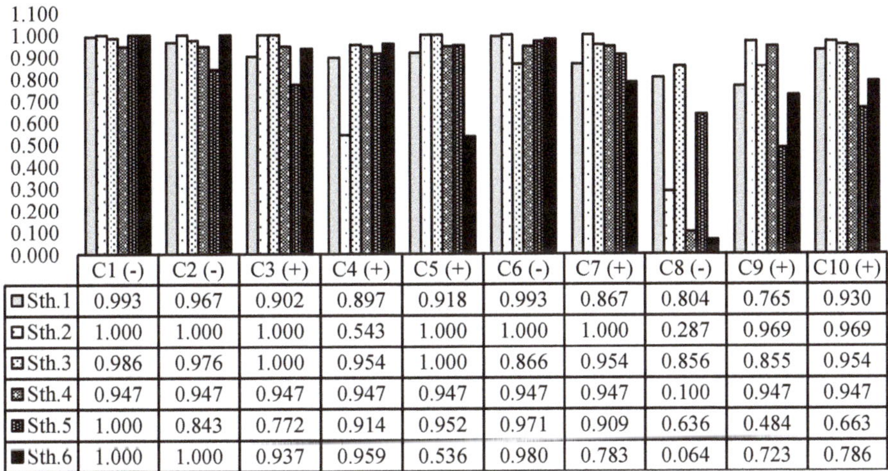

	C1 (-)	C2 (-)	C3 (+)	C4 (+)	C5 (+)	C6 (-)	C7 (+)	C8 (-)	C9 (+)	C10 (+)
Sth.1	0.993	0.967	0.902	0.897	0.918	0.993	0.867	0.804	0.765	0.930
Sth.2	1.000	1.000	1.000	0.543	1.000	1.000	1.000	0.287	0.969	0.969
Sth.3	0.986	0.976	1.000	0.954	1.000	0.866	0.954	0.856	0.855	0.954
Sth.4	0.947	0.947	0.947	0.947	0.947	0.947	0.947	0.100	0.947	0.947
Sth.5	1.000	0.843	0.772	0.914	0.952	0.971	0.909	0.636	0.484	0.663
Sth.6	1.000	1.000	0.937	0.959	0.536	0.980	0.783	0.064	0.723	0.786

Figure 7. The defuzzified weights of the selected criteria in each individual stakeholder's viewpoint.

Finally, the criteria are weighted based on the entropy method. According to Step 4, the stakeholders' power weights are determined by using linguistic variables, which are finally defuzzified. Consequently, the final criteria weights in the individual stakeholders' viewpoints are presented in Figure 8.

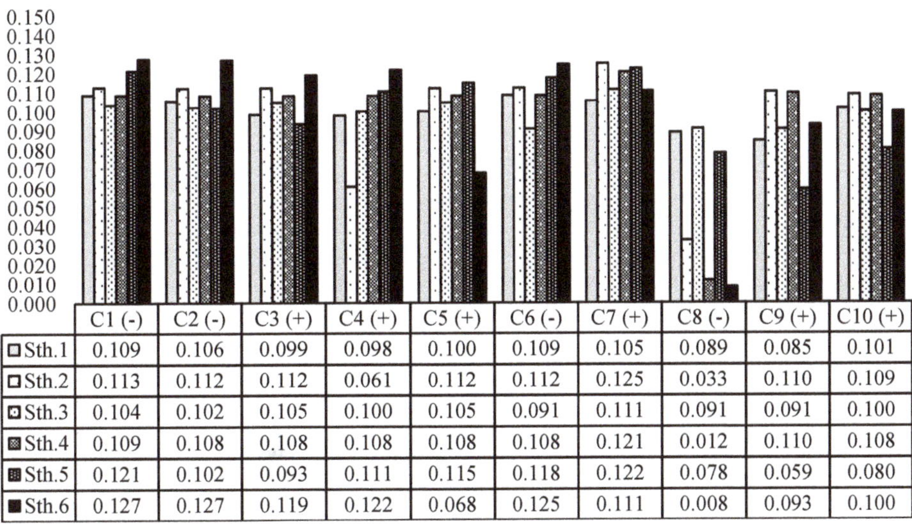

	C1 (-)	C2 (-)	C3 (+)	C4 (+)	C5 (+)	C6 (-)	C7 (+)	C8 (-)	C9 (+)	C10 (+)
Sth.1	0.109	0.106	0.099	0.098	0.100	0.109	0.105	0.089	0.085	0.101
Sth.2	0.113	0.112	0.112	0.061	0.112	0.112	0.125	0.033	0.110	0.109
Sth.3	0.104	0.102	0.105	0.100	0.105	0.091	0.111	0.091	0.091	0.100
Sth.4	0.109	0.108	0.108	0.108	0.108	0.108	0.121	0.012	0.110	0.108
Sth.5	0.121	0.102	0.093	0.111	0.115	0.118	0.122	0.078	0.059	0.080
Sth.6	0.127	0.127	0.119	0.122	0.068	0.125	0.111	0.008	0.093	0.100

Figure 8. The final weights of the selected criteria in each individual stakeholder's viewpoint.

The final criteria weights in the group viewpoint are represented in Figure 9.

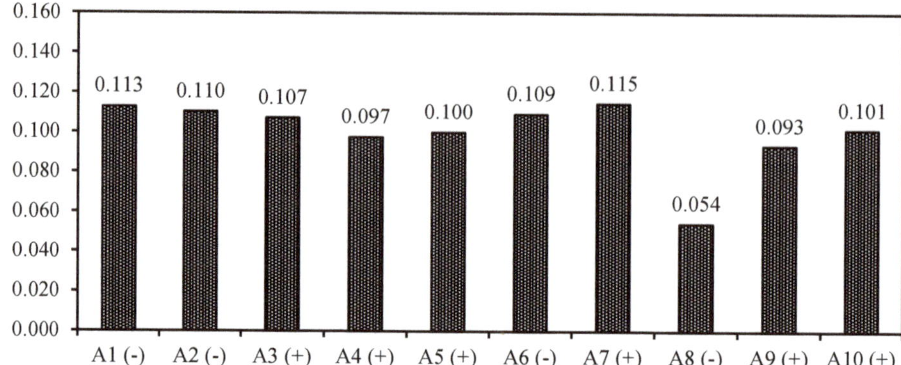

Figure 9. The final weights of the selected criteria in the group viewpoint.

2.2.7. Decision Matrix (Evaluation Matrix)

Regarding Step 5, the decision matrix is formed for evaluating the five water strategies with respect to the 10 selected sustainable criteria (see Table 6). The decision matrix elements are common among the six stakeholders. In the decision matrix, each evaluation value is the influence of implementing each water strategy on each criterion, which is obtained from MODSIM modeling outputs report [81], the data related to the hydrologic report [75], the hydrogeologic and budget reports [76,86], the socio-economic report [77], and the several demands reports [71,78–80] (see the data source in Table 6). Accordingly, the evaluation matrix of the water strategies with respect to the sustainable development criteria for the Kashafroud watershed is presented in Table 7 (see Appendix C, Table A6).

Table 7. Evaluation matrix of the water strategies with respect to the criteria for the Kashafroud watershed.

Selected Criteria	Dimension	S_1	S_2	S_3	S_4	S_5
$C_1^{(-)}$	Nondimensional	0.958	0.973	1.136	0.919	1.012
$C_2^{(-)}$	Nondimensional	0.753	0.742	0.704	0.728	0.742
$C_3^{(+)}$	Nondimensional	0.798	0.818	0.890	0.828	0.869
$C_4^{(+)}$	m^3/Person	269.8	269.8	251.9	253.4	245.0
$C_5^{(+)}$	Nondimensional	0.943	0.948	0.966	0.950	0.961
$C_6^{(-)}$	Nondimensional	1.027	1.027	1.133	0.980	1.081
$C_7^{(+)}$	Nondimensional	0.248	0.248	0.555	0.241	0.541
$C_8^{(-)}$	Nondimensional	0.232	0.232	0.232	0.171	0.171
$C_9^{(+)}$	Nondimensional	1.101	1.103	1.081	1.418	0.941
$C_{10}^{(+)}$	Nondimensional	0.915	0.921	0.944	0.925	0.938

It is noticeable that, in the decision matrix, some of the criteria, including C_3, C_4, C_5, C_7, C_9, and C_{10}, are positive (C^+), and the other criteria, including C_1, C_2, C_6, and C_8, are negative (C^-).

3. Results

3.1. Risk Analysis-Based Scores of the Water Strategies

According to Step 6, the decision matrix is first normalized; then, the weighted normalized decision matrix is formed for each of the stakeholders. Regarding Step 8, the weighted normalized decision matrix associated with each stakeholder is applied to implement the external risk analysis-based aggregation process. The scores of strategies in each stakeholder's viewpoint are calculated in several risk-taking cases. The results are presented in Figures 10 and 11 for the two risk-taking cases of completely risk-averse and completely risk-prone standpoints.

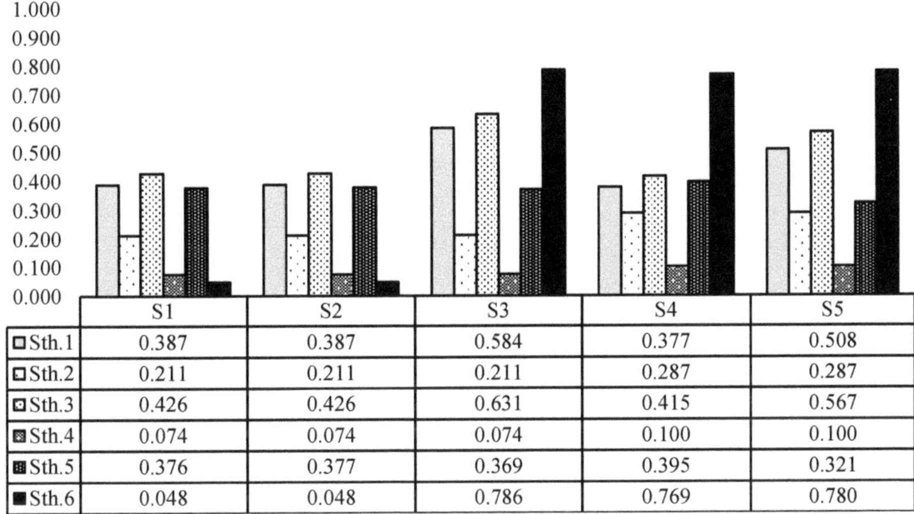

	S1	S2	S3	S4	S5
☐ Sth.1	0.387	0.387	0.584	0.377	0.508
☐ Sth.2	0.211	0.211	0.211	0.287	0.287
☐ Sth.3	0.426	0.426	0.631	0.415	0.567
▨ Sth.4	0.074	0.074	0.074	0.100	0.100
▦ Sth.5	0.376	0.377	0.369	0.395	0.321
■ Sth.6	0.048	0.048	0.786	0.769	0.780

Figure 10. Scores of the water strategies in each stakeholder's viewpoint (satisfying all criteria).

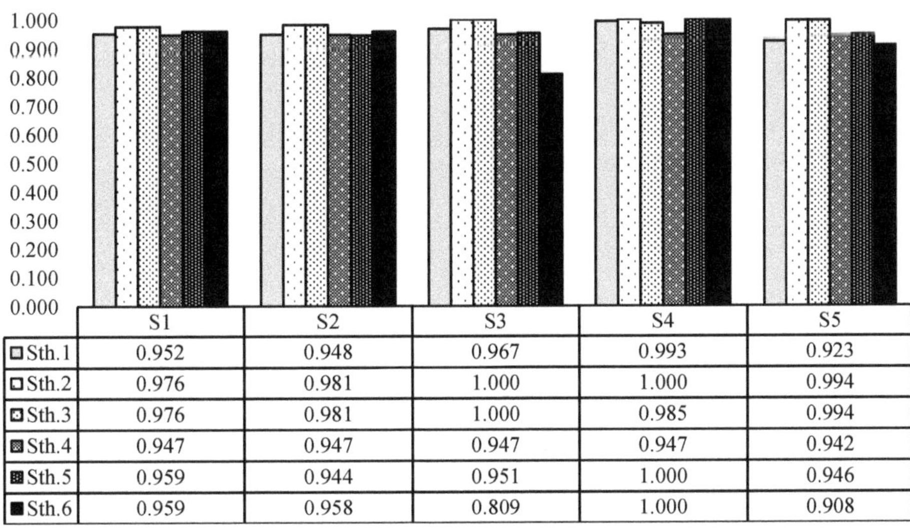

	S1	S2	S3	S4	S5
☐ Sth.1	0.952	0.948	0.967	0.993	0.923
☐ Sth.2	0.976	0.981	1.000	1.000	0.994
☐ Sth.3	0.976	0.981	1.000	0.985	0.994
▨ Sth.4	0.947	0.947	0.947	0.947	0.942
▦ Sth.5	0.959	0.944	0.951	1.000	0.946
■ Sth.6	0.959	0.958	0.809	1.000	0.908

Figure 11. Scores of the water strategies in each stakeholder's viewpoint (satisfying at least one criterion).

Following Step 9, the scores of water strategies in the viewpoint of a group of stakeholders are calculated based on the two types of EMOWA and IMOWA operators. Figures 12 and 13 represent the scores of strategies in each group viewpoint in several risk-taking attitudes, based on the EMOWA and IMOWA operators, respectively:

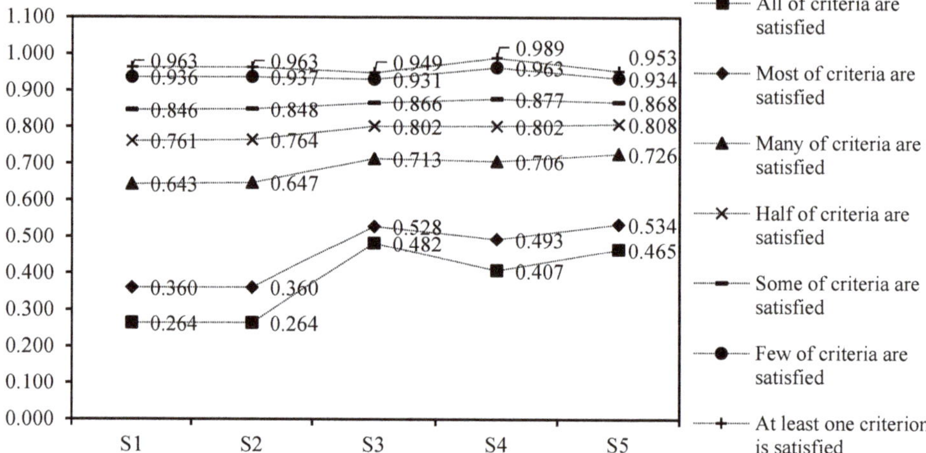

Figure 12. The external modified ordered weighted averaging (EMOWA) scores of the Kashafroud water strategies in the group viewpoint for the risk cases.

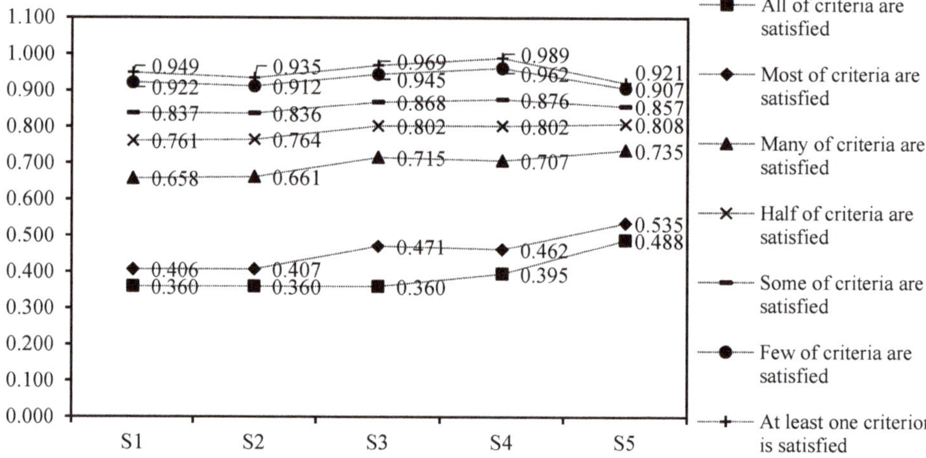

Figure 13. The internal modified ordered weighted averaging (IMOWA) scores of the Kashafroud water strategies in the group viewpoint for the risk cases.

3.2. Group Consensus Measurements of the Water Strategies

Regarding Step 10, the consensus measurements of water strategies in the viewpoint of a group of stakeholders are calculated based on the Euclidean Minkowski's distance-based method two types of EMOWA and IMOWA operators. According to the group decision-making amongst the stakeholders of the Kashafroud watershed, the TLA index is selected as the linguistic variable of "slightly high", which equals a numerical value of 0.800. Accordingly, the consensus measurement of each water strategy is compared with the numerical value of TLA.

Figures 14 and 15 represent the consensus measurements of water strategies in a group viewpoint in the several risk-taking attitudes, based on the EMOWA and IMOWA operators, respectively. The TLA of 0.800 is represented by the dashed line.

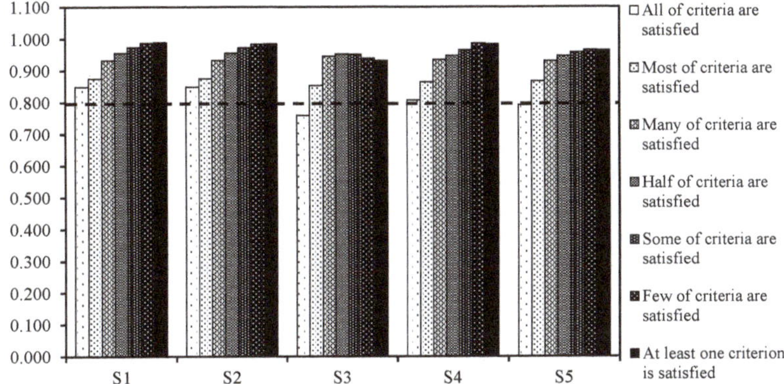

Figure 14. The EMOWA group consensus measurements of water strategies in the risk cases.

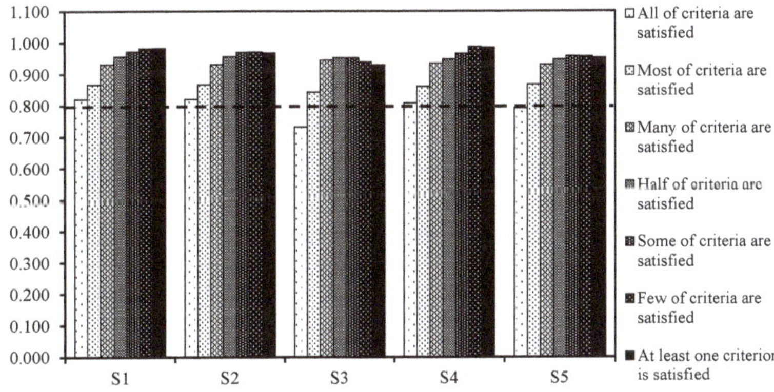

Figure 15. The IMOWA group consensus measurements of water strategies in the risk cases.

3.3. The Final Ranking of the Water Strategies

Finally, according to Step 11, the watershed strategies are ranked based on the group scores calculated by the two types of EMOWA and IMOWA operators in the several risk-taking cases, presented in Table 8. Additionally, the number of criteria that are satisfied in each case are specified [13].

Table 8. The final ranking of water strategies for the Kashafroud watershed in the risk-taking cases.

Type No. 1: EMOWA Operator						
Satisfaction of criteria by strategies	Satisfies at least one (1) criterion	Satisfies few of (2) the criteria	Satisfies some of (3) the criteria	Satisfies half of (5) the criteria	Satisfies many of (7) the criteria	Satisfies most of (9) the criteria	Satisfies all of (10) the criteria
S_1	2	3	5	5	5	5	5
S_2	3	2	4	4	4	4	4
S_3	5	5	3	3	2	2	1
S_4	1	1	1	2	3	3	3
S_5	4	4	2	1	1	1	2
Type No. 2: IMOWA Operator						
Satisfaction of criteria by strategies	Satisfies at least one (1) criterion	Satisfies few of (2) the criteria	Satisfies some of (3) the criteria	Satisfies half of (5) the criteria	Satisfies many of (7) the criteria	Satisfies most of (9) the criteria	Satisfies all of (10) the criteria
S_1	3	3	4	5	5	5	5
S_2	4	4	5	4	4	4	4
S_3	2	2	2	3	2	2	3
S_4	1	1	1	2	3	3	2
S_5	5	5	3	1	1	1	1

4. Discussion

Discussion about the results of the risk-based consensus-based GDSS modeling for the Kashafroud urban watershed is illustrated in four subjects, including 1—importance degrees of the criteria in the viewpoints of the group of stakeholders, 2—scores of water strategies in each stakeholder's viewpoint and the group of stakeholders' opinions, 3—group consensus measurements for the strategies, and 4—final ranking of the water strategies.

According to the results of criteria weights in group viewpoint (Figure 9), the stakeholders' group assigned the most weight to the criteria C_7 and C_1, respectively. In the viewpoint of the group, the priority of the criterion C_7 (purified sewerage ratio) in comparison with the other criteria shows that utilization of purified wastewater for some agricultural demands could reduce its withdrawal from groundwater resources, which instead be used to supply the increasing urban potable demand. Additionally, the relative priority of C_1 (water stress) emphasizes the importance of a close ratio between water withdrawal and renewable water resources in a semi-arid climate in order to control the withdrawal from other water resources. On the other hand, the group of stakeholders assigns the least weight for the criterion C_8 (potable water losses), because the criterion C_8 has no significant effect on water stress in comparison with the other factors.

Regarding the results related to the scores of water strategies in each stakeholder's viewpoint (Figures 10 and 11), in the completely risk-averse case, each stakeholder desires to select the strategy that satisfies all criteria. In this conservative viewpoint, half of the stakeholders choose the strategy S_3 as the more desirable strategy. These stakeholders have the supply management approach with an emphasis on utilization of purified wastewater for agricultural irrigation and dependency on water transfer from the Doosti Dam. Vice versa, in the completely risk-prone standpoint, each stakeholder desires to select the strategy that satisfies at least one criterion. Therefore, in this nonconservative standpoint, half of the stakeholders choose the strategy S_4 as the more desirable strategy. These stakeholders have just the demand management approach while considering the water transfer from the Doosti Dam. As it is expected from the risk analysis results, the scores of strategies in each stakeholder's viewpoint in the completely risk-prone viewpoint (completely optimistic viewpoint) are greater than the scores in the completely risk-averse viewpoint (completely pessimistic viewpoint). The completely optimistic viewpoint emphasizes on a fully positive and fully nonconservative approach of each stakeholder, while the completely pessimistic standpoint emphasizes on a fully negative and a fully conservative approach of each stakeholder.

With respect to the results of the group scores of water strategies (Figures 12 and 13), the group scores of strategies are increased from the completely risk-averse viewpoint to the completely risk-prone standpoint. Risk-averse cases have a conservative viewpoint and emphasize a pessimistic approach from stakeholders in the GDSS process, while the risk-prone cases have a nonconservative standpoint and emphasize an optimistic approach from stakeholders. For several risk-taking cases, the trend of changes for EMOWA scores is almost the same as the trend of changes for IMOWA scores, except for the completely risk-averse case. According to the EMOWA results, in the completely risk-averse viewpoint, strategy S_3 is selected as the more desirable strategy by the group. On the other hand, in the completely risk-averse viewpoint of the IMOWA results, strategy S_5 is chosen as the more desirable strategy by the group. It means that, for the Kashafroud watershed, the completely risk-averse viewpoint of the EMOWA operator emphasizes a supply management approach with dependency on water transfer from the Doosti Dam, whereas the completely risk-averse viewpoint of the IMOWA operator emphasizes a combined supply-demand management approach with no dependency on water transfer from the Doosti Dam. On the other hand, in accordance with the EMOWA and IMOWA results, in the completely risk-prone standpoint, strategy S_4 is chosen as the more desirable strategy by the group. For this watershed, the completely risk-prone viewpoint of the EMOWA and IMOWA operators emphasizes a demand management approach with dependency on the water transfer from the Doosti Dam.

Following the results of group consensus measurements (Figures 14 and 15), the consensus measurements of all water strategies in several risk-taking cases are higher than the selected TLA, except for the strategy S_3 in the completely risk-averse viewpoint (which has a consensus measurement with a really small distance to the selected TLA of 0.800). Therefore, a final group agreement amongst the stakeholders was reached. Additionally, according to the results of Figures 14 and 15, it is observed that the group consensus measurements have an increasing trend from a completely risk-averse viewpoint to a completely risk-prone standpoint. It is therefore more difficult to achieve group consensus by satisfying all criteria by the water strategies in the completely risk-averse viewpoint than accomplishing a group consensus by satisfying just one criterion by the strategies in the completely risk-prone standpoint. In addition, changing the Minkowski's parameter of q from 1 to infinity, the deviation and conflict between the individual and group viewpoints about the water strategies increased, which caused a decrease of group consensus measurements on strategies. After the achievement of the group consensus, the final ranking of water strategies can be implemented to determine the more desirable strategy in several risk-taking cases in both the EMOWA and IMOWA operators.

Consequently, according to the results of group scores achieved by the EMOWA and IMOWA operators, the final ranking of water strategies is determined in several risk-taking cases (Table 8). Regarding the EMOWA results, in the three risk-prone cases, the strategy S_4 is selected as the more desirable water strategy. In the neutral risk and the two risk-averse cases, the strategy S_5 is chosen as the more desirable strategy. Additionally, in the completely risk-averse case, the strategy S_3 is selected as the more desirable strategy. In accordance with the IMOWA results, in the three risk-prone cases, the strategy S_4 is selected as the more desirable water strategy, while, in the neutral risk and the three risk-averse cases, the strategy S_5 is chosen as the more desirable strategy.

5. Conclusions

In modeling the GDSS for effective urban watershed management, there are numerous stakeholders and beneficiaries with several opinions and preferences that should be used to evaluate water strategies with respect to sustainable development criteria for selecting the more desirable water strategy. The stakeholders' group may have several risk-taking attitudes, each of which risk-taking cases is related to satisfying the number of criteria by water strategies. The risk-taking attitudes vary from a completely risk-averse viewpoint to a completely risk-prone standpoint. The completely risk-averse viewpoint (completely conservative opinion) believes that all criteria should be satisfied by water strategies, while the completely risk-prone standpoint (completely nonconservative opinion) believes that at least one criterion can be satisfied by strategies. The other risk-taking attitudes are expressed between these two limited risk-taking cases. Accordingly, for analyzing the effect of risk-taking cases on the selection of the more desirable water strategy, the risk-based consensus-based GDSS model should be developed for effective urban watershed management.

In this research, in order to select the more desirable water strategy for the Kashafroud watershed, the risk-based EMOWA and IMOWA operators were proposed in the two types of external and internal aggregations to calculate the group scores of water strategies with respect to the criteria. These operators consider the importance degrees of criteria, the risk-taking degrees of the stakeholders' group, and the stakeholders' power weights simultaneously. Additionally, the group consensus-seeking process was implemented based on the weighted Minkowski's method, in which the group consensus measurements for strategies have been calculated using the squared mean deviation between the individual and group viewpoints of stakeholders. Finally, the ranking of the water strategies was determined in several risk-taking attitudes of the group of stakeholders with respect to the EMOWA and IMOWA scores for the strategies.

Therefore, the proposed methodology, including the main phases of water strategies' scoring, group consensus measuring, and the water strategies' ranking, was successfully developed for the study area of the Kashafroud watershed. The scoring results related to the EMOWA and IMOWA operators represents that the group scores of the water strategies are dependent on the risk-taking attitudes of the stakeholders within the watershed. Accordingly, for each strategy, the group scores in the risk-prone cases (at least one, a few, and some of the criteria satisfied by the strategies) are greater than the group scores in the risk-averse situations (many, most, and all of the criteria satisfied by the strategies). In addition, the group consensus measuring results shows that the final agreement among the stakeholders for all strategies was almost fully achieved. According to the findings of each strategy, the group consensus measurements in the risk-prone cases are greater than the group consensus measurements in the risk-averse situations. Finally, regarding the ranking results of strategies, for the risk-averse viewpoint in the EMOWA results, the group of stakeholders has a conservative approach and tend to select the strategy of S_3 as a supply management strategy, which satisfies all sustainable development criteria, while, in the IMOWA results with the risk-averse viewpoint, the group of stakeholders tends to choose the strategy of S_5 as a combined supply-demand management strategy. For the risk-prone standpoint in both EMOWA and IMOWA results, the group of stakeholders have a nonconservative approach and like to select the strategy of S_4 as a demand management strategy, which satisfies at least one sustainable development criteria.

Besides the advantages of the proposed risk-based consensus-based GDSS model in this study, there are some issues that should be improved in future studies, which include:

- Improving the GDSS model for use of the other input variables in the MCDM process, including the combination of crisp and linguistic data, as well as fuzzy interval valued data.
- Considering the alternative generation process during the GDSS modeling by use of the design theory, such as the K-C and K-C-P methodologies.
- Modeling the other probable water strategies such as climate changes strategies; additionally, changes in the percentage of water supply for the agricultural demand with respect to more several criteria.
- Resolving probable conflicts among stakeholders within the GDSS model using the game-theoretical Nash Bargaining solution.

For future studies, it is suggested to develop this proposed risk-based consensus-based GDSS model for any other watershed management by generating several water strategies based on the stakeholders' group consensus, which considers the combination of agricultural, industrial, and environmental demands and climate changes conditions. Furthermore, a conflict resolution process among stakeholders within the risk-based consensus-based GDSS process for resolving the probable conflicts of preferences among the watershed stakeholders should be analyzed. Additionally, an analysis of the varieties of the Minkowski's parameter and its effect on the group consensus measurement should be studied for future research.

Author Contributions: Conceptualization, R.J.S. and A.P.N.; methodology, R.J.S.; software, R.J.S.; validation, R.J.S. and A.P.N.; formal analysis, R.J.S.; investigation, R.J.S. and A.P.N.; resources, R.J.S.; data curation, R.J.S. and A.P.N.; writing—original draft preparation, R.J.S.; writing—review and editing, A.P.N.; visualization, R.J.S. and A.P.N.; supervision, A.P.N.; project administration, R.J.S. All authors have read and agreed to the published version of the manuscript.

Funding: This research received no external funding.

Acknowledgments: The authors would like to acknowledge the collaboration between the researchers of the Civil Engineering Department at Hakim Sabzevari University and the Department of Biosystems and Agricultural Engineering at Michigan State University for developing this paper. Additionally, the first author would like to thank the support from the Khorasan Razavi Regional Water Company for this research. In addition, the authors would like to thank Toossab Consultant Company for providing the real data related to the study area of the Kashafroud watershed.

Conflicts of Interest: The authors declare no conflicts of interest. In addition, the funder had no role in the design of the study; in the collection, analysis, or interpretation of data; in the writing of the manuscript; or in the decision to publish the results.

Appendix A

A.1. The Average 40-Year Long-Term Hydrological and Hydrogeological Budget of Kashafroud (1975–2015)

The summary of the results related to the average 40-year long-term hydrological and hydrogeological budget for the Kashafroud watershed (1975–2015) are presented in Table A1 [74–77].

Table A1. Long-term hydrological and hydrogeological budget of the Kashafroud watershed [74–77].

Hydrological Budget							
Mean annual precipitation (MCM)		Mean annual runoff (MCM)		Mean annual soil moisture (MCM)			
Plain	Highland	Plain	Highland	Plain		Highland	
				Evaporation	Infiltration	Evaporation	Infiltration
896	1953	21	342	758	117	1279	332

Hydrogeological Budget						
Mean annual groundwater recharge from (MCM)				Mean annual groundwater discharge by (MCM)		Mean annual changes in reservoir (MCM)
Rainfall	Surface runoff	Agricultural reversible water	Urban and Industrial reversible water	Withdraw	Outflow from aquifer	
449	207	181	180	1105	31	−119

A.2. The Estimated Water Consumptions and Predicted Water Demands for Kashafroud by the 2040 Vision

Additionally, for several urban, agricultural, industrial, and environmental water demands of the Kashafroud watershed, the current water consumptions have been specified, and the water demands by the 2040 vision have been predicted, which are presented in Table A2 [71,78–80].

Table A2. Water consumptions and water demands for the Kashafroud watershed by 2040 [71,78–80].

Current Water Consumptions					
Annual consumptions from surface water resources (MCM)			Annual consumptions from groundwater resources (MCM)		
Agricultural water	Urban water	Industrial water	Agricultural water	Urban water	Industrial water
173	173.5	1.5	806	227	32

Predicted Water Demands by the 2040 Vision						
Urban demand (MCM)	Equivalent Produced sewerage (MCM)	Agricultural demand (MCM)	Allocable sewerage for demand (MCM)	Industrial demand (MCM)	Equivalent Produced sewerage (MCM)	Environmental water demand (MCM)
489	382	806	404	90	52	43

Appendix B

B.1. Sample Questionnaire for the Selection of the Final Criteria

(1) Firstly, please overview the definitions of the initial criteria. After that, overview Table A3 containing the sustainable development objectives and the relevant initial criteria. Ultimately, give your viewpoint about the importance degree of each of the following criteria for the decision-making process and water resources planning and management in the study area of the Kashafroud urban watershed. (Please mark √ as a linguistic importance degree for each criterion within just one of the 4th to 12th columns of the table, according to the name of the criterion and the description of that corresponding criterion.)

- Please note that, in Table A3, the 21 initial criteria (taking into account the sustainability objectives including water resources sustainability, environmental sustainability, economic sustainability, and social sustainability) are specified and defined. Choose your priorities so that you can ultimately choose from all four objectives to be included in the final decision-making process.

Definitions of criteria:

- C'_1: (Urban water withdrawal + Agricultural water withdrawal + Industrial water withdrawal)/ (Total water storage);
- C'_2: (Exploitation from groundwater resources) / (Urban water withdrawal + Agricultural water withdrawal + Industrial water withdrawal + Environmental water withdrawal);
- C'_3: (Adjustable water potential from surface water resources—adjusted surface water resources by hydraulic structures) / (Surface water resources);
- C'_4: (Groundwater withdrawal) / (Renewable groundwater resources);
- C'_5: (Supplied agricultural water demand) / (Agricultural water demand);
- C'_6: (Supplied potable urban water demand) / (Potable urban water demand);
- C'_7: (Supplied industrial water demand) / (Industrial water demand);
- C'_8: (Supplied environmental water demand) / (Environmental water demand);
- C'_9: (Surface water resources + Renewable groundwater resources)/(Population);
- C'_{10}: (Sold urban water to urban water consumer) / (Urban water population);
- C'_{11}: (Industrial water withdrawal) / (Industrial employed population);
- C'_{12}: 0.25 × {(Supplied agricultural water demand) / (Agricultural water demand) + (Supplied potable urban water demand) / (Potable urban water demand) + (Supplied industrial water demand) / (Industrial water demand) + (Supplied environmental water demand) / (Environmental water demand)};
- C'_{13} (Qualitative): Balancing between the supply and withdrawal from surface water and groundwater resources;
- C'_{14}: (Groundwater discharge) / (Groundwater recharge);
- C'_{15}: (Entranced surface water + Transferred surface water) / (Surface water resources + Entranced surface water + Transferred surface water);
- C'_{16}: (Purified sewerage) / (Urban sewerage + Industrial sewerage);
- C'_{17}: (Total urban water withdrawal for water distribution network—sold urban water to urban water consumer) / (Total urban water withdrawal for water distribution network);
- C'_{18}: (Total benefit of implementation of water strategy) / (Total cost of implementation of water strategy);
- C'_{19} (Qualitative): Resolving conflicts among stakeholders in agricultural water, potable water, industrial water, and environmental water;
- C'_{20} (Qualitative): Creating job opportunities in agricultural, industrial, and service sectors during the implementation and operation periods of the strategies; and
- C'_{21}: 0.25 × {0.970 × (Supplied agricultural water demand) / (Agricultural water demand) + 1 × (Supplied potable urban water demand) / (Potable urban water demand) + 0.970 × (Supplied industrial water demand) / (Industrial water demand)}.

Table A3. Sample questionnaire for the linguistic importance degrees of the initial criteria.

1	2	3	4	5	6	7	8	9	10	11	12
Objective	Criterion	Definition	Linguistic Importance Degree								
			NI	VLI	LI	SLI	MI	SHI	HI	VHI	PI
Water resources sustainability	C'_1: Water stress										
	C'_2: Groundwater dependency										
	C'_3: Adjustable protentional of Surface Water Resources										
	C'_4: Development of groundwater										
	C'_5: Percentage of water supply for agricultural demand										
	C'_6: Percentage of water supply for potable urban demand										
	C'_7: Percentage of water supply for industrial demand										
	C'_8: Percentage of water supply for environmental demand										
	C'_9: Renewable water resources per capita										
	C'_{10}: Potable water consumption per capita										
	C'_{11}: Industrial water consumption per capita										
	C'_{12}: Reliability for water supply										
	C'_{13}: Balancing between using surface water and groundwater										
	C'_{14}: Groundwater unsustainability										
	C'_{15}: Surface water dependency on other watersheds										
Environmental sustainability	C'_{16}: Purified sewerage ratio										
Economic sustainability	C'_{17}: Potable water losses										
	C'_{18}: Benefit per cost ratio										
Social sustainability	C'_{19}: Conflict resolution amongst water stakeholders										
	C'_{20}: Creating job opportunities										
	C'_{21}: Social equity										

NI: no importance, VLI: very low importance, LI: low importance, SLI: slightly low importance, MI: moderate importance, SHI: slightly high importance, HI: high importance, VHI: very high importance, and PI: perfect importance.

B.2. Stakeholders' Individual Viewpoints about the Importance Degree of the Initial Criteria

Table A4. The linguistic importance degrees of the initial criteria in the individual's viewpoints.

Member's ID	C'_1	C'_2	C'_3	C'_4	C'_5	C'_6	C'_7	C'_8	C'_9	C'_{10}	C'_{11}	C'_{12}	C'_{13}	C'_{14}	C'_{15}	C'_{16}	C'_{17}	C'_{18}	C'_{19}	C'_{20}	C'_{21}
$Sth._{11}$	P	VH	N	VH	P	SH	SL	H	P	SH	SL	P	VH	P	SL	VH	H	SL	H	H	H
$Sth._{12}$	P	P	M	VH	H	SL	L	L	SH	SL	H	M	SH	P	VL	M	M	VH	M	M	SH
$Sth._{13}$	P	P	VL	VH	M	VH	VH	VH	H	M	VL	P	VH	P	SH	P	P	M	H	SL	VH
$Sth._{14}$	P	VH	L	VH	H	VH	H	H	M	L	SL	VH	VH	P	H	M	SH	M	L	VL	H
$Sth._{15}$	P	P	SL	H	P	VH	VH	SH	P	VH	VH	P	VH	P	VH	P	VH	VH	VH	M	P
$Sth._{16}$	P	VH	H	VH	H	VH	VH	VH	P	VH	VH	VH	VH	P	H	VH	VH	H	SH	M	H
$Sth._{17}$	VH	H	SH	VH	VH	VH	VH	VH	P	H	VH	VH	VH	VH	H	H	H	VH	SH	VH	P
$Sth._{18}$	P	P	M	VH	P	VH	VH	VH	VH	VH	VH	P	VH	P	SL	P	P	P	VH	VH	P
$Sth._{19}$	P	P	L	VH	P	L	M	VL	H	L	M	VH	VH	P	H	P	SL	VH	SL	H	VH
$Sth._{21}$	P	P	M	VL	P	SL	SL	VH	L	L	SL	P	VH	P	M	P	M	P	H	VH	P
$Sth._{22}$	P	P	M	VL	P	SL	M	M	P	L	SH	P	VH	P	M	P	VL	VH	VH	H	VH
$Sth._{31}$	P	P	M	VH	P	VH	VH	M	P	VH	VH	P	VH	P	VL	P	P	VH	VH	H	P
$Sth._{32}$	P	P	VH	VH	P	VH	VH	VH	P	VH	VH	P	VH	P	VH	P	P	P	VH	VH	P
$Sth._{33}$	P	P	VL	M	P	VL	VH	H	P	VH	SH	P	VH	M	VH	P	M	VH	VH	M	P
$Sth._{34}$	VH	H	VL	M	P	SH	M	SL	SH	H	M	P	VH	VH	VH	SH	H	M	SH	H	SH
$Sth._{41}$	P	P	M	N	P	SL	VL	SL	P	L	VH	P	SL	P	SH	P	L	P	M	M	P
$Sth._{51}$	P	H	H	M	M	L	VH	SL	VH	H	VH	H	M	VH	H	SH	SL	L	M	SH	M
$Sth._{52}$	P	SH	H	VH	P	VH	SH	VH	H	VH	SH	P	VH	P	SH	P	P	SH	H	M	SH
$Sth._{61}$	P	P	N	L	VH	H	VH	H	P	H	VH	SL	VH	P	SL	H	N	VH	VH	SL	M
$Sth._{62}$	P	P	N	L	VH	H	VH	H	VH	H	VH	M	VH	P	SL	VH	N	P	VH	SL	VH
$Sth._{63}$	P	P	H	VH	VH	M	M	M	VH	H	M	VH	M	VH	SH	M	SL	SL	SL	H	VH

B.3. Sample Questionnaire for Weighting the Final Criteria

(2) Please overview the Table A5 containing the final selected criteria. Give your viewpoint about the importance degree (linguistic weight) of each of the following criteria for the decision-making process and water resources planning and management in the study area of the Kashafroud urban watershed. (Please mark $\sqrt{}$ as a linguistic importance degree for each criterion within just one of the 4th to 12th columns of the table, according to the name of the criterion and the description of that corresponding criterion.)

Table A5. Sample questionnaire for the linguistic importance degrees of the final criteria.

1	2	3	4	5	6	7	8	9	10	11	12
Objective	Criterion	Definition	Linguistic Importance Degree								
			NI	VLI	LI	SLI	MI	SHI	HI	VHI	PI
Water resources sustainability	C'_1: Water stress										
	C'_2: Groundwater dependency										
	C'_5: Percentage of water supply for agricultural demand										
	C'_9: Renewable water resources per capita										
	C'_{12}: Reliability for water supply										
	C'_{14}: Groundwater unsustainability										
Environmental sustainability	C'_{16}: Purified sewerage ratio										
Economic sustainability	C'_{17}: Potable water losses										
	C'_{18}: Benefit per cost ratio										
Social sustainability	C'_{21}: Social equity										

Appendix C

Determination of the Water Strategies' Evaluation Values

In order to select the evaluation values of each water strategy with respect to the final sustainable development criteria for the risk-based GDSS model, the related variables have been extracted from the relevant IWRM project reports [71,74–77] and MODSIM modeling report [81], which have been provided by the Iran Water Resources Management Company and ToossAb Water Engineering Consultant Company and approved by the Iran Ministry of Energy. Indeed, the variables of the GDSS modeling for the Kashafroud watershed are related to the outputs of the IWRM project and MODSIM modeling project, which have been already analyzed for this study area.

The input data that should be entered into the MODSIM modeling includes the following:

1. Monthly data for the simulation of the model, including the data of the hydrometric stations related to the surface water, as well as the data of the aquifer unit hydrograph associated with the groundwater resources.
2. Monthly evaporation from the reservoir of each dam.
3. Monthly water withdrawal from the aquifer.
4. Monthly existing water consumptions for the base strategy, including urban, agricultural industrial, and environmental waters.
5. Estimated infiltration fraction (return flow) from urban, agricultural, and industrial consumptions.

Accordingly, the output data that are taken out from the MODSIM modeling includes the simulated results in hydrometric stations and aquifer unit hydrograph, which are compared with the observed data, based on an iterative calibration-validation process.

The parameters that are involved in the model calibration include the return water to aquifers from agricultural, urban, and industrial consumptions, the surface runoff infiltration values into the groundwater reservoirs, the outflow groundwater, and, if necessary, the efficiency of used water.

Additionally, the model calibration criterion for surface flows is primarily the hydrometric stations and the water resources budget data of the study area.

In addition, the model calibration criterion for the aquifers and groundwater reservoir is primarily the unit hydrograph of the aquifer and then the water resources budget data of the study area.

Accordingly, the model calibration is done in two parts: surface water and groundwater. In the surface water calibration, the output data from the station in the model is compared with the hydrometric station data within the watershed.

To calibrate and validate the outputs of the groundwater reservoir, it is done by calculating the changes in groundwater volume from the unit hydrograph of the aquifer. This is implemented for the modeling period. Then, the changes in the volume of the aquifer are compared with the changes in the volume of the groundwater reservoir in the simulated model. It should be noted that the calibration of the surface and groundwater due to the dependence of the parameters on each other should be performed simultaneously.

Ultimately, in order to determine the evaluation values of each water strategy with respect to each sustainable development criteria, the related data (see Table 6, Appendix B—Definitions of criteria) are extracted from the IWRM project reports [70,71,75–80,86], as well as the relevant variables are obtained from the outputs of the MODSIM modeling project report [81]. Accordingly, the detailed variables used for calculation of the evaluation values of the five water strategies with respect to the 10 final selected sustainable development criteria are presented in Table A6:

Table A6. The detailed variables for the determination of the evaluation matrix of the water strategies.

	S_1	S_2	S_3	S_4	S_5
C_1	$\frac{(488.2+880.3+71.5)}{(1503.3)}$	$\frac{(488.2+902.3+71.5)}{(1503.3)}$	$\frac{(488.2+981.9+71.5)}{(1356.7)}$	$\frac{(437.2+792.7+65.8)}{(1410.3)}$	$\frac{(437.2+831.7+65.8)}{(1319.4)}$
C_2	$\frac{(1092.45)}{(488.2+880.3+71.5+10.9)}$	$\frac{(1092.45)}{(488.2+902.3+71.5+10.9)}$	$\frac{(1092.45)}{(488.2+981.9+71.5+10.9)}$	$\frac{(951.19)}{(437.2+792.7+65.8+10.9)}$	$\frac{(999.09)}{(437.2+831.7+65.8+10.9)}$
C_3	$\frac{(880.3)}{(1103.4)}$	$\frac{(902.3)}{(1103.4)}$	$\frac{(981.9)}{(1103.4)}$	$\frac{(792.7)}{(956.8)}$	$\frac{(831.7)}{(956.8)}$
C_4	$\frac{(1533.8\times10^6)}{(5686046)}$	$\frac{(1533.8\times10^6)}{(5686046)}$	$\frac{(1432.2\times10^6)}{(5686046)}$	$\frac{(1440.8\times10^6)}{(5686046)}$	$\frac{(1392.9\times10^6)}{(5686046)}$
C_5	$0.25\times(0.798+1+1+0.973)$	$0.25\times(0.818+1+1+0.973)$	$0.25\times(0.890+1+1+0.973)$	$0.25\times(0.828+1+1+0.973)$	$0.25\times(0.869+1+1+0.973)$
C_6	$\frac{(1110.95)}{(1082)}$	$\frac{(1110.95)}{(1082)}$	$\frac{(1110.95)}{(980.4)}$	$\frac{(969.69)}{(989)}$	$\frac{(1017.59)}{(941.1)}$
C_7	$\frac{(105.858)}{(426.667)}$	$\frac{(105.858)}{(426.667)}$	$\frac{(236.976)}{(426.667)}$	$\frac{(94.445)}{(391.41)}$	$\frac{(211.775)}{(391.41)}$
C_8	0.232	0.232	0.232	0.171	0.171
C_9	1.101	1.103	1.081	1.418	0.941
C_{10}	$0.333\times$ $(0.97\times0.798+1\times1+0.97\times1)$	$0.333\times$ $(0.97\times0.818+1\times1+0.97\times1)$	$0.333\times$ $(0.97\times0.890+1\times1+0.97\times1)$	$0.333\times$ $(0.97\times0.828+1\times1+0.97\times1)$	$0.333\times$ $(0.97\times0.869+1\times1+0.97\times1)$

References

1. United Nations, Department of Economic Social Affairs, Population Division 2019. *World Population Prospects 2019: Highlights*, 1st ed.; United Nations: New York, NY, USA, 2019; pp. 1–2.
2. Bahri, A. *Global Water Partnership: Integrated Urban Water Management*, 1st ed.; Global Water Partnership: Stockholm, Sweden, 2012; pp. 11–12.
3. United Nations Development Programme 2006. *Global Partnership for Development: Annual Report 2006*, 1st ed.; United Nations: New York, NY, USA, 2006; pp. 1–2.
4. Simonovic, S.P. *Managing Water Resources: Methods and Tools for a Systems Approach*, 1st ed.; UNESCO Publishing: Paris, France; Earthscan James & James: London, UK, 2009; pp. 172–240.
5. Bender, M.J.; Simonovic, S.P. Consensus as the Measure of Sustainability. *Hydrol. Sci. J.* **1997**, *42*, 493–500. [CrossRef]
6. Parkinson, J.N.; Goldenfum, J.A.; Tucci, C. *Integrated Urban Water Management: Humid Tropics (Urban Water Series-UNESCO-IHP)*, 1st ed.; CRC Press: Boca Raton, FL, USA, 2009; p. 2.
7. Tahmasebi Birgani, Y.; Yazdandoost, F. An Integrated Framework to Evaluate Resilient-Sustainable Urban Drainage Management Plans Using a Combined-Adaptive MCDM Technique. *Water Resour. Manag.* **2018**, *32*, 2817–2835. [CrossRef]
8. Mianabadi, H.; Sheikhmohammady, M.; Mostert, E.; Van de Giesen, N. Application of the Ordered Weighted Averaging (OWA) Method to the Caspian Sea Conflict. *Stoch. Environ. Res. Risk Assess.* **2014**, *28*, 1359–1372. [CrossRef]
9. Minatour, Y.; Bonakdari, H.; Zarghami, M.; Ali Bakhshi, M. Water Supply Management Using an Extended Group Fuzzy Decision-Making Method: A Case Study in North-Eastern Iran. *Appl. Water Sci.* **2015**, *5*, 291–304. [CrossRef]
10. Prodanovic, P.; Simonovic, S.P. Comparison of Fuzzy Set Ranking Methods for Implementation in Water Resources Decision-Making. *Can. J. Civil. Eng.* **2002**, *29*, 692–701. [CrossRef]
11. Chitsaz, N.; Azarnivand, A. Water Scarcity Management in Arid Regions based on an Extended Multiple Criteria Technique. *Water Resour. Manag.* **2017**, *31*, 233–250. [CrossRef]
12. Anbari, M.J.; Tabesh, M.; Roozbahani, A. Risk Assessment Model to Prioritize Sewer Pipes Inspection in Wastewater Collection Networks. *J Environ. Manag.* **2017**, *190*, 91–101. [CrossRef]
13. Emrouznejad, A.; Marra, M. Ordered Weighted Averaging Operators 1988–2014: A Citation-based Literature Survey. *Int. J. Intell. Syst.* **2014**, *29*, 994–1014. [CrossRef]
14. Bouzarour-Amokrane, Y.; Tchangani, A.; Peres, F. A Bipolar Consensus Approach for Group Decision Making Problems. *Expert Syst Appl* **2015**, *42*, 1759–1772. [CrossRef]
15. Simonovic, S.P.; Bender, M.J. Collaborative Planning-Support System: An Approach for Determining Evaluation Criteria. *J. Hydrol.* **1996**, *177*, 237–251. [CrossRef]
16. Bender, M.J.; Simonovic, S.P. A System Approach for Collaborative Decision Support in Water Resources Planning. *Int. J. Technol. Manag.* **2000**, *19*, 546–556. [CrossRef]

17. Urtiga, M.M.; Morais, D.C.; Hipel, K.W.; Kilgour, D.M. Group Decision Methodology to Support Watershed Committees in Choosing Among Combinations of Alternatives. *Group Decis. Negot.* **2017**, *26*, 729–752. [CrossRef]

18. Dweiri, F.; Ahmed Khan, S.; Almulla, A. A Multi-Criteria Decision Support System to Rank Sustainable Desalination Plant Location Criteria. *Desalination* **2018**, *444*, 26–34. [CrossRef]

19. Kumar, A.; Sah, B.; Singh, A.R.; Deng, Y.; He, X.; Kumar, P.; Bansal, R.C. A Review of Multi Criteria Decision Making (MCDM) towards Sustainable Renewable Energy Development. *Renew Sust. Energ. Rev.* **2017**, *69*, 596–609. [CrossRef]

20. Roozbahani, A.; Ebrahimi, E.; Banihabib, M.E. A Framework for Ground Water Management based on Bayesian Network and MCDM Techniques. *Water Resour. Manag.* **2018**, *32*, 4985–5005. [CrossRef]

21. Tscheikner-Gratl, F.; Egger, P.; Rauch, W.; Kleidorfer, M. Comparison of Multi-Criteria Decision Support Methods for Integrated Rehabilitation Prioritization. *Water* **2017**, *9*, 68. [CrossRef]

22. Golfam, P.; Ashofteh, P.; Rajaee, T.; Chu, X. Prioritization of Water Allocation for Adaptation to Climate Change Using Multi-Criteria Decision Making (MCDM). *Water Resour. Manag.* **2019**, *33*, 3401–3416. [CrossRef]

23. Prodanovic, P.; Simonovic, S.P. Fuzzy Compromise Programming for Group Decision Making. *IEEE Trans. Syst. Man Cybern.-A* **2003**, *33*, 358–365. [CrossRef]

24. Bender, M.J.; Simonovic, S.P. A Fuzzy Compromise Approach to Water Resource Systems Planning Under Uncertainty. *Fuzzy Sets Syst.* **2000**, *115*, 35–44. [CrossRef]

25. Alhumaid, M.; Ghumman, A.R.; Haider, H.; Al-Salamah, I.S.; Ghazaw, Y.M. Sustainability Evaluation Framework of Urban Stormwater Drainage Options for Arid Environments Using Hydraulic Modeling and Multicriteria Decision-Making. *Water* **2018**, *10*, 581. [CrossRef]

26. Sapkota, M.; Arora, M.; Malano, H.; Sharma, A.; Moglia, M. Integrated Evaluation of Hybrid Water Supply Systems Using a PROMETHEE–GAIA Approach. *Water* **2018**, *10*, 610. [CrossRef]

27. Gigovic, L.; Pamucar, D.; Bajic, Z.; Drobnjak, S. Application of GIS-Interval Rough AHP Methodology for Flood Hazard Mapping in Urban Areas. *Water* **2017**, *9*, 360. [CrossRef]

28. Aschilean, I.; Giurca, I. Choosing a Water Distribution Pipe Rehabilitation Solution Using the Analytical Network. *Water* **2018**, *10*, 484. [CrossRef]

29. Fattahi, P.; Fayyaz, S. A Compromise Programming Model to Integrated Urban Water Management. *Water Resour. Manag.* **2010**, *24*, 1211–1227. [CrossRef]

30. Takahashi, R.H.C.; Deb, K.; Wanner, E.F.; Greco, S. A MCDM Model for Urban Water Conservation Strategies Adapting Simos Procedure for Evaluating Alternatives Intra-criteria. In Proceedings of the 6th International Conference on Evolutionary Multi-Criterion Optimization (EMO 2011), Ouro Preto, Brazil, 5–8 April 2011; Springer: Berlin, Heidelberg, 2011.

31. Roozbahani, A.; Zahraie, B.; Tabesh, M. PROMETHEE with Precedence Order in the Criteria (PPOC) as a New Group Decision Making Aid: An Application in Urban Water Supply Management. *Water Resour. Manag.* **2012**, *26*, 3581–3599. [CrossRef]

32. Roozbahani, A.; Zahraie, B. Tabesh, Integrated Risk Assessment of Urban Water Supply Systems from Source to Tap. *Stoch. Environ. Res. Risk Assess.* **2013**, *27*, 923–944. [CrossRef]

33. Abrishamchi, A.; Ebrahimian, A.; Tajrishi, M.; Marino, M.A. Case Study: Application of Multicriteria Decision Making to Urban Water Supply. *J. Water Res. Plan. Man.* **2005**, *131*, 326–335. [CrossRef]

34. Jesiya, N.P.; Girish, G. Groundwater Suitability Zonation with Synchronized GIS and MCDM approach for Urban and Peri-Urban Phreatic Aquifer Ensemble of Southern India. *Urban Water J.* **2018**, *15*, 801–811.

35. Qin, X.S.; Huang, G.H.; Chakma, A.; Nie, X.H.; Lin, Q.G. A MCDM-based Expert System for Climate-Change Impact Assessment and Adaptation Planning—A Case Study for the Georgia Basin, Canada. *Expert Syst. Appl.* **2008**, *34*, 2164–2179. [CrossRef]

36. Kim, Y.; Chung, E.-S.; Jun, S.-M.; Kim, S.-U. Prioritizing the Best Sites for Treated Wastewater Instream Use in an Urban Watershed Using Fuzzy TOPSIS. *Resour. Conserv. Recy.* **2013**, *73*, 23–32. [CrossRef]

37. Kumar, A.; Pandey, A.C. Geoinformatics based groundwater potential assessment in hard rock terrain of Ranchi urban environment, Jharkhand state (India) using MCDM–AHP techniques. *Groundw. Sustain. Dev.* **2016**, *2–3*, 27–41. [CrossRef]

38. Yu, L.; Lai, K.K. A Distance-based Group Decision-Making Methodology for Multi-Person Multi-Criteria Emergency Decision Support. *Decis. Support Syst.* **2011**, *51*, 307–315. [CrossRef]

39. Singh, R.K.; Choudhury, A.K.; Tiwari, M.K.; Shankar, R. Improved Decision Neural Network (IDNN) Based Consensus Method to Solve a Multi-Objective Group Decision Making Problem. *Adv. Eng. Inform.* **2007**, *21*, 335–348. [CrossRef]

40. Pluchinotta, I.; Kazakçi, A.O.; Giordano, R.; Tsoukiàs, A. Design Theory for Generating Alternatives in Public Decision Making Processes. *Group Decis. Negot.* **2019**, *28*, 341–375. [CrossRef]

41. Ullah, A.; Rashid, M.M.; Tamaki, J. On some Unique Features of C–K Theory of Design. *CIRP J. Manufact. Sci. Technol.* **2012**, *5*, 55–66. [CrossRef]

42. Agogué, M.; Kazakçi, A.O. 10 Years of C–K Theory: A Survey on the Academic and Industrial Impacts of a Design Theory. In *An Anthology of Theories and Models of Design; Philosophy, Approaches and Empirical Explorations*, 1st ed.; Chakrabarti, A., Blessing, L.T.M., Eds.; Springer: London, UK, 2014; pp. 219–235.

43. Hooge, S.; Béjean, M.; Arnoux, F. Organising for Radical Innovation: The Benefits of the Interplay between Cognitive and Organisational Processes in Kcp Workshops. *Int. J. Innov. Manag.* **2016**, *20*, 1–33. [CrossRef]

44. Vinogradova, I.; Podvezko, V.; Zavadskas, E.K. The Recalculation of the Weights of Criteria in MCDM Methods Using the Bayes Approach. *Symmetry* **2018**, *10*, 205. [CrossRef]

45. Zardari, N.H.; Ahmed, K.; Shirazi, S.M.; Bin Yusop, Z. *Weighting Methods and their Effects on Multi-Criteria Decision Making Model Outcomes in Water Resources Management*, 1st ed.; Springer: Cham, Germany, 2015; pp. 69–100.

46. Kandel, A. *Fuzzy Techniques in Pattern Recognition*, 1st ed.; John Wiley and Sons: New York, NY, USA, 1982; pp. 170–182.

47. Akter, T.; Simonovic, S.P.; Salonga, J. Aggregation of Inputs from Stakeholders for Flood Management Decision-Making in the Red River Basin. *Can. Water Resour. J.* **2004**, *29*, 251–266. [CrossRef]

48. Simonovic, S.P.; Akter, T. Participatory Floodplain Management in the Red River Basin, Canada. *Annu. Rev. Control.* **2006**, *30*, 183–192. [CrossRef]

49. Herrera, F.; Herrera-Viedma, E.; Verdegay, J.L. Direct Approach Processes in Group Decision-Making Using Linguistic OWA Operators. *Fuzzy Set Syst.* **1996**, *79*, 175–190. [CrossRef]

50. Mosadeghi, R.; Warnken, J.; Tomlinson, R.; Mirfenderesk, H. Comparison of Fuzzy-AHP and AHP in a Spatial Multi-Criteria Decision Making Model for Urban Land-Use Planning. *Comput. Environ. Urban Syst.* **2015**, *49*, 54–65. [CrossRef]

51. Zekai, S. *Fuzzy Logic and Hydrological Modeling*, 1st ed.; Taylor & Francis Group: New York, NY, USA, 2010; pp. 60–100.

52. Akter, T.; Simonovic, S.P. Aggregation of Fuzzy Views of a Large Number of Stakeholders for Multi-Objective Flood Management Decision Making. *J. Environ. Manag.* **2005**, *77*, 133–143. [CrossRef] [PubMed]

53. Ross, T.J. *Fuzzy Logic with Engineering Applications*, 3rd ed.; Wiley: Hoboken, NJ, USA, 2009; pp. 89–116.

54. Azarnivand, A.; Hashemi-Madani, F.S.; Banihabib, M.E. Extended Fuzzy Analytic Hierarchy Process Approach in Water and Environmental Management (Case Study: Lake Urmia Basin, Iran). *Environ. Earth Sci.* **2015**, *73*, 13–26. [CrossRef]

55. Shih, H.S.; Shyur, H.J.; Lee, E.S. An Extension of TOPSIS for Group Decision-Making. *Math. Comp. Model.* **2007**, *45*, 801–813. [CrossRef]

56. Llopis-Albert, C.; Merigo, J.M.; Liao, H.; Xu, Y.; Grima-Olmedo, J.; Grima-Olmedo, C. Water Policies and Conflict Resolution of Public Participation Decision-Making Processes Using Prioritized Ordered Weighted Averaging (OWA) Operators. *Water Resour. Manag.* **2018**, *32*, 497–510. [CrossRef]

57. Javidi Sabbaghian, R.; Zarghami, M.; Nejadhashemi, A.P.; Sharifi, M.B.; Herman, M.R.; Daneshvar, F. Application of Risk-based Multiple Criteria Decision Analysis for Selection of the Best Agricultural Scenario for Effective Watershed Management. *J. Environ. Manag.* **2016**, *168*, 260–272. [CrossRef]

58. Khakzad, H. OWA Operators with Different Orness Levels for Sediment Management Alternative Selection Problem. *Water Supp.* **2020**, *20*, 173–185. [CrossRef]

59. Zadeh, L.A. A computational approach to fuzzy quantifiers in natural languages. *Comput. Math. Appl.* **1983**, *9*, 149–184. [CrossRef]

60. Ben-Arieh, D.; Chen, Z. Linguistic-Labels Aggregation and Consensus Measure for Autocratic Decision Making Using Group Recommendations. *IEEE Trans. Syst. Man Cybern.-A* **2006**, *36*, 558–568. [CrossRef]

61. Bouzarour-Amokrane, Y.; Tchangani, A.P.; Peres, F. Defining and Measuring Risk and Opportunity in BOCR Framework for Decision Analysis. In Proceedings of the 9th International Conference on Modeling, Optimization and Simulation (MOSIM 2012), Bordeaux, France, 6–8 June 2012.

62. Tchangani, A.P.; Bouzarour-Amokrane, Y.; Peres, F. Evaluation Model in Decision Analysis: Bipolar Approach. *Informatica* **2012**, *23*, 461–485. [CrossRef]

63. Chiclana, F.; Taipa-Garcia, J.M.; del Moral, M.J.; Herrera-Viedma, E. A Statistical Comparative Study of Different Similarity Measures of Consensus in Group Decision Making. *Inform. Sci.* **2013**, *221*, 110–123. [CrossRef]

64. Wang, J.; Meng, X.; Dong, Z.; Lu, H.; Sun, J. A Consensus Degree based Multiple Attribute Group Decision Making Method. In Proceedings of the International Conference on Information and Automation (ICIA 2014), Hailar, China, 28–30 July 2014.

65. Dong, Y.; Zha, Q.; Zhang, H.; Kou, G.; Fujita, H.; Chiclana, F.; Herrera-Viedma, E. Consensus Reaching in Social Network Group Decision Making: Research Paradigms and Challenges. *Knowl.-Based Syst.* **2018**, *162*, 3–13. [CrossRef]

66. Kacprzyk, J.; Fedrizzi, M. A 'Soft' Measure of Consensus in the Setting of Partial (Fuzzy) Preferences. *Eur. J. Oper. Res.* **1988**, *34*, 316–325. [CrossRef]

67. Verma, D. *Decision Making Style: Social and Creative Dimensions.*, 1st ed.; Global India Publications: New Delhi, India, 2009; pp. 205–232.

68. Zarghami, M.; Szidarovszky, F. On the Relation between Compromise Programming and Ordered Weighted Averaging Operator. *Inform. Sci.* **2010**, *180*, 2239–2248. [CrossRef]

69. Javidi Sabbaghian, R.; Zarghami, M.; Sharifi, M.B.; Mianabadi, H. Developing a Distance-Based Group Consensus Model under Risk Assessment for Effective Watershed Management. In Proceedings of the 23rd International Conference on Multiple Criteria Decision Making (MCDM 2015), Hamburg, Germany, 2–7 August 2015.

70. Tooss Ab Consultant, Co. *Updating of Integrated Water Resources Management of Iran Project, The Eastern Basins of Iran: Report of Population Studies, Quaraqum Watershed*, 1st ed.; Tooss Ab Co.: Mashhad, Iran, 2012.

71. Tooss Ab Consultant, Co. *Updating of Integrated Water Resources Management of Iran Project, The Eastern Basins of Iran: Report of Urban and Rural Water Consumptions and Demands, Quaraqum Watershed*, 1st ed.; Tooss Ab Co.: Mashhad, Iran, 2012.

72. Davari, K.; Omranian Khorasani, H.; Shafi'ee, M.; Salarian, M. *The Process of Water Management, Revised Version for Kashafroud Watershed: Water and Sustainable Development*, 1st ed.; Khorasan Razavi Regional Water Co.: Mashhad, Iran, 2013; pp. 1–33.

73. Mesgaran, M.; Azadi, P. *A National Adaptation Plan for Water Scarcity in Iran: Stanford Iran 2040 Project*, 1st ed.; Stanford University: Stanford, CA, USA, 2018; pp. 1–36.

74. Tooss Ab Consultant, Co. *Updating of Integrated Water Resources Management of Iran Project, The Eastern Basins of Iran: Report of Meteorology and Climatology Studies, Quaraqum Watershed*, 1st ed.; Tooss Ab Co.: Mashhad, Iran, 2012.

75. Tooss Ab Consultant, Co. *Updating of Integrated Water Resources Management of Iran Project, The Eastern Basins of Iran: Report of Quantitative and Qualitative Surface Water, Quaraqum Watershed*, 1st ed.; Tooss Ab Co.: Mashhad, Iran, 2013.

76. Tooss Ab Consultant, Co. *Updating of Integrated Water Resources Management of Iran Project, The Eastern Basins of Iran: Report of Quantitative and Qualitative Groundwater Studies, Quaraqum Watershed*, 1st ed.; Tooss Ab Co.: Mashhad, Iran, 2012.

77. Tooss Ab Consultant, Co. *Updating of Integrated Water Resources Management of Iran Project, The Eastern Basins of Iran: Report of Socio-economic Studies, Quaraqum Watershed*, 1st ed.; Tooss Ab Co.: Mashhad, Iran, 2012.

78. Tooss Ab Consultant, Co. *Updating of Integrated Water Resources Management of Iran Project, The Eastern Basins of Iran: Report of Agricultural Consumptions and Demands Studies, Quaraqum Watershed*, 1st ed.; Tooss Ab Co.: Mashhad, Iran, 2012.

79. Tooss Ab Consultant, Co. *Updating of Integrated Water Resources Management of Iran Project, The Eastern Basins of Iran: Report of Industrial Consumptions and Demands Studies, Quaraqum Watershed*, 1st ed.; Tooss Ab Co.: Mashhad, Iran, 2012.

80. Tooss Ab Consultant, Co. *Updating of Integrated Water Resources Management of Iran Project, The Eastern Basins of Iran: Report of Environmental Studies, Quaraqum Watershed*, 1st ed.; Tooss Ab Co.: Mashhad, Iran, 2013.

81. Tooss Ab Consultant, Co. *Updating of Integrated Water Resources Management of Iran Project, The Eastern Basins of Iran: Report of Modeling of the Water Resources and Consumptions, Quaraqum Watershed*, 1st ed.; Tooss Ab Co.: Mashhad, Iran, 2013.

82. Vrba, J.; Lipponen, A. *Groundwater Resources Sustainability Indicators: Groundwater Working Group UNESCO, IAEA, IAH*, 14th ed.; UNESCO: Paris, France, 2007; pp. 7–27.

83. DiSano, J. *Indicators of Sustainable Development: Guidelines and Methodologies.*, 3rd ed.; United Nations Publications: New York, NY, USA, 2007; pp. 69–72.

84. Tooss Ab Consultant, Co. *Updating of Integrated Water Resources Management of Iran Project, The Eastern Basins of Iran: Report of Identifying and Determining the Criteria of the Water Resources and Consumptions, Quaraqum Watershed*, 1st ed.; Tooss Ab Co.: Mashhad, Iran, 2013.

85. United Nations Sustainable Development Solutions Network. *Indicators and a Monitoring Framework for the Sustainable Development Goals, A Report to the Secretary-General of the United Nations by the Leadership Council of the Sustainable Development Solutions Network*, 1st ed.; United Nations: New York, NY, USA, 2015; pp. 1–233.

86. Tooss Ab Consultant, Co. *Updating of Integrated Water Resources Management of Iran Project, The Eastern Basins of Iran: Report of hydrological-hydrogeological budget, Quaraqum Watershed*, 1st ed.; Tooss Ab Co.: Mashhad, Iran, 2015.

Article

System Dynamics Approach for Assessing the Behaviour of the Lim Reservoir System (Serbia) under Changing Climate Conditions

Milan Stojkovic * and Slobodan P. Simonovic

Department of Civil and Environmental Engineering, University of Western Ontario, London, ON N6A 3K7, Canada
* Correspondence: mstojko@uwo.ca; Tel.: +1-519-661-2139

Received: 9 July 2019; Accepted: 1 August 2019; Published: 6 August 2019

Abstract: Investigating the impact of climate change on the management of a complex multipurpose water system is a critical issue. The presented study focuses on different steps of the climate change impact analysis process: (i) Use of three regional climate models (RCMs), (ii) use of four bias correction methods (BCMs), (iii) use of three concentration scenarios (CSs), (iv) use of two model averaging procedures, (v) use of the hydrological model and (vi) use of the system dynamics simulation model (SDSM). The analyses are performed for a future period, from 2006 to 2055 and the reference period, from 1971 to 2000. As a case study area, the Lim water system in Serbia (southeast Europe) is used. The Lim river system consists of four hydraulically connected reservoirs (Uvac, Kokin Brod, Radojnja, Potpec) with a primary purpose of hydropower generation. The results of the climate change impact analyses indicate change in the future hydropower generation at the annual level from −3.5% to +17.9%. The change has a seasonal variation with an increase for the winter season up to +20.3% and decrease for the summer season up to −33.6%. Furthermore, the study analyzes the uncertainty in the SDSM outputs introduced by different steps of the modelling process. The most dominant source of uncertainty in power production is the choice of BCMs (54%), followed by the selection of RCMs (41%). The least significant source of uncertainty is the choice of CSs (6%). The uncertainty in the inflows and outflows is equally dominated by the choice of BCM (49%) and RCM (45%).

Keywords: system dynamics; system analysis; complex water system; uncertainty assessment; climate change; regional climate models; averaging procedures; HEC-HMS; Lim river; Lim water systems

---

1. Introduction

Water resources management relies on the application of a systems approach to deal with complex problems [1]. This approach uses a systems analysis in finding solutions for complex water problems trying to balance between conflicting social, ecological and economic concerns that affect the decision-making process [2]. This study uses one of the two main techniques of systems analysis -simulation. In particular, the system dynamics simulation based on the causal, stock and flow diagraming is implemented. System dynamics simulation is an appropriate approach for the analyses of interconnecting processes and functional relationships of the water resource system components [3]. Using systems analysis, the water resources professionals are able to define plans, design and define reservoir operations for complex water systems under present and future climates.

1.1. System Dynamics Simulation Approach for the Climate Change Impacts Assessment

Water resource management is affected by changing climate conditions combined with accelerated environmental and social changes [4]. An increase in temperature, changes in the precipitation pattern, an increase in the frequency and magnitude of extreme climatic events and a decrease of the snow cover are some of the climate change implications with significant impacts on water resources [5]. The intensification of the hydrological cycle under a changing climate will result in higher flows during the winter season and lower flows for the summer and autumn months. The demand for freshwater, water for irrigation, hydropower generation and instream flows to sustain river health are likely to increase due to a rapid increase in the population. Hence, it is evident that the existing reservoir operations will need to address the challenges of the imbalance between water supply and demand [6,7]. The changing climate is therefore expected to affect the management of complex water resource systems significantly [8]. The implications of climate change on water resource management will affect the planning and real-time operation stages of water systems [6]. Under the changing climate, population growth and water technology deployments, the water systems in the planning stage should explicitly consider the trade-offs between releases from the reservoirs to maintain normal operational levels, environmental flow, water demand for industry and households, agricultural irrigation and hydropower generation. For the case of existing reservoirs, the design capacity of the reservoir storage and hydropower plant (HPP) characteristics need to be re-evaluated together with the real-time reservoir operations in the changing climate conditions. The revisions may end by suggesting the use of both, non-structural and structural measures [9]. For example, the addition of a turbine to an existing power plant due to higher flows in winter is a structural measure, while the change of reservoir operating rules is considered as a non-structural adaptation measure. The latter adjusts levels in reservoirs over the winter and spring months, increasing the releases through the turbines, and consequently increasing the hydropower generation [10]. Applying adaptive real-time reservoir operation rules and water management policy for the river basins impacted by climate change can reduce the vulnerability associated with the hydropower generation and should be an effective non-structural measure capable of responding to changing climate conditions [10].

1.2. System Dynamics Simulation Modelling Processes and Its Uncertainty

Several system dynamics simulation modelling studies have been performed in the past focusing on the reservoir operations and adaptation measures under the changing climate conditions [9–13]. The climate change impact studies use the following steps to estimate the impacts on the reservoir system operations: (1) Selection of the global climate model/s (GCMs) and/or regional climate model/s (RCMs), (2) selection of the concentration scenarios, (3) correction of raw climatic data from the climate models using bias correction methods (BCMs), (4) application of averaging procedures to combine the outputs from the climate models, (5) application of the hydrological model to estimate the inflows for the reservoir system simulation model, (6) development of the system dynamics model and operational rules to transform available water resources in space and time considering the constraints of the system and needs of the users.

To close the science-practice gap, it is required to provide a clear understanding of the uncertainty within the climate change impact analysis processes [4]. The uncertainty associated with future climate variations and natural hydrologic variability represents a great challenge for the water resource system management [14]. The reservoir inflows are the most significant contributor of uncertainty to water resources management. The sources of uncertainty in the reservoir inflows originate in model parameters and the model structure [14].

Furthermore, inflows can be uncertain due to differences in the space and time distribution scales [15]. However, the greatest uncertainty in the hydrological outputs under the present and future climate conditions stems from the uncertainty in the model structure, not the parameters [16]. The significant uncertainty in the inflows, and consequently in the outflows, is associated with the choice of the climate model, rather than the concentration scenario and choice of the hydrological model [17,18]. The statistical post-processing tools contribute to the highest level of uncertainty in the reservoir inflows, propagating it to the hydropower generation [8]. To provide scientifically based advice to decision-makers, it is highly recommended to use different approaches for each step of the climate change impact modelling process. Integrating uncertainties into the decision-making process of reservoir management can increase the utility of the decision support tools [4].

1.3. The Goals of the Study

This study is carried out for the Lim water system that includes the Lim and Uvac rivers (Serbia, southeast Europe), Figure 1. The primary purpose of the system is the hydropower generation. In addition, the Lim water system mitigates adverse hazardous events and improves the downstream water quality [19]. Variation of the inflows and hydropower generation mostly depend on the climate conditions. Therefore, the primary goal of the presented study is to assess the climate-related impacts on the Lim river basin using the simulation model of hydraulically connected reservoirs (Figure 1).

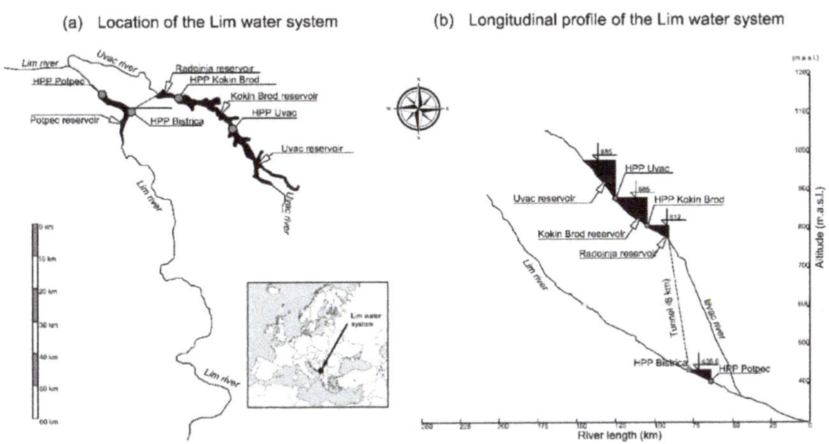

Figure 1. Location of the Lim water system reservoirs (**a**) together with the longitudinal profile (**b**).

The assessment of the climate-related impacts on the hydropower generation of the Lim water system has been performed in [19]. A traditional engineering approach (flow duration analyses under present and future climates) was used to estimate the impacts of the changing climate on the hydropower generation [20]. This study extends the World Bank analyses using an approach based on the system analysis and simulation of climate-related impacts on the Lim water system [1]. It employs the climate change impact analysis processes and corresponding steps implemented in this study are illustrated in Figure 2. The impact assessment is obtained by using three RCMs, three CSs, four BCMs, two averaging techniques, a single hydrological model and system dynamics simulation model (SDSM). The BCMs transform raw precipitation from the climate models. Subsequently, averaging procedures are used to identify the processes that contribute to the uncertainty in the SDSM outputs. Next, the hydrological model is used to convey the climate change signal to the watershed response. Finally, the SDSM is used to capture the complex system structure and assess the climate change-related impacts on the outputs of SDSM within the Lim river basin.

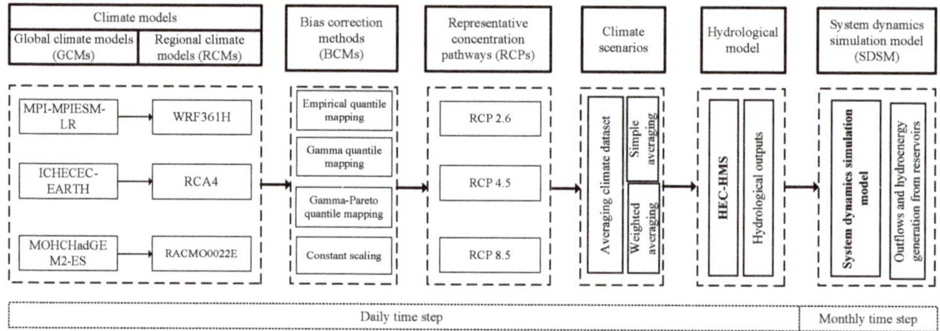

Figure 2. The climate change impact analysis process used in the study.

The goals of the study are therefore multifold: (1) To develop the SDSM and to propose adaptive operational rules for the existing system reservoirs within the Lim river basin, (2) to assess the impacts of the changing climate on the SDSM outputs, (3) to quantify the contribution of the uncertainty of each individual climate change impact analysis steps on the system performance using SDSM.

The rest of the paper is organized as follows. The description of the Lim water system is presented in Section 2. Section 3 describes the SDSM of the Lim water system alongside the operational rules of the reservoir system. The methodology for the uncertainty assessment within the different steps of the impact assessment process is presented in Section 4. The assessment results and their discussion are in Section 5, while conclusions end the paper in Section 6.

2. Lim Water System

The Lim water system is located in Serbia (southeast Europe) and extends over an area of around 3600 km^2. It represents a multipurpose complex system including four hydraulically connected reservoirs (the Potpec, Uvac, Kokin Brod and Radojnja). The locations of the Lim water system reservoirs alongside the longitudinal profiles of the Lim and Uvac rivers are depicted in Figure 1. The Potpec reservoir lies on the Lim river, while the remaining three reservoirs, namely the Uvac, Kokin Brod and Radojnja, are located on the Uvac river. The Lim and Uvac rivers are part of the international trans-boundary Drina river system [19]. The Lim river and Uvac river (the biggest tributary of the Lim river), are respectively 220 km and 115 km long. The specific water yield of the Uvac river is 9.9 l/s/km^2, while the contribution of the Lim river is much greater, ranging from 26.0 l/s/km^2 to 43.8 l/s/km^2 [21,22].

The primary purpose of the Lim water system is the hydropower generation (Table 1). In addition, the Lim water system is used for mitigation of floods, and downstream water quality control by regulation of the outflows over the low-flow season. The management of the Lim water system depends on the actual volume of water stored in the reservoirs, inflows and energy demand.

Table 1. The characteristics of reservoirs in the Lim water system.

Reservoirs	Year Built	Drainage Area (km^2)	Annual Inflows (m^3/s)	Active Volume (10^{6}m^3)	Maximal Operational Level (m.a.s.l.)	Minimal Operational Level (m.a.s.l.)	Spillway Capacity (m^3/s)	Spillway Crest Elevation (m.a.s.l.)
Potpec	1967	3605	79.9	19.8	437	423.6	3000	439
Uvac	1979	920	9.5	160	988	940	1050	986
Kokin Brod	1962	1170	13	209	888	845	1400	885.5
Radojnja	1959	1331	13.5	4.1	815	805	1400	816.2

The concrete gravity dam of the Potpec reservoir is 46 m in height and 215 m in length (Figure 1a). The power plant has three turbines (51 MW) for power generation with the maximal and minimal discharges of 165 m^3/s and 18.5 m^3/s, respectively. Active storage of the Potpec reservoir is 19.8 $\times$ 10^6 m^3. Three gated spillways are located at the middle part of the dam with the capacity of 3000 m^3/s. The Uvac, Kokin Brod and Radojnja reservoirs are hydraulically connected. They satisfy the demand for the hydropower generation and provide storage for attenuation of flood waves (Figure 1). The Uvac reservoir is the second largest reservoir within the Lim water system with active storage of 160 $\times$ 10^6 m^3. The Kokin Brod and Radojnja reservoirs are located on the downstream river section (Figure 1a,b). The Kokin Brod reservoir has the largest active storage of 209 $\times$ 10^6 m^3. Downstream of the Kokin Brod reservoir is situated the Radojnja reservoir with the active storage of 4.1 $\times$ 10^6 m^3. The lateral flow between the Uvac and Kokin Brod reservoirs is equal to 2.5 m^3/s, while the lateral flow which contributes to the Radojnja reservoirs is 0.5 m^3/s. The dams of the Kokin Brod and Radojnja reservoirs are rockfill dams, while the Uvac dam is an earthfill dam. The dams of the Uvac, Kokin Brod and Radojnja reservoirs are 110 m, 82 m and 42 m high and 307 m, 1220 m and 361 m long, respectively. The Uvac and Kokin Brod power plants have turbines with the maximal discharge of 43 m^3/s (36 MW) and 36 m^3/s (2 $\times$ 10.7 MW), respectively. The Radojnja reservoir is hydraulically connected with diversion-type turbines at the Bistrica power plant by the 8 km pressure tunnel (Figure 1a,b). The Bistrica power plant has two turbines (2 $\times$ 18 m^3/s, 2 $\times$ 51.5 MW) and is located at the Lim river nearby the Potpec reservoir. The pressure tunnel conveys the water from the Uvac river to the Lim river providing a significant contribution to the total annual flow at the Potpec reservoir (77.6 m^3/s). Note that the total annual inflow of the Radojnja reservoir is equal to 14.4 m^3/s. The maximal spillway capacities of the Uvac, Kokin Brod and Radojnja reservoirs are 1050 m^3/s, 1400 m^3/s and 1400 m^3/s, respectively.

3. System Dynamics Simulation Model of the Lim Water System

3.1. General Approach

The presented climate change impact analysis study organization is shown in Figure 2. Three RCMs from the EURO-CORDEX initiative are selected [23]: WRF361H (DWD—Deutscher Wetterdienst), RCA4 (SMHI—Swedish Meteorological and Hydrological Institute) and RACMO 22E (Royal Netherlands Meteorological Institute). The selected RCMs driven by the boundary conditions from GCMs are illustrated in Figure 2. Climate simulations cover the case study area with the datasets at a high spatial resolution of 0.11 degrees (~2.5 km). Such RCM resolution is selected since it can reproduce extreme precipitation behavior. Furthermore, the availability of the simulated climate for the future period under the RCP 2.6, RCP 4.5 and RCP 8.5 emission trajectories is used. Selected climate options cover a wide range of future socioeconomic scenarios and projected CS. Furthermore, the EURO-CORDEX collection also offers climate simulations from many RCMs with different spatial resolutions. For example, HIRHAM5 and REMO2009, REGCM3 and HadRM3Q16, RACMO2 and RCA provide the simulated precipitation and temperature with spatial resolutions of 12.5 km^2, 25 km^2, and 50 km^2, respectively.

Then, BCMs are applied to statistically correct climate model outputs. Selected BCMs include empirical quantile mapping—EQM [24], gamma quantile mapping—GQM [25], gamma-Pareto quantile mapping—GPQM [13] and constant scaling—CS [26]. Next, the averaging methods are used to estimate average climate signals from multiple RCMs and BCMs. The simple model averaging [27] is introduced to analyze uncertainty in the climate models, statistical post-processing tools and concentration scenarios (CSs). The Bates-Granger averaging [28] is used to form the median of the ensemble using the weighted realizations from each RCM and BCM. Using the averaging procedures, a set of climate scenarios are formed. Then, the hydrological modelling system (HEC-HMS) version 4.2.1 is used to derive a watershed response under the climate scenarios [29]. The model hydrological structure includes six modelling components: Meteorological input, snow, precipitation loss, direct runoff, baseflow, and river routing. The applied structure of the hydrological model is capable of performing continuous hydrological simulation under present and future climates [30,31]. Average climate outputs are used by the hydrological model to obtain the inflows for the Lim water system reservoirs (Figure 1a). The inflows are then used by SDSM to simulate the Lim water system performance. The SDSM is developed using the Vensim system dynamics simulation software [32]. The SDSM transforms (in space and time) the outputs of the hydrological model through the hydraulically connected system of reservoirs resulting in the spillway, turbine flow and environmental reservoir releases. A description of the SDSM is provided in the following section.

3.2. The System Dynamics Simulation Model

3.2.1. Reservoir Operations

The Lim water system reservoirs regulate downstream flows in time and space to meet the demand for the hydropower generation, attenuate flood waves and improve the downstream water quality. The reservoir capacity of the Lim water system is divided into three parts [33]: Active storage, dead storage and flood storage. The active storage capacity of the reservoirs is used for the hydropower generation and environmental flow management, while the flood storage capacity is reserved to reduce the flooding downstream. The dead storage capacity of the reservoirs is required for the sediment deposition. The existence of large reservoirs on the Uvac river enables the active storage capacity of the Uvac and Kokin Brod reservoirs to be separated into two parts: Over-year storage capacity, and within-year storage volume. The Radojnja and Potpec reservoirs participate only in daily flow regulation due to the smaller reservoir capacity.

Operations of the Lim water system are driven by technically experienced staff of the public power utility "Elektroprivreda Serbia" to meet the needs of the Serbian power system over the peak demand hours. However, the change of climatic conditions will intensify the hydrological cycle, leading to changes in the annual and seasonal streamflow distribution. Since reservoirs are built to alter the natural spatial and temporal distribution of streamflow, flexible operational rules, associated with the use of the active storage capacity, are required to deal with the annual and interannual changes of the inflows. The climate change impacts on the operational rules of the Uvac and Kokin Brod reservoirs, which can regulate the seasonal and annual flow variations, are much more important than the impacts on the Radojnja and Potpec reservoirs operating at a daily basis. Accordingly, the over-year storage capacity and within-year storage volume of the Uvac and Kokin Brod reservoirs is re-optimized. The operational rules at the monthly time scale are developed using the yield model solved by the sequent peak method [33]. This method enables the adaptation of reservoir operations according to the changes in the multi-annual and seasonal flow distributions [10]. In this way, the release policy is adapted to the climate signal changes because the climate drivers are the main forcing factor of flow variations. For the Radojnja and Potpec reservoirs, multi-annual and interannual flow distribution does not affect their regular operations and, therefore, the standard operational policy is applied [33].

The operational rules for the reservoirs shown in Figure 3 identify the storage volume zones associated with a particular release policy. The Uvac and Kokin Brod reservoirs have the within-year storage capacity below zone 2 because the distribution of the within-year inflows requires additional reservoir capacity (Figure 3a,b). For this purpose, monthly hydropower releases with 50% reliability are used to define the optimal yields and the corresponding reservoir release rules. Zone 1 indicates the release policy to satisfy the variations between the annual inflow distribution and annual yield estimated as a median of the actual annual hydropower releases (Figure 3a,b). Maximal operational levels and the upper bound of zone 2 correspond to zone 3 (Figure 3a,b). If at any time t the storage volume is in zone 3, releases from the Uvac and Kokin Brod reservoirs cannot be greater than the maximal turbine discharge given at 43 m³/s and 37.4 m³/s, respectively. Within zone 2, releases from reservoirs at any time t should not be higher than the optimal yield with a probability of 50% for each month. Releases in zone 1 can be also made to satisfy any demand lower or equal to the annual yield for the Uvac (9.3 m³/s) and Kokin Brod reservoirs (11.3 m³/s). If the water levels in the reservoirs are beyond the limits of zone 3, there are no hydropower releases. Consequently, spillway releases from the reservoirs can be made to reduce the actual storage in the reservoirs to spillway crest elevations

(Table 2). The storage above these elevations is considered as the flood storage volume for the Uvac and Kokin Brod reservoirs.

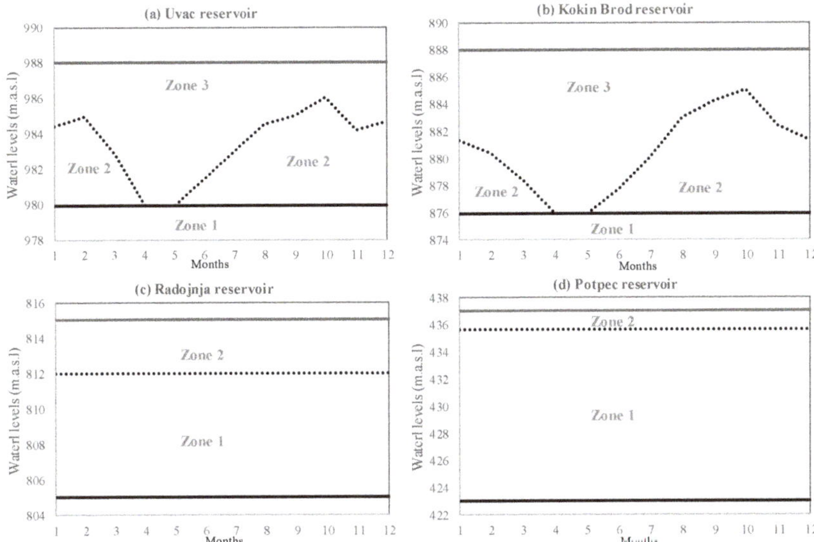

Figure 3. Reservoir release rules for the Lim water system: (**a**) Uvac reservoir, (**b**) Kokin Brod reservoir, (**c**) Radojnja reservoir, (**d**) Potpec reservoir.

The Radojnja and Potpec reservoirs use the standard operational policy. The active storage capacity is divided in accordance with their minimal, normal and maximal operation levels. In the first zone, minimal and normal operational levels are used to set operational rules related to the first part of their active storage (Figure 3c,d). The release policy made within the first zone is to satisfy any demand limited with the maximal and minimal turbine discharges (Table 1). The maximal release policy is made in the second part zone 2 corresponding to the installed turbine capacity of the Bistrica (36 m^3/s) and Potpec (165 m^3/s) HPPs. Note that the upper limits of zone 2 represent the maximal operational levels of the powerplants. If at any time *t* the water levels rise above zone 2, spillways are activated reducing the water levels below the maximal operational levels. In addition to operational flood rules, releases from these reservoirs have to be made to improve the downstream water quality. The release has to meet environmental flows standing at 1.2 m^3/s and 13.9 m^3/s for the Radojnja and Potpec reservoirs, respectively.

Table 2. The characteristics of hydropower plants in the Lim water system.

Hydropower Plant	Type	Number of Turbines	Maximal Discharge (m^3/s)	Instaled Power (MW)	Annual Energy Generation (GWh)	Maximal Water Head (m)	Minimal Water Head (m)
Potpec	Non-diversion	3	165	52	216	38.4	25.6
Uvac	Non-diversion	1	43	36	72	100	55
Kokin Brod	Non-diversion	2	37.4	21.4	60	73	-
Radojnja	Diversion	2	36	103	370	378	345

3.2.2. System Dynamics Simulation Model of the Lim Water System

The SDSM of the complex Lim water system is developed in the Vensim software [32]. The SDSM uses a stock and flow diagraming to capture the system structure. The stock and flow diagrams use four graphical objects to represent a complex system structure: (1) Stocks, (2) flows, (3) auxiliary variables and (4) arrows. The reservoirs of the Lim water system are modelled as stocks because they represent state variables accumulating over time. Inflows to and releases from the reservoirs are modelled as flows. They are attached to stocks and change the state of the accumulated water in reservoirs. Other variables in the Lim water system SDSM are represented using auxiliaries. Arrows are connecting stocks, flows and auxiliary variables to close the system structure. They convey information from one variable to another. The stock and flow diagram of the Lim water system is presented in Figure 4. The model simulation uses monthly time step and hydrological model outputs under current conditions. Three reservoirs on the Uvac river, namely the Uvac, Kokin Brod and Radojnja, are in series. The hydropower and spillway releases of the Uvac reservoir are the inflows for the Kokin Brod reservoir together with an additional contribution of lateral flows. Similarly, the outflow from the Kokin Brod reservoir is the major inflow of the Radojnja reservoir, which represents the last reservoir in the series. The hydropower releases of the Radojnja reservoir (the Bistrica HPP) contribute to the inflow of the Potpec reservoir transferring water from the Uvac river to the Lim river (Figure 3). Note that the spillway and environmental releases from the Uvac reservoir continue to flow downstream the Uvac river until its confluence with the Lim river (Figure 1). Alongside outflows from the Radojnja reservoir, the Potpec reservoir at the Lim river receives more significant inflow from the major course-the Lim river. The SDSM is developed to follow the release policy described in Section 3.2.1. As any reservoirs, the Lim water system reservoirs, accumulate their flows in the following way [1]:

$$Reservoir(t) = \int_{t_0}^{t} [Inflow(t) - Outflow(t) - Losses(t)]\mathrm{d}t + Reservoir(t_0), \tag{1}$$

where $Inflow(t)$ and $Outflow(t)$ are the values of the reservoir flows at any time t between the initial time t_0 and current time t, while $Losses(t)$ are the cumulative losses (evaporation and seepage) from the reservoirs over time t_0 and t. $Reservoir(t)$ and $Reservoir(t_0)$ denote the actual volume of reservoirs in m^3 at time steps t and t_0, respectively.

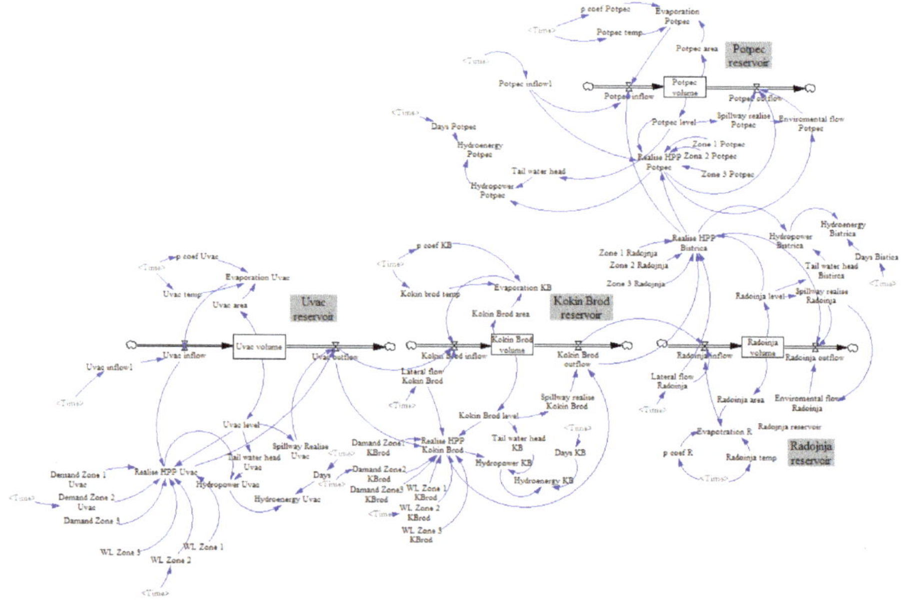

Figure 4. Stock and flow diagram of the Lim water system consisting of four reservoirs (Uvac, Kokin Brod, Radojnja and Potpec).

Evaporation losses in Equation (1) are estimated due to their importance, especially for reservoirs with large storage capacity (Uvac and Kokin Brod reservoirs). For the monthly evaporation rates from the surface of reservoirs, the Blaney–Criddle formula is applied [34]:

$$Losses(t) = \frac{p(0.457{\cdot}T + 8.128) \times Surface\_area(t)}{1000}, \tag{2}$$

where T is the mean monthly temperature in degrees Celsius at time t, p is the percentage of mean annual daytime hours for each month over the year and $Surface\_area$ denotes the area of reservoirs in square metres at time t.

Although seepage losses contribute toward the total releases from the Lim water system, there are no available information on driving factors (e.g., seepage rates, geological formation, permeability, soil moisture conditions) to estimate the seepage rates from the reservoirs. Therefore, the SDSM model does not account for seepage losses.

Inflows in Equation (1) are natural streamflows simulated by the HEC-HMS (Section 3.1). They are transformed over time and space in accordance with the release policy described in Section 3.2.1. The outflows in Equation (1) account for the hydropower (*Release_hydropower*) and spillway (*Release_spillway*) releases as well as environmental flow releases (*Release_env*). The hydropower releases for reservoirs are expressed as a function of time depending on inflows in the system (*Inflow*) and a random term (*RND*):

$$Release\_hydropower(t) = Inflow(t-2) + RND\left(0, \sigma^2\right). \tag{3}$$

The discrepancy in the seasonal pattern between the inflows and hydropower demand is simulated with a time delay function in the Vensim software. The time delay of two months is used considering the observed inflows and turbine releases. In addition, uncertainties in hydropower releases are simulated by the random term $RND\left(0, \sigma^2\right)$ under the assumption that it follows the normal distribution. The differences among the observed inflows and hydropower releases serve as a basis to estimate the variance of the random term (σ^2) with the median having the value zero.

Based on the simulated turbine flows, the following equation is used for the hydropower calculation:

$$Hydropower(t) = \frac{9.81 \times 10^6 Release\_hydropower(t) \times H(t) \times \mu}{3600 \times 1000}, \tag{4}$$

where $Hydropower(t)$ is the megawatts of power produced in time t, $H(t)$ is the net head in metres of water available for the hydropower generation during time t and μ is the turbine efficiency.

Spillway releases (*Release_spillway*) are solved simultaneously until the water level in reservoirs reaches the spillway crest elevation. The spillway release curves are linearly interpolated to determine *Release_spillway* at time step t. Moreover, if the outflows on the downstream sections of the Uvac and Lim river are below the values required to maintain the river health, an additional release from the Radojnja reservoir (*Release_env* = 1.2 m^3/s) and Potpec reservoir (*Release_env* = 13.9 m^3/s) are calculated to meet these requirements.

4. Uncertainty Assessment within the Climate Change Impact Analysis Process

The outflows from the Lim system of the reservoirs and corresponding hydropower generation reflect the release policy acting to modify the natural hydrological regime to satisfy the multiple water needs (e.g., demand for hydropower generation, flood mitigation, environmental flow releases). Inflows under the changing climate, as the main inputs into the system dynamics simulations, increase the total level of uncertainty due to uncertainty in climate modelling (choice of climate model, bias correction tools and concentration scenarios) and statistical post-processing (selection of the averaging method). The system dynamics modelling and model's ability to capture the complex system structure accurately can add extra uncertainty to the SDSM outputs. There are several uncertainty assessment approaches (e.g., Monte-Carlo simulation [35], GLUE [36], direct variance method [37]). Since this study aims at quantifying how different steps of the impact analyses process influence the SDSM simulated variables, the direct variance method is selected. It is a straightforward approach capable of assessing the contribution of each step to the total uncertainty by varying the samples and calculating the variance of the ensemble. As a robust statistical method, the variance method represents the proxy measure for determining the uncertainty in the climate change impact analysis process. The drawback lies in the fact that it considers individual variations of each step as a proxy measure of uncertainty, rather than using the hindcast process accuracy during the recorded period. The mean variances $\bar{\sigma}^2$ of particular variables (e.g., hydropower generation, inflows, outflows) with respect to the selection of RCMs, BCMs and CSs are estimated in the following way:

$$\bar{\sigma}\,(RCM)^2 = \frac{1}{N_{RCM}-1}\sum_{i=1}^{N}\left(\bar{Y}_{RCM}(i)-\bar{Y}(i)\right)^2,\tag{5}$$

$$\bar{\sigma}\,(BCM)^2 = \frac{1}{N_{BCM}-1}\sum_{i=1}^{N}\left(\bar{Y}_{BCM}(i)-\bar{Y}(i)\right)^2,\tag{6}$$

$$\bar{\sigma}\,(CS)^2 = \frac{1}{N_{ES}-1}\sum_{i=1}^{N}\left(\bar{Y}_{CS}(i)-\bar{Y}(i)\right)^2.\tag{7}$$

A simple model averaging method is used with equally weighted variables $\bar{Y}_{RCM}$, $\bar{Y}_{BCM}$ and $\bar{Y}_{CS}$ from each RCM, BCM and CS [27]:

$$Y(RCM) = \frac{1}{N_{BCM}+N_{CS}}\sum_{j}\sum_{k}Y_{j,k},\; i\;=\;WRF;\;RCA4;\;RACMO,\tag{8}$$

$$Y(BCM) = \frac{1}{N_{RCM}+N_{CS}}\sum_{i}\sum_{k}Y_{i,k},\; j\;=\;EQM;\;GQM;\;GPQM,\tag{9}$$

$$Y(CS) = \frac{1}{N_{RCM}+N_{BCM}}\sum_{i}\sum_{j}Y_{i,j},\; k\;=\;RCP\;2.6;\;RCP\;4.5;\;RCP\;8.5,\tag{10}$$

where the numbers of selected RCMs, BCMs and CSs are denoted as N_{RCM}, N_{BCM} and N_{CS}, respectively.

Next, the Bates-Granger averaging is applied to estimate the average values of the ensemble $\bar{Y}$ with all individual realizations of RCMs, BCMs and CSs [28]:

$$\bar{Y} = \frac{1}{N_{ES}}\sum_{i}\sum_{j}\sum_{k}\beta_{i,j}\cdot Y_{i,j,k},\tag{11}$$

where $\beta_{i,j}$ are the weights of each RCM and BCM obtained with respect to their hindcast accuracy over the reference period. The contributions of each climate change impact analysis step (RCMs, CSs, BCMs) to the total uncertainty are estimated as follows:

$$\Delta_{RCM} = \frac{\bar{\sigma}^2_{RCM}}{\bar{\sigma}^2_{RCM}+\bar{\sigma}^2_{BCM}+\bar{\sigma}^2_{CS}} \times 100,\tag{12}$$

$$\Delta_{BCM} = \frac{\bar{\sigma}^2_{BCM}}{\bar{\sigma}^2_{RCM}+\bar{\sigma}^2_{BCM}+\bar{\sigma}^2_{CS}} \times 100,\tag{13}$$

$$\Delta_{CS} = \frac{\bar{\sigma}^2_{ES}}{\bar{\sigma}^2_{RCM}+\bar{\sigma}^2_{BCM}+\bar{\sigma}^2_{CS}} \times 100,\tag{14}$$

where Δ_{RCM}, Δ_{BCM} and Δ_{CS} denote the percent contribution of uncertainty introduced by RCMs, BCMs and CSs, respectively.

5. Results and Discussion

5.1. Model Verification

The SDSM is developed for four reservoirs of the Lim water system (Figure 1). Simulations are performed using the monthly time step. The Lim water system model (Figure 4) is implemented in the Vensim software which allows an easy modification of the system structure (by the manipulation of four graphical objects) and model simulations using different dataset. In this study, the outputs of the hydrological model are used as inputs to system dynamics simulations. Hydrological simulations are conducted with a variety of climatic datasets including the observed and simulated climate (Section 3.1). The HEC-HMS deterministic hydrological model is set for the Lim river basin and Lim water system.

The SDSM structure of the Lim water system is verified over the referenced period. The SDSM simulation results are compared with the observed values of the inflows, outflows and hydropower generation from 1971 to 2000. Note that the simulated climate dataset considered seven realizations from three RCMs (WRF, RCA4, RACMO), and four BCMs (EQM, GQM, GPQM, CS) by applying the simple model averaging method (Section 4). In addition, the average ensemble value is estimated with the Bates-Granger averaging. In this manner, eight time series for each variable of the SDSM are determined for the reference period of 1971–2000. The simulated inflows and outflows from the system of reservoirs at the Lim and Uvac rivers are shown in Figure 5, while Figure 6 shows the hydropower generation for each powerplant within the system. The median values of inflows/outflows for each reservoir well match the observed values (Uvac—9.5 m^3/s; Kokin Brod—13.5 m^3/s; Radojnja—13.5 m^3/s; Potpec—79.9 m^3/s). The simulated medians of monthly inflows/outflows depend on the climate data analyzed. They are in the range of 8.2–10.4 m^3/s, 11.7–15.0 m^3/s, 11.8–15.4 m^3/s and 74.3–91.3 m^3/s for the Uvac, Kokin Brod, Radojnja and Potpec reservoirs, respectively. Similarly, the simulated annual hydropower generation for most HPPs reasonably fit the observed long-term values (Uvac—72 GWh, Kokin Brod—60 GWh, Bistrica—370 GWh, Potpec—216 GWh). This is highlighted for the Uvac and Kokin Brod power plants where the medians of the annual hydropower generation are equal to 63.9–76.2 GWh and 60.9–72.8 GWh, respectively. The annual hydropower generation for the Bistrica (281.6–340.6 GWh) and Potpec (170.8–210.5 GWh) powerplants are somewhat overestimated, perhaps due to uncertainty in the reservoir elevation–volume rating curves available for these reservoirs.

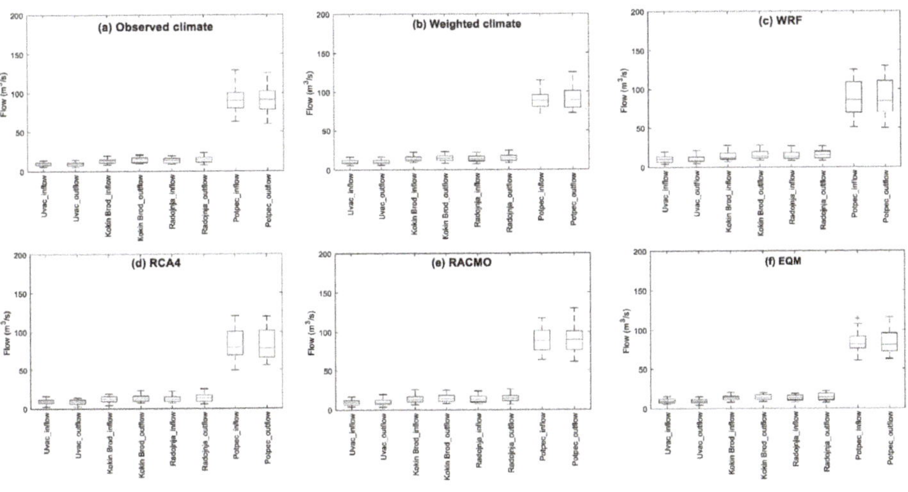

Figure 5. *Cont.*

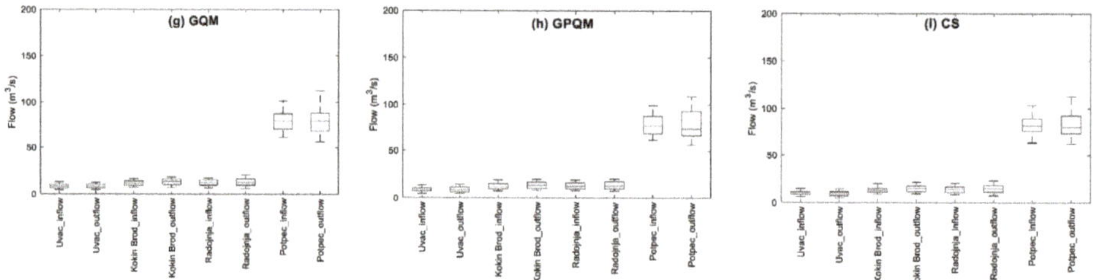

Figure 5. Monthly inflows to and outflows from the Lim water system reservoirs (Uvac, Kokin Brod, Radojnja and Potpec) estimated with the following climate datasets: (**a**) Observed climate, (**b**) weighted climate simulations, (**c–e**) regional climate model simulations—WRF, RCA4, RACMO, (**f–i**) bias correction methods—EQM (empirical quantile mapping), GQM (Gamma quantile mapping), GPQM Gamma Pareto quantile mapping), CS (constant scaling).

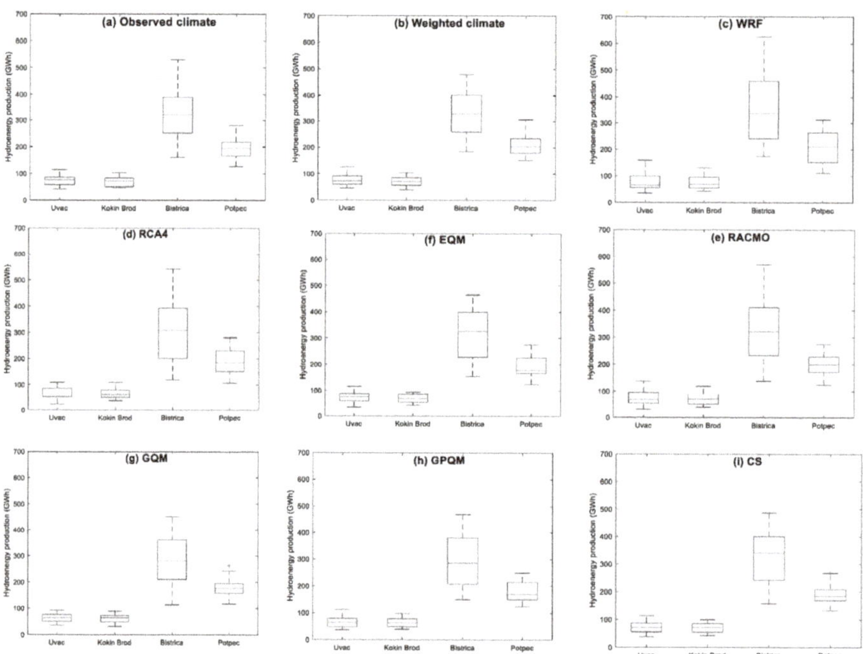

Figure 6. Annual hydropower generation of the Lim water system estimated with the following climate datasets: (**a**) Observed climate, (**b**) weighted climate simulations, (**c–e**) regional climate model simulations—WRF, RCA4, RACMO, (**f–i**) bias correction methods—EQM, GQM, GPQM, CS.

5.2. Future Projections

The verified SDSM model of the Lim water system is used to obtain the long-term prediction of system outflows and hydropower generation. The selected RCMs, BCMs and averaging procedures (Figure 2) are used to derive the precipitation and temperature for the Lim and Uvac river basins under RCP 2.6, RCP 4.5, RCP 8.5 concentration scenarios. The combinations of the different climate change modelling steps are shown in Figure 7a, while Figure 7b illustrates realisations of each climate dataset for the observed period of 1971–2000 and future period of 2006–2055 under the RCP options. Climate realisations under the aforementioned options cover the synchronous period with a length of 50 years.

Therefore, the available climate outputs for the entire time period are used in the study since they provide more uncertain estimates of the future climate.

Based on the different climate data set realizations, the HEC-HMS is used to estimate inflows into the reservoir system for the future period of 2006–2055. The simulation model is then utilized to obtain 24 time series for each system variable under different concentration scenarios for the future period of 2006–2055. These time series include 21 realizations from each RCM/BCM combination and three realizations for the entire ensemble. The latter uses the weighted precipitation and corresponding hydrological response, derived from the Bates-Granger averaging, to obtain reservoir outflows and hydropower generation. This averaging method weights the selected RCMs and BCMs based on their hindcast accuracy. Therefore, the members of the ensemble that reproduce better climate drivers are heavily weighted providing the minimum possible uncertainty in the projected inflows used for system dynamics simulations.

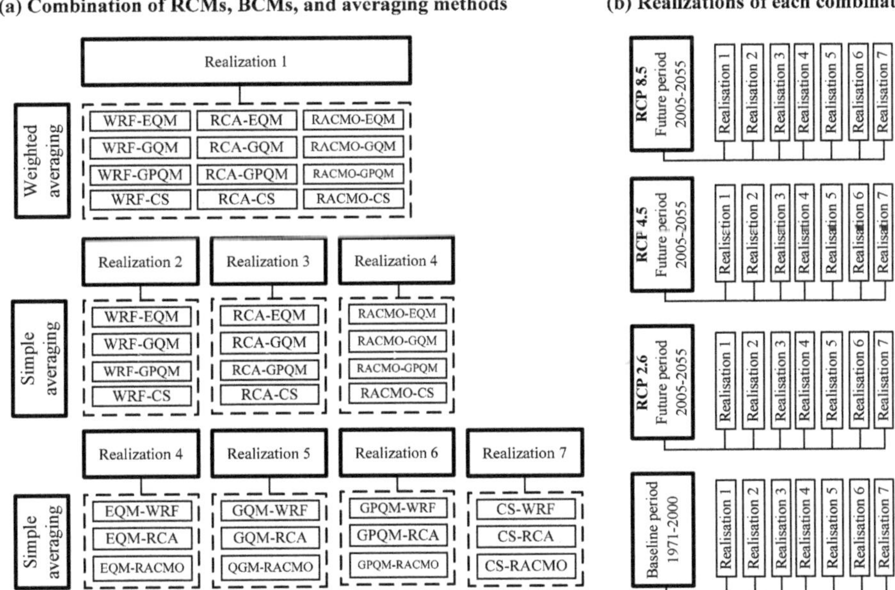

Figure 7. Climate scenarios within the Lim river basin: (**a**) Combination of RCMs (regional climate models), BCMs (bias correction methods), and averaging procedures, (**b**) realisations of each climate dataset combination for the baseline time period of 1971–2000 and future time period of 2006–2055 under RCP (representative concentration pathway) 2.6, RCP 4.5, and RCP 8.5.

The projected changes in annual inflows of the Uvac river reservoirs are in the range from −3.5% to 12.5%. In contrast, the Lim river shows less significant annual change in the inflows ranging from −5.9% to 4.9%. Variation in the hydrological inputs has an impact on the hydropower generation shown as relative change in the electricity production with respect to the reference period of 1971–2000 (Table 3). The expected changes in the annual hydropower generation are from −3.5% to +17.9% and from −2.7% to +7.9% for the Uvac and Lim reservoirs, respectively. The projected annual change in the outputs of the SDSM significantly depends on the concentration scenarios. The most severe option is RCP 8.5 with an annual decrease of the hydropower generation of −7.9%, while the RCP 4.5 scenarios shows an increase in the hydropower generation for the analyzed reservoirs (+9.6%). The annual decrease of the hydropower generation for the Uvac river reservoirs is equal to −2.3% under the RCP 2.6 scenario.

The impacts of climate change on the hydropower generation for the Lim water system are also analyzed by the study given in the literature [19]. This analysis encompasses the future period from 2011 to 2040 with regard to the reference period from 1961 to 1990. Two RCMs are used for further hydroenergetic analysis under RCP 4.5 and RCP 8.5 concentration scenarios, namely CCLM 4-8-19 and ALADIN 5.2 [19]. The study uses the traditional engineering approach based on the flow duration curve under climate change [20]. This approach allows the estimation of the percentage of corresponding turbines working time. The findings suggest that the changes in annual electricity production for the Lim river reservoirs are in the range from −6.3% to +9.4% [19]. In spite of the different modelling approach used in the study presented in this paper, the results show a significant agreement. For example, the estimated annual change in the electricity production for the Potpec reservoir is from −7.9% to +2.7% (Table 3).

Table 3. The relative annual change (%) in the hydropower generation of the Lim water system (future period of 2006–2055 production compared to the baseline period of 1971–2000 production).

Emissions Scenarios	Uvac HPP	Kokin Brod HPP	Bistrica HPP	Potpec HPP
RCP 2.6	−3.5	−1.6	−1.8	+2.7
RCP 4.5	+8.1	+17.9	+15.3	−3.0
RCP 8.5	−0.7	−1.0	−2.4	−7.9

Since the seasonality adds an important variation in the hydropower generation, the traditional approach based on the flow duration curve is not capable of capturing the changes in the seasonal power generation [20]. Higher temperature over the year will modify the seasonality of the inflows available for the hydropower generation [38]. The natural climate variation is also an important factor which affects the variation in seasonal flows [39]. In this study, the system dynamics simulation approach is employed to assess the changes in the seasonal electricity generation which depends on the inflows available for power generation and demand for electricity in Serbia. In the SDSM the turbine flows are simulated as a time delay function of the inflows and random terms (Equation (3)) to ensure that the changes in the hydropower generation follow the seasonal variations of the inflows. Figure 8 shows the hydropower generation for both, the reference period of 1971–2000 and the future period of 2006–2055 under RCP 2.6, RCP 4.5, RCP 8.5 scenarios. The simulated values are estimated by the Bates–Granger averaging of the climate dataset (Section 4). The most prominent change in the seasonal hydropower generation pattern can be seen for the Uvac HPP (Figure 8a) and Potpec HPP (Figure 8d). These reservoirs receive unregulated inflows with the modified seasonal distribution due to the changing climate. Therefore, an increase in the hydropower generation within the winter months (January–March) ranging from +1.6% to +20.3% can be expected. A rapid decrease in monthly flows over the summer and autumn months brings about the lower level of electricity generation from August to September (from −2.8% to −33.6%). Large variations in the seasonal power generation are noted for the downstream HPPs at the Uvac river, Kokin Brod HPP and Bistrica HPP (Figure 8). These variations can be attributed to the dynamics of the multiple reservoir operations since the operations of the upstream reservoir (Uvac reservoir) has an effect on the downstream Kokin Brod and Radojnja reservoirs [3].

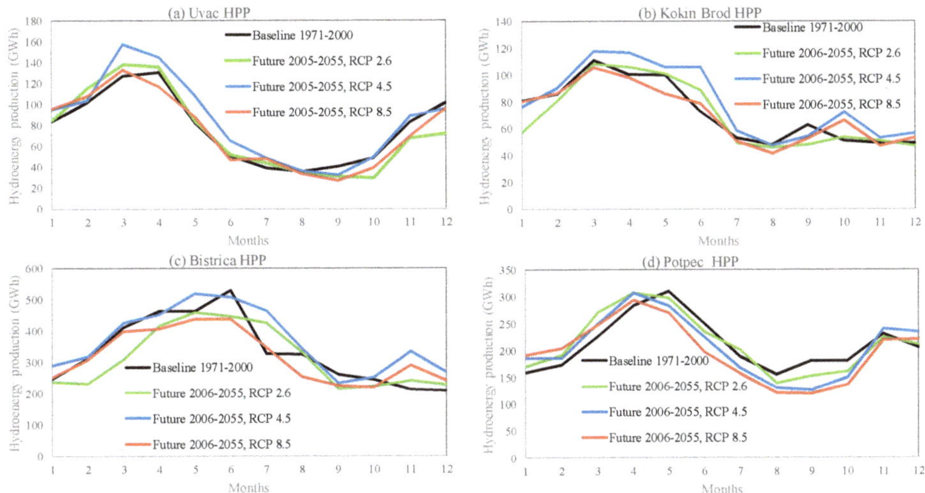

Figure 8. Seasonal distribution of the hydropower generation (gigawatt hours (GWh)) for the hydropower plants (HPPs) of the Lim water system: Bold line—baseline period 1971–2000, thin lines—future period 2006–2055 under RCP 2.6, RCP 4.5, RCP 8.5. ((**a**) Uvac HPP, (**b**) Kokin Brod HPP, (**c**) Bistrica HPP, (**d**) Potpec HPP).

Monthly values in the hydropower generation are also used to analyze the impact of the the the changing climate on the simulation outputs of SDSM. The box plots of monthly predicted inflows, outflows and hydropower generation for the reference period (1971–2000) and future period (2006–2055) are shown, respectively in Figures 9–11. Based on the monthly values, the duration curves for each realization of SDSM are calculated to examine the changes in the lower and higher values of the monthly hydropower generation. A decrease in the lower values of the hydropower generation are between −9.8% and −5.6% for the RCP 2.6 and RCP 8.5 scenarios, respectively. However, an increase in the hydropower generation for the case of the RCP 4.5 scenario equal to 10% is observed. Higher values of the hydropower generation suggest an overall increase of 3% for the RCP 2.6 and RCP 8.5 scenarios, while the RCP 4.5 shows that the change in the hydropower generation is negligible.

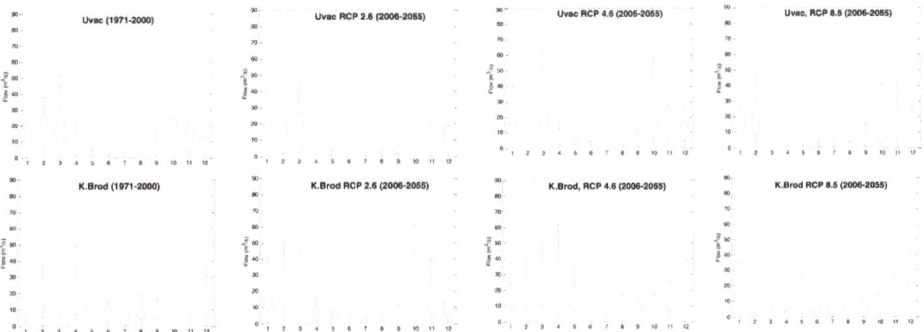

Figure 9. *Cont.*

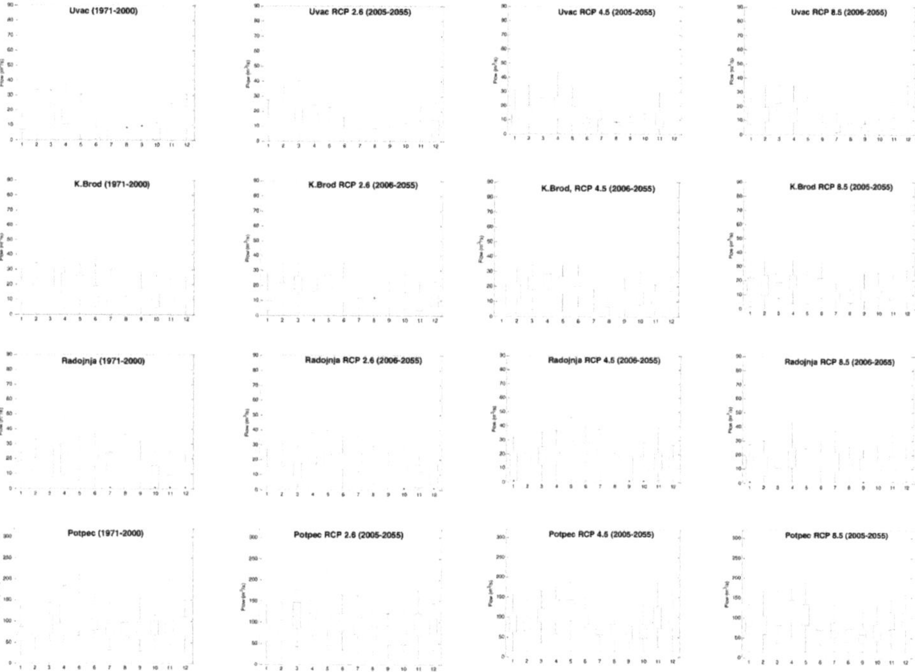

Figure 9. Box plots of the monthly inflows into the Lim water system (Uvac, Kokin Brod, Radojnja and Potpec).

Figure 10. Box plots of the monthly outflows from the Lim water system reservoirs (Uvac, Kokin Brod, Radojnja and Potpec).

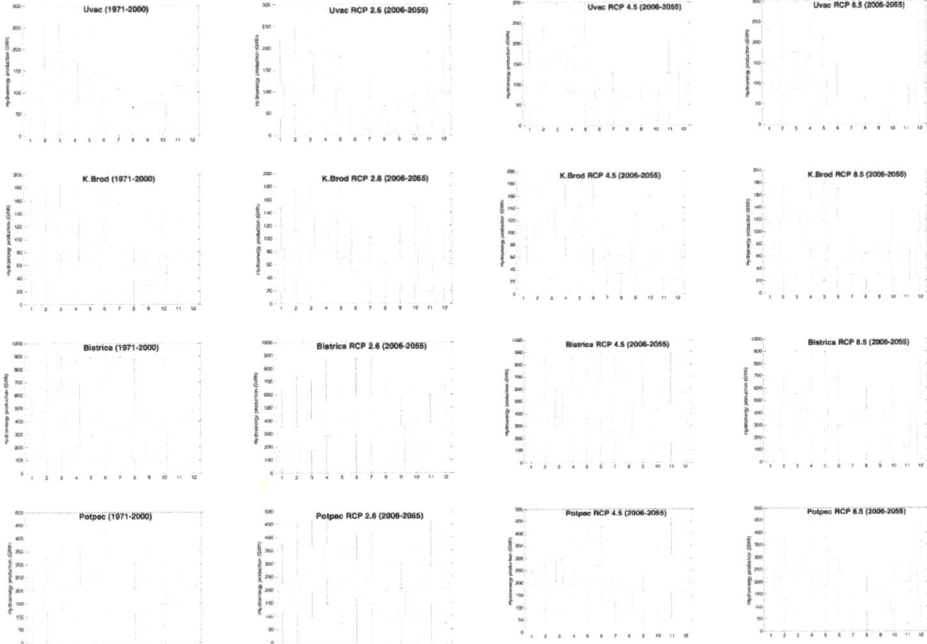

Figure 11. Box plots of the monthly hydropower generation of the Lim water system power plants (Uvac, Kokin Brod, Bistrica and Potpec).

5.3. Uncertainty Assessment

Sources of uncertainty within the climate change impact studies of the Lim water system have not yet been evaluated, mainly because of the oversimplified method used to assess the climate change impacts on the hydropower generation [19]. However, the unawareness of the uncertainty in the SDSM outputs may introduce high risk into the system planning and management activities.

This study employs the different climate change impact analysis process (Figure 2). It should be noted that some of the steps in the assessment process are limited to single modeling approaches. For instance, in this study only one hydrological model is used as well as only one system dynamics model. Therefore, the assessment of the uncertainties introduced by these approaches requires a model ensemble. Significantly more attention is devoted to the assessment of the uncertainty introduced by the selection of climate models, statistical post-processing tools and concentration scenarios. A spread of variance in the monthly time series of inflows, hydropower generation and outflows from the system of reservoirs are analyzed as a proxy measure to quantify the contribution of uncertainties introduced by the choice of RCMs, BCMs and CSs to the total uncertainty. The variance of each step is calculated by Equations (5)–(7). A higher spread of variance corresponds to a higher uncertainty in the process analyzed and vice versa. The entire ensemble is used to assess the spread of variance of the individual climate change impact analysis process steps. The Bates–Granger averaging estimates the weights of different process steps based on their ability to reproduce the variables within the reference period of 1971–2000. Subsequently, the estimated weights are applied for the future period of 2006–2055. The simple averaging of the SDSM outputs is performed by Equations (8)–(10). Finally, the application of Equations (12)–(14) provides the contribution of each step uncertainty (RCM, BCM, ES) to the total uncertainty in the SDSM outputs.

The inflows and outflows from the reservoirs of the Lim water system, as well as the hydropower generation of HPPs, are selected to quantify the contribution of different modelling steps to the total uncertainty. The results presented in Table 4 suggest that the largest source of uncertainty in the SDSM variables is contributed by the choice of BCMs followed by the contribution introduced by the choice of RCMs. The least significant source of uncertainty is obviously the choice of CSs. These findings agree with the results from the previous study done for Canada [8]. The results of the broader literature indicate that the climate models are the predominant source of uncertainty in the projection of monthly flows during most of the annual cycle, while the statistical post-processing methods are the most important source of uncertainty in the extreme flows [37].

The inflows and outflows from the Lim water system reservoirs shows a similar contribution of uncertainty arising from different modelling steps. Uncertainty introduced by BCMs is slightly higher (~49%) than uncertainty introduced by RCMs (~45%). It is clear that the lowest source of uncertainty is introduced by CSs, accounting only for 5%. However, the uncertainty contribution in the hydropower generation has a quite different distribution (Table 4). Note that the highest level of uncertainty is introduced by the choice of BCMs (~54%), followed by the choice of RCMs (about 41%) and emissions scenarios contributing only ~6%.

Table 4. The contribution of the different climate change impact assessment steps uncertainty (RCMs, BCMs, CSs) to the total uncertainty in the predicted inflows, outflows and hydropower generation of the Lim water system for the future period (2006–2055).

Modelling Steps	Uncertainty within the Different Processes (%)		
	Inflows	Outflows	Hydropower Generation
RCMs	44.7	45.7	40.6
BCMs	49.0	49.2	53.5
CSs	6.3	5.0	5.9

The different distribution between the contribution of uncertainty in the SDSM outputs can be attributed to their nature. Namely, the hydropower generation is calculated using Equation (4) accumulating the uncertainties from the inflows and reservoir water levels. It is, therefore, evident that multiplying the individual sources of uncertainty increases the uncertainty in the selection of BCMs. In contrast, the uncertainty in the hydropower generation introduced by the selection of RCMs is lower than the uncertainty in the inflows/outflows (~41%) and the uncertainty contribution from the choice of concentration scenarios (~5–6%). This finding suggests that the different RCMs and BCMs exhibit similar tendencies in each concentration scenario.

6. Conclusions

The study focuses on the development of the system dynamics simulation model of the complex water resources system alongside the operational rules under present and future climates. The main objective of the study is to estimate the climate change impacts on the performance of the Lim water system represented by the outputs of the simulation model using different climate change impact analysis process steps. In addition, the study assesses the uncertainty contribution of each process step to the total uncertainty of the system performance.

The Lim water system in Serbia (southeast Europe) is selected as a case study. This water system includes four hydraulically connected reservoirs at the Lim (Potpec reservoir) and Uvac rivers (Uvac, Kokin Brod, Radojnja reservoirs). The primary purpose of the Lim water system is the hydropower generation, followed by flood mitigation and downstream water quality management. Different climate drivers introduce variation in the inflows of the Lim water system resulting in different hydropower generations. Therefore, addressing the impacts of the changing climate on the Lim reservoir operations is of high importance for the region.

The study uses the following process steps to assess the climate change-related impacts: Three RCMs, four BCMs, three CSs, two averaging procedures, single hydrological model and one SDSM. The raw precipitation data from the RCMs under the RCP 2.6, RCP 4.5 and RCP 8.5 concentration scenarios are corrected using the multiple BCMs. Then, the simple and weighted averaging methods are applied to the precipitation datasets to quantify the contribution of process uncertainty to the total uncertainty in the system performance. Prior to the system dynamics modelling, the HEC-HMS hydrological model is applied to transform the climate data information into the hydrological response (flow). The SDSM is built using the Vensim system dynamics simulation software and uses the inflows derived by the HEC-HMS hydrological model. The system dynamics simulation model of the Lim water system is developed in a holistic manner to evaluate the climate-related impacts on the system outflows and hydropower generation. For the reservoirs with multi-annual and seasonal flow regulation (Uvac and Kokin Brod reservoirs), the reservoir operations are developed using the yield model solved by the sequent peak method. Reservoirs with daily regulation (Potpec and Radojnja reservoirs) use the standard operational policy. Note that the proposed operational rules can be adjusted in accordance with the changes in the seasonal flow distribution.

The projected changes in the hydropower generation of the Lim water system depend on the inflows into the reservoirs and their available water volume. The findings suggest that the change in annual power production can be expected in the range from −3.5% to +17.9%. More severe change is expected at the seasonal level with the decrease in annual production ranging from −2.8% to −33.6% for the summer and autumn seasons. A rapid increase in power production is reported for the winter months, from +1.6% to +20.3%. The study also provides the quantification of the uncertainty contribution for each process step using the variance method. The most dominant source of uncertainty in the hydropower generation is introduced by the choice of BCMs (53.5%) followed by the choice of RCMs (40.6%). The least contribution of uncertainty comes from the choice of CSs (5.9%). However, the contributions of uncertainty of different process steps to the uncertainty in the inflows into and outflows from the Lim system reservoirs have different distributions. The choice of BCM and RCM contribute to 45.7–44.7% and 49.0–49.2%, respectively. The ES contributes the least to the uncertainty in the predicted inflows/outflows (5.0–6.3%). In contrast to the inflows or outflows, the hydropower generation accumulates the uncertainties in the inflows and reservoir water levels, resulting in the different distribution of uncertainty contributions.

The findings point out the importance of analyzing the operation of the Lim system reservoirs under changing climate conditions to support a more efficient decision-making. Future research should concentrate on investigating additional sources of uncertainty, particularly those stemming from the reservoir operation and hydrological model structure. That will require the use of multiple operational rules obtained by different reservoir operations modelling approaches [33] and the use of different hydrological models [40,41].

Author Contributions: M.S. and S.P.S. designed the research framework; M.S. conducted simulations and wrote the paper; S.P.S. supervised the paper.

Funding: This research was funded by the Natural Sciences and Engineering Research Council of Canada and Chaucer Syndicates Ltd.

Acknowledgments: We acknowledge the financial support provided to the second author by the Natural Sciences and Engineering Research Council of Canada and Chaucer Syndicates Ltd.

Conflicts of Interest: The authors declare no conflict of interest.

References

1. Simonovic, S.P. *Managing Water Resources: Methods and Tools for a Systems Approach*; UNSECO Publishing: Paris, France; Earthscan James & James: London, UK, 2009.
2. Oni, S.K.; Dillon, P.J.; Metcalfe, R.A.; Futter, N.N. Dynamic Modelling of the Impact of Climate Change and Power Flow Management Options using STELLA: Application to the Steephill Falls Reservoir, Ontario, Canada. *Can. Water Resour. J.* **2012**, *37*, 125–148. [CrossRef]
3. Teegavarapu, R.; Simonovic, S.P. Object Oriented Simulation of Multiple Hydropower Reservoir Operations using System Dynamics Approach. *Water Resour. Manag.* **2014**, *28*, 1937–1958. [CrossRef]

4. Hollermann, B.; Evers, M. Coping with uncertainty in water management: Qualitative system analysis as a vehicle to visualize the plurality of practitioners' uncertainty handling routines. *J. Environ. Manag.* **2019**, *235*, 213–223. [CrossRef] [PubMed]

5. *EEA Climate Impacts on Water Resources*; European Environment Agency: Copenhagen, Denmark, 2017.

6. Ehsani, N.; Vorosmarty, C.J.; Fekete, B.M.; Stakhiv, E.Z. Reservoir operations under climate change: Storage capacity options to mitigate risk. *J. Hydrol.* **2017**, *555*, 435–446. [CrossRef]

7. International Commission for the Protection of the Danube River (ICPDR). *Danube Study—Climate Change Adaptation, Final Report*; International Commission for the Protection of the Danube River: Vienna, Austria, 2012.

8. Mandal, S.; Arunkumar, R.; Breach, P.A.; Simonovic, S.P. Reservoir Operations under Changing Climate Conditions: Hydropower-Production Perspective. *J. Water Resour. Plan. Manag.* **2019**, *145*, 04019016. [CrossRef]

9. Arsenault, R.; Brissette, F.; Malo, J.S.; Minville, M.; Leconte, R. Structural and Non-Structural Climate Change Adaptation Strategies for the Péribonka Water Resource System. *Water Resour. Manag.* **2013**, *27*, 2075–2087. [CrossRef]

10. Haguma, D.; Leconte, R.; Côté, P.; Krau, S.; Brissette, F. Optimal Hydropower Generation Under Climate Change Conditions for a Northern Water Resources System. *Water Resour. Manag.* **2017**, *28*, 4631–4644. [CrossRef]

11. Ahmadi, M.; Haddad, O.B.; Loáiciga, H.A. Adaptive Reservoir Operation Rules Under Climatic Change. *Water Resour. Manag.* **2015**, *29*, 1247–1266. [CrossRef]

12. Burn, H.D.; Simonovic, S.P. Sensitivity Of Reservoir Operation Performance To Climatic Change. *Water Resour. Manag.* **1996**, *6*, 463–478. [CrossRef]

13. Mandal, S. Uncertainty Modeling in The Assessment of Climate Change Impacts on Water Resources Management. Ph.D. Thesis, University of Western Ontario, London, ON, Canada, 2017.

14. Goharian, E.; Zahmatkesh, Z.; Sandoval-Solis, S. Uncertainty Propagation of Hydrologic Modeling in Water Supply System Performance: Application of Markov Chain Monte Carlo Method. *J. Hydrol. Eng.* **2018**, *23*, 04018013. [CrossRef]

15. Gaur, A.; Simonovic, S.P. Towards Reducing Climate Change Impact Assessment Process Uncertainty. *Environ. Process. J.* **2015**, *2*, 275–290. [CrossRef]

16. Poulin, A.; Brissette, F.; Leconte, R.; Arsenault, R.; Malo, J.S. Uncertainty of hydrological modelling in climate change impact studies in a Canadian, snow-dominated river basin. *J. Hydrol.* **2011**, *409*, 626–636. [CrossRef]

17. Das, J.; Treesa, A.; Umamahesh, N. Modelling Impacts of Climate Change on a River Basin: Analysis of Uncertainty Using REA & Possibilistic Approach. *Water Resour. Manag.* **2018**, *32*, 4833–4852.

18. Joseph, J.; Ghosh, S.; Pathak, A.; Sahai, A.K. Hydrologic impacts of climate change: Comparisons between hydrological parameter uncertainty and climate model uncertainty. *J. Hydrol.* **2018**, *566*, 1–12. [CrossRef]

19. *Support to Water Resources Management in the Drina River Basin*; World Bank: Washington, DC, USA, 2019. Available online: http://www.wb-drinaproject.com/index.php/en/ (accessed on 4 August 2019).

20. Reichl, F.; Hack, J. Derivation of Flow Duration Curves to Estimate Hydropower Generation Potential in Data-Scarce Regions Data-Scarce Regions. *Water* **2017**, *9*, 572. [CrossRef]

21. Institute for the Development of Water Resources Jaroslav Cerni. *Water Resource Management Plan of Montenegro WRMPM*; Institute for the Development of Water Resources Jaroslav Cerni: Belgrade, Serbia, 2001.

22. Institute for the Development of Water Resources Jaroslav Cerni. *Water Resource Management Plan of Serbia WRMPS*; Institute for the Development of Water Resources Jaroslav Cerni: Belgrade, Serbia, 2010.

23. Jacob, D.; Petersen, J.; Eggert, B.; Alias, A.; Christensen, O.B.; Bouwer, L.M.; Braun, A.; Colette, A.; Déqué, M.; Georgievski, G.; et al. EURO-CORDEX: New high-resolution climate change projections for European impact research. *Reg. Environ. Chang.* **2014**, *14*, 563–578. [CrossRef]

24. Teutschbeina, C.; Seibert, J. Bias correction of regional climate model simulations for hydrological climate-change impact studies: Review and evaluation of different methods. *J. Hydrol.* **2012**, *456*, 12–29. [CrossRef]

25. Piani, C.; Haerter, J.O.; Coppola, E. Statistical bias correction for daily precipitation in regional climate models over Europe. *Theor. Appl. Climatol.* **2010**, *99*, 187–192. [CrossRef]

26. Cannon, A.J.; Sobie, S.R.; Murdock, T.Q. Bias Correction of GCM Precipitation by Quantile Mapping: How Well Do Methods Preserve Changes in Quantiles and Extremes? *J. Clim.* **2015**, *28*, 6938–6959. [CrossRef]

27. Mani, A.; Tsai, F.T.C. Ensemble Averaging Methods for Quantifying Uncertainty Sources in Modeling Climate Change Impact on Runoff Projection. *J. Hydrol. Eng.* **2016**, *22*, 04016067. [CrossRef]
28. Bates, J.M.; Granger, C.W.J. The combination of forecasts. *Oper. Res. Q.* **1969**, *20*, 451–468. [CrossRef]
29. Scharffenberg, W. *Hydrological Modelling System*; HEC-HMS: US Army Corps of Engineers, Institute for Water Resources, Hydrologic Engineering Centre: Davis, CA, USA, 2016.
30. Azmat, M.; Qamar, M.U.; Huggel, C.; Hussain, E. Future climate and cryosphere impacts on the hydrology of a scarcely gauged catchment on the Jhelum river basin, Northern Pakistan. *Sci. Total Environ.* **2018**, *639*, 961–976. [CrossRef] [PubMed]
31. Annex 1. Development of the Hydrologic Model for the Sava River Basin. In *Water Climate and Adaptation Plan for the Sava River Basin*; Final Report; World Bank: Washington, DC, USA, 2015. Available online: http://documents.worldbank.org/curated/en/111101468188370674/Annex-one-development-of-the-hydrologic-model-for-the-Sava-river-basin (accessed on 4 August 2019).
32. *Vensim Reference Manual Version: 7.3*; VENTANA System Inc.: Tucson, AZ, USA. Available online: https://vensim.com/docs/ (accessed on 9 July 2019).
33. Loucks, D.P.; Beek, E. *Water Resources Systems Planning and Management: An Introduction to Methods, Models and Applications*; Deltares and UNESCO-IHE: Delft, The Netherlands, 2017.
34. Brouwer, C.; Heibloem, M. *Irrigation Water Management: Irrigation Water Needs*; The International Institute for Land Reclamation and Improvement: Rome, Italy, 1986.
35. Wilby, R.L.; Harris, I. A framework for assessing uncertainties in climate change impacts: Low-flow scenarios for the River Thames, U.K. *Water Resour. Res.* **2006**, *42*. [CrossRef]
36. Bastola, S.; Murphy, C.; Fealy, R. Generating probabilistic estimates of hydrological response for Irish catchments using a weather generator and probabilistic climate change scenarios. *Hydrol. Process.* **2012**, *26*, 2307–2321. [CrossRef]
37. Yuan, F.; Chun, K.P.; Sapriza-Azuri, G.; Bruen, M.; Wheater, H.S. Assessing the relative importance of parameter and forcing uncertainty and their interactions in conceptual hydrological model simulations. *J. Hydrol.* **2017**, *554*, 434–450. [CrossRef]
38. Ranzani, A.; Bonato, M.; Patro, E.R.; Gaudard, L.; Michele, C.D. Hydropower Future: Between Climate Change, Renewable Deployment, Carbon and Fuel Prices. *Water* **2017**, *10*, 1197. [CrossRef]
39. Chai, Y.; Li, Y.T.; Yang, Y.P.; Zhu, B.Y.; Li, S.X.; Xu, C.; Liu, C.C. Influence of Climate Variability and Reservoir Operation on Streamflow in the Yangtze River. *Sci. Rep.* **2019**, *9*, 5060. [CrossRef]
40. Markstrom, S.L.; Regan, R.S.; Hay, L.E.; Viger, R.J.; Webb, R.M.T.; Payn, R.A.; LaFontaine, J.H. *PRMS-IV, the Precipitation-Runoff Modeling System*; Version 4; Geological Survey: Reston, VA, USA, 2015.
41. SWAT: Soil & Water Assessment Tool. Available online: https://swat.tamu.edu/ (accessed on 2 June 2019).

Article

Aggregation–Decomposition-Based Multi-Agent Reinforcement Learning for Multi-Reservoir Operations Optimization

Milad Hooshyar [1], S. Jamshid Mousavi [2], Masoud Mahootchi [3] and Kumaraswamy Ponnambalam [4],*

[1] Princeton Institute for International and Regional Studies (PIIRS), Princeton University, Princeton, NJ 08544, USA; hooshyar@princeton.edu

[2] School of Petroleum, Civil and Mining Engineering, Amirkabir University of Technology, Tehran 1591634311, Iran; jmosavi@aut.ac.ir

[3] School of Industrial Engineering, Amirkabir University of Technology, Tehran 1591634311, Iran; mmahootchi@gmail.com

[4] Department of Systems Design Engineering, University of Waterloo, Waterloo, ON N2L 3G1, Canada

* Correspondence: ponnu@uwaterloo.ca

Received: 13 June 2020; Accepted: 23 September 2020; Published: 25 September 2020

Abstract: Stochastic dynamic programming (SDP) is a widely-used method for reservoir operations optimization under uncertainty but suffers from the dual curses of dimensionality and modeling. Reinforcement learning (RL), a simulation based stochastic optimization approach, can nullify the curse of modeling that arises from the need for calculating a very large transition probability matrix. RL mitigates the curse of the dimensionality problem, but cannot solve it completely as it remains computationally intensive in complex multi-reservoir systems. This paper presents a multi-agent RL approach combined with an aggregation/decomposition (AD-RL) method for reducing the curse of dimensionality in multi-reservoir operation optimization problems. In this model, each reservoir is individually managed by a specific operator (agent) while co-operating with other agents systematically on finding a near-optimal operating policy for the whole system. Each agent makes a decision (release) based on its current state and the feedback it receives from the states of all upstream and downstream reservoirs. The method, along with an efficient artificial neural network-based robust procedure for the task of tuning Q-learning parameters, has been applied to a real-world five-reservoir problem, i.e., the Parambikulam–Aliyar Project (PAP) in India. We demonstrate that the proposed AD-RL approach helps to derive operating policies that are better than or comparable with the policies obtained by other stochastic optimization methods with less computational burden.

Keywords: multireservoir operations; optimization; multi-agent reinforcement learning; aggregation–decomposition; neural networks

---

1. Introduction

Multi-reservoir optimization models are generally non-linear, non-convex, and large-scale in terms of the number of variables and constraints. Moreover, uncertainties in stochastic variables such as inflows, evaporation, and demands make it difficult to find even a sub-optimal operating policy. In general, two types of stochastic programming approach are used to optimize multi-reservoir systems operations under uncertainty, i.e., implicit and explicit. In implicit stochastic optimization (ISO), a large number of historical or synthetically generated sequences of random variables such as streamflow are generated as the input for a deterministic optimization model. The generated results represent different aspects of the underlying stochastic process, such as spatial or temporal correlations among random variables involved in the process. Optimal operation policies are then acquired by performing post-processing analysis on the outputs of the deterministic optimization model solved for different input sequences (samples).

Different optimization models have been used in water reservoir problems through ISO, including linear programming (LP) [1,2], successive LP [3], successive quadratic programming [4], method of multipliers [5], and dynamic programming (DP) [6]. ISO is computationally intensive in large multi-reservoir systems, especially the highly non-linear ones, because solving a large-scale non-linear optimization is often tedious and time-consuming. More importantly, this approach does not directly lead to optimal operating rules as functions of hydrological variables available at the time of decision making; so a post-processing technique such as multiple linear regression [7], neural nets [8], and fuzzy rule-based modeling [9–11] may be used in deriving operating policies.

In the explicit stochastic optimization (ESO), probability distributions of random variables are used to derive a transition probability matrix (TPM) required for modeling dynamics of the underlying stochastic process. Several stochastic optimization methods have been applied to optimal multi-reservoir operations problems such as chance-constrained programming [12,13], reliability programming [14–16], the Fletcher–Ponnambalam (FP) method that does not require discretization or inflow scenarios [17–19], stochastic LP [20], stochastic dynamic programming (SDP), with some approximations for dimensionality reduction [21–35], stochastic dual DP (SDDP) [36–40], sampling SDP (SSDP) [41], and reinforcement learning (RL) [6,42,43].

SDP is one of the most widely-used explicit stochastic methods in reservoir operations which requires models to derive the TPM, and the optimal steady-state policy is guaranteed to be global for a given discretization. However, the application of SDP is limited since perfect knowledge about the underlying stochastic model is needed, and the computational effort increases exponentially by the size of the system or the number of state variables (the curse of dimensionality). More details on SDP is presented later in the corresponding section.

In order to address the drawback of deriving TPM for SDP, RL can be applied in a similar framework as SDP but in a simulation-based setup. In RL, an agent (operator) learns to take proper actions through independent interaction with a stochastic environment. In one of the first RL application in water management in literature, Bhattacharya et al. [44] developed a controller for a complex water system in the Netherlands using coupled artificial neural networks (ANN) and RL model, where RL is just used to mitigate the error of the ANN. They also stated that RL and its combinations could be very helpful in water management problems such as reservoir operations, but such works did not appear in literature until 2007. Lee and Labadie [6] compared the performance of RL with some other optimization techniques in a 2-reservoir case study. They used all possible actions as a set of admissible actions. This choice can make the applicability of the learning method inefficient as some actions-state pairs are practically impossible (infeasible), an issue that is dealt with in detail later. Mahootchi et al. [42] applied a method called opposition-based RL (OBRL) in a reservoir case study where the agent takes an action and its opposite action simultaneously in order to speed up the convergence of RL. Castelletti et al. [45] developed a new method based on RL and tree-based regression called Tree-based RL (TBRL), which uses operational data through the learning process. They applied TBRL to a system that consisted of one reservoir and nine run-of-the-river hydro plants. In comparison with SDP, TBRL is computationally more tractable and has better performance, especially during floods. More recent applications of RL to multireservoir operations optimization can be seen in [46,47].

Although RL is capable of eliminating the need for prior perfect knowledge of the underlying stochastic model of the environment, as the learning can be performed using historical operational data, and also provides some computational advantages, it still suffers from the curse of dimensionality for large-scale problems. As a remedy, this paper presents an RL-based model combined with an aggregation–decomposition (AD) approach (referred to as AD-RL) that reduces the dimensionality problem to efficiently solve a stochastic multi-reservoir operation optimization problem.

2. Aggregation–Decomposition Methods and Reinforcement Learning

2.1. Reservoir Operation Optimization Model

In the following, the objective function and constraints for the optimization model of the reservoir operations are described.

2.1.1. Objective Function

In reservoir problems, releases and/or storage values of reservoirs are the decision variables and the objective function may be defined as the maximization of net benefit or minimization of a penalty function. A general form of the objective function in a multi-reservoir application can be written as:

$$f\left(R^t\right) = \sum_{i=1}^{N}\sum_{j=1}^{N}\sum_{t=1}^{T} f_{ij}^t\left(s_i^t, R_{ij}^t, D_i^t\right) \tag{1}$$

where f is the cost or revenue function which could be a function of s_i^t (the storage volume of reservoir i at the beginning of time period t), R_{ij}^t (release from reservoir i to reservoir j in time period t), D_i^t (the demand to be met by reservoir i in time period t), and N denotes the number of reservoirs.

2.1.2. Constraints

There are three types of constraint in reservoir operations. The first type is balance equations which are in fact equality constraints referring to the conservation of mass (in appropriate units) with respect to inflow to and outflow from the reservoirs as expressed below:

$$s_i^{t+1} = s_i^t - \sum_{l=1,l\neq i}^{N_D} R_{il}^t \times \delta_{il} + I_i^t - v_i^t + \sum_{l=1,l\neq i}^{N_U} R_{il}^t \times \delta_{li} \quad \forall i = 1 \ldots N \,\&\, t = 1 \ldots T \tag{2}$$

where I_i^t is the amount of incremental natural inflow to every reservoir i in period t, v_i^t is the loss (e.g., evaporation) from reservoir i in period t, R_{il}^t is the amount of inflow from upstream reservoir i to reservoir l in period t, N_D is the number of downstream reservoirs, N_U is the number of upstream reservoirs. δ_{ij} is an element of the routing matrix that is 1 if ith reservoir is physically connected to jth reservoir and is zero otherwise.

The second set of constraints are upper and lower bounds on reservoir storage variables. The upper bounds may consider the flood control objective or physical reservoirs storage capacities, while the lower bounds may take the objectives of sedimentation, recreation, and functionality of power generation into account. The constraints can be expressed as:

$$Smin_i^t \leq s_i^t \leq Smax_i^t \tag{3}$$

where $Smin_i^t$ and $Smax_i^t$ are the maximum and minimum storage volumes of reservoir i in time period t, respectively.

The last set of constraints are upper and lower bounds of reservoir releases. The purpose of these constraints is to provide a minimum instream flow for water quality and ecosystem services and to supply water to meet different demands while considering capacities of outlet works and preventing downstream flooding. These constraints are often modeled as:

$$Rmin_i^t \leq \sum_{l=1}^{N} R_{il}^t \times \delta_{il} \leq Rmax_i^t \tag{4}$$

where $Rmin_i^t$ and $Rmax_i^t$ are the maximum and minimum releases from reservoir i, respectively.

2.2. Stochastic Dynamic Programming (SDP)

Discrete SDP is a popular method for stochastic optimization of reservoir operations [45]. The solution of SDP for long-term operations is typically a steady-state operating policy representing optimal decisions (e.g., the volume of releases) for all possible combinations of states (e.g., storages of reservoirs, etc.) which is obtained through solving the Bellman's optimality equation iteratively. The state of the system is usually divided into some specific discrete values and the recursive function (a function of the state and the decision vectors) is updated in every iteration based on the estimated value function, $V_t(i)$, defined as the maximum accumulated reward from period t to a termination point in time for a given state i. The value function is obtained based on the optimality equation proposed by Bellman [48] as:

$$V^t(i) = \max_{a \in A(i)} \sum_j P^t_{ij}(a) \times \left[r^t_{ij}(a) + \gamma V^t(j) \right] \tag{5}$$

where $P^t_{ij}(a)$ is the probability of transition from state i to state j when an action a in period t is taken, $r^t_{ij}(a)$ is the reward function pertinent to action a for that transition in period t, $A(i)$ is the set of admissible actions for state i, and γ is the discount factor. Equation (5) is solved backward, i.e., $t : T, T-1, \cdots, 1$ where T is the last time period.

SDP suffers from a dual curse which makes it unsuitable to cope with large-scale problems [45]. First, the dimension of the optimization problem grows exponentially with the number of the state and decision variables (the curse of dimensionality). Therefore, the SDP algorithm is not computationally tractable in systems where the number of reservoirs is more than a few. Second, prior knowledge about the underlying Markov decision process (explicit model) of inflow variables, including state TPM and rewards, is required (curse of modeling). This prior knowledge may be difficult to access due to the complexities of multi-reservoir systems or insufficient data [6], considering the spatial and temporal dependence structure of inflow stochastic variables.

The attempts to overcome the curse of dimensionality can be categorized into two main classes, namely, methods based on function approximation and aggregation/decomposition. In methods based on function approximation, the combination of a coarse discretization size and approximation of value functions is used in order to save the quality of the extracted policy. Different methods are applied to approximate value function, including Hermitian polynomials [27], cubic piecewise polynomial [24], and ANN [22].

In the aggregation/decomposition-based methods, the original problem is broken down into some tractable sub-problems solvable by SDP. Each sub-problem is related to one specific reservoir (might be more than one reservoir in some cases) which is connected to other reservoirs based on the designed configuration. The main goal in each sub-problem is to find the best release policy for that reservoir based on two different state variables: actual and virtual states. Turgeon [34] developed a simple aggregation/decomposition method for a serial or parallel configuration called one-at-a-time where an N-reservoir problem changed into N one-reservoir sub-problems. The sub-problems are solved successively by SDP. Turgeon [35] also modified their method by considering only the potential hydropower energy of the downstream reservoirs in addition to the corresponding state of the actual reservoir.

The technique of state aggregation may also be performed in a different way in which some unimportant action-state pairs are systemically eliminated. For instance, Mousavi and Karamouz [26] considered eliminating infeasible action-state pairs in order to speed up the convergence of DP-based methods in multi-reservoir problems. Saad and Turgeon [29] also developed a method to eliminate some components of state vector based on principal component analysis which was modified by Saad et al. [31] through applying censored data algorithm.

Some researchers also combined both aggregation and function approximation techniques to alleviate the curse of dimensionality in SDP. Saad et al. [30] aggregated a 5-reservoir problem into one-reservoir problems which were solved then using SDP.

In this study, we use the aggregation/decomposition methods proposed by Archibald, et al. [21], and Ponnambalam and Adams [28]. The detailed description of these methods is presented in the following sections.

2.3. Aggregation–Decomposition Dynamic Programming (AD-DP)

The aggregation-decomposition method, proposed by Archibald et al. [21] and referred to as aggregation–decomposition-dynamic programming (AD-DP), decomposes the original problem into some sub-problems equivalent to the number of reservoirs. All sub-problems could be solved in parallel using SDP. The set of state variables for each sub-problem can be defined as the beginning storage for actual (focus) reservoir, the summation of beginning storages of all upstream reservoirs (virtual up-stream reservoir), and the summation of beginning storages of all down-stream reservoirs (virtual down-stream reservoir). The decisions at each period are the next state of the upstream reservoirs, released from the focus (actual) reservoir and the next state of the non-upstream reservoirs. The main limitation of AD-DP is that the total storage of the virtual upstream and non-upstream reservoirs in each iteration of SDP should be proportionally distributed among the respective reservoirs based on their capacities. Given the end-of-period storage for the virtual reservoir, one could analogously find the end-of-period storages for these upstream reservoirs as well. Archibald et al. [21] note that the proposed aggregation/decomposition method in a multi-reservoir system would end up with a near-optimal release policy, however, it is still more restrictive than the method that we describe next.

2.4. Multilevel Approximation-Dynamic Programming (MAM-DP)

Ponnambalam and Adams [28] proposed a decomposition method to overcome the main restriction of Turgeon's model [35] in which only the serial or parallel configurations can be tackled. The number of reservoirs considered was only two (one virtual), and the objective function was separable by the reservoir. In multilevel approximation-dynamic programming (MAM-DP), the state variables for each reservoir can be defined similarly to AD-DP [21]; however, the virtual reservoir only includes the summation of all characteristics (inflow, capacity, minimum storage, etc.) of the rest of the reservoirs (e.g., the capacity of this virtual reservoir is the total summation of capacities of all corresponding reservoirs). The way of decomposition generally depends on the presented configuration (i.e., decomposition might be different from one problem to another; in Ponnambalam and Adams [49] the algorithm proceeded from upstream to downstream, so the virtual reservoir only considered downstream reservoirs).

2.5. Reinforcement Learning (RL)

RL [50] is a computational method based on learning through interaction with the environment. Despite the SDP, RL does not presume knowledge about the underlying model as the knowledge about the environment is gained through real-world experience (on-line) or simulation (off-line). Using simulation, RL is able to overcome the curse of modeling; however, RL only mitigates the curse of dimensionality to some extent as it searches the feasible action-state pairs heuristically.

The basic idea of RL can simply be described as a learning agent interacting with its environment to achieve a goal [50]. In reservoir operations, the agent is the operator of a reservoir who makes the decisions over the release (as the action) based on a policy while the state space consists of a set of discrete values of storage.

Beyond the agent and the environment, there are four main sub-elements in RL: a policy, a reward function, a value function, and optionally a model of the environment [50]. The model components of RL determine the next state and the reward of the environment based on a mathematical function. Figure 1 illustrates a schematic perspective of RL.

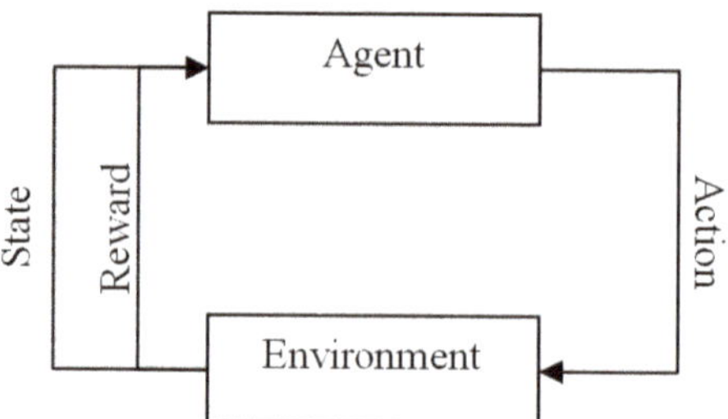

Figure 1. The schematic view of reinforcement learning.

2.5.1. Action-Taking Policy

The chance of taking an action in a specific state is actually a trade-off between exploration (taking an action randomly) and exploitation (taking the best action), which leads to four common action policies in the literature: random, greedy, ε-greedy and Softmax. In the random policy, there is no action preference. In the ε-greedy policy, greedy action (a^*) in each state is chosen most of the time; however, once in a while, the agent tries to choose an action in the set of admissible actions with the probability of $\frac{\varepsilon}{|A(i)|}$ where ε is the probability of taking non-greedy actions and $A(i)$ is the set of admissible actions for state i. In the greedy policy, the agent picks the best one among all admissible actions in each iteration, with respect to the last estimate of the action-value function for all admissible actions in the respective state. In contrast to the greedy policy, the Softmax policy derived from the Boltzmann's distribution can be defined in which the proportion of exploration versus exploitation changes as the process of learning continues.

2.5.2. Admissible Actions

The agent should choose an action among candidate actions, which are called possible actions. Evaluation of all state-action pairs is computationally expensive and makes the learning process inefficient. According to the stochastic environment of reservoir operations, some actions might be infeasible to take in some conditions of stochastic variables (e.g., minimum inflow and maximum evaporation) and some are always infeasible. One may define the set of admissible actions based on the worst condition of stochastic variables. However, adopting a such pessimistic approach [42] limits the number of admissible actions, eliminating some actions which might be infeasible in rare conditions; so it is not efficient in terms of finding the optimal policy. Considering the best possible conditions of stochastic variables, the set of admissible actions can be defined using an optimistic approach.

2.5.3. Q-Learning

Q-learning [51] is an RL formulation which has been derived from the formulation of SDP. The value function in SDP is substituted with an action-value function, which is a value defined for every pair of action-states, instead of every state in the value function. Since the learning process is implemented by direct interaction with the environment, value functions have to be updated after each interaction. To find an updated value function based on the Bellman equation, all admissible actions in the respective state and period must be tested with respect to this equation, and the best value is chosen as a new value function. Therefore, we can introduce another terminology called action-value function, $Q(i, a)$, demonstrating the expected accumulated reward when a decision maker starts from state i and takes action a. Using these new values, the formulation of SDP in the Bellman equation can be written as follows:

$$Q^t(i,a) = \sum_j p_{ij}^t(a)\left[r_{ij}^t(a) + \gamma \max_{b \in A(j)} Q^{t+1}(j,b)\right] \tag{6}$$

where $r^t_{ij}(a)$ is the immediate reward in period t when the action a is taken for the transition from state i to state j, and b and b^* are admissible actions and the best one with respect to the next state j, respectively.

As demonstrated in Equation (6), the action-value function is the expected value of the sequence of data in the form of $\left[r^t_{ij}(a) + \gamma \max_{b \in A(j)} Q^{t+1}(j,b) \right]$. Using the Robbins–Monro algorithm [52], it is simple to find a new formulation as follows:

$$Q^t(i,a) \leftarrow Q^t(i,a) + \alpha \times \left[r^t_{ij}(a) + \gamma \max_{b \in A(j)} Q^{t+1}(j,b) - Q^t(i,a) \right] \tag{7}$$

where α is called the learning parameter. Q-learning is a model-free algorithm in which the transition probabilities are not used for updating the action values.

2.6. Aggregation/Decomposition Reinforcement Learning (—RL)

In this section, a new method, namely aggregation/decomposition RL (AD-RL) is proposed in which Q-learning, is used jointly with AD-DP method. Furthermore, the way of using aggregation/decomposition has been derived from [21] which was generalized in [25].

As mentioned above, Archibald et al. [21] proposed an aggregation/decomposition method in which the original problem should be decomposed to n sub-problems where n is the number of reservoirs. All these sub-problems can be individually solved using SDP or Q-learning. The release of each actual reservoir in each sub-problem is a function of states: the beginning storage of focus (actual) reservoir, the total beginning storage of upstream and non-upstream reservoirs. This means that two virtual reservoirs are assumed to be connected to the actual reservoir in which one is stated in its upstream and the other is in its downstream. As has been explained in Archibald's technique [21], to decompose the original problem, the upstream and non-upstream should be properly defined. In this decomposition technique, the upstream reservoirs are initially specified. The rest of the reservoirs are, therefore, non-upstream reservoirs. Based on his definition, the upstream reservoirs of an actual reservoir in a sub-problem are those whose releases directly or indirectly reach the focus (actual) reservoir. We defined the non-upstream reservoirs as those which directly or indirectly receive the release from the focus (actual) reservoir. The rest of the reservoirs are upstream ones. Based on some numerical examples, we found that the new definition of virtual reservoirs would end up with a superior release policy in SDP or Q-learning.

Indeed, there are different interrelated sub-problems in AD-RL that are connected to each other through the releases from actual reservoirs. That means that action (release) taken for each focus (actual) reservoir in a sub-problem in one period is used as an input to the actual reservoir in the next sub-problem in the same period. Therefore, it can be assumed that every sub-problem is conducted by a specific agent and all these agents could share their information through their releases in every interaction of the learning process in AD-RL. Therefore, AD-RL might be interpreted as a multi-agent RL algorithm in which every agent should make a proper decision using its own information and the information received from other agents, including releases and the beginning storages for their respective actual reservoirs. Moreover, the feedback for each agent in the next stage (e.g., the next period) comes from the respective agent and other agents after making their decisions. This feedback includes the next end-of-period storages and the immediate reward for all actual reservoirs in the sub-problems. In other words, each agent should be individually trained until it converges in a steady-state situation in which an optimal policy for that agent can be obtained. The schematic way of training is illustrated in Figure 2.

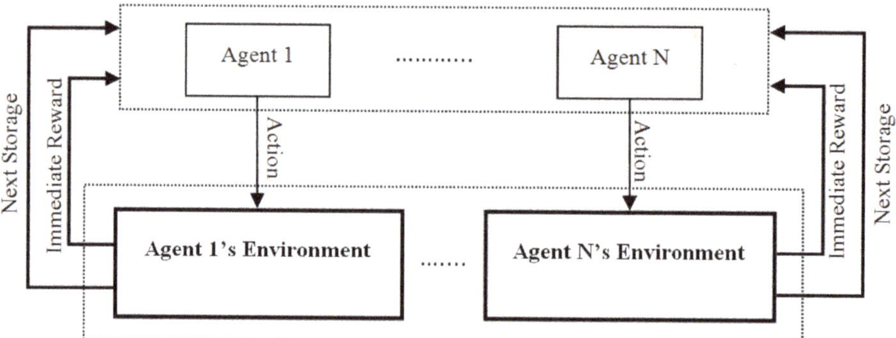

Figure 2. Schematic of aggregation/decomposition reinforcement learning (AD-RL) algorithm.

As illustrated in Figure 2, each agent should interact directly with a specific environment; however, it has indirect interaction with other environments pertinent to other agents. In other words, AD-RL can be projected to a traditional RL in which there are multiple agents (the upper dotted-line box in Figure 2) and multiple environments (the lower dotted-line box in Figure 2) instead of having a single agent and single environment. Therefore, at the time of decision making, each agent should access the beginning storages of all reservoirs. The training part for each agent (updating the action-value functions) should be individually accomplished. The immediate rewards used in the training part for one specific agent come from the reservoir which is related to that agent and from the non-upstream reservoirs that are related to other agents. The developed algorithm (AD-RL) can be summarized as the following steps:

Step 1—The original problem is decomposed to some sub-problems in which the first sub-problem starts for the most upstream reservoir where no releases flow to that reservoir. The last sup-problem is, therefore, one with no downstream reservoirs. As previously mentioned, for each sub-problem, there are two virtual reservoirs: upstream and non-upstream reservoirs. The non-upstream reservoirs should be initially defined. The rest of the reservoirs are upstream reservoirs.

Step 2—For each sub-problem, the states for the respective agent should be defined as follows:

- The beginning storage of focus (actual) reservoir (s_i^t)
- The summation of beginning storage of all non-upstream reservoirs (Snu_i^t)
- The summation of beginning storage of all upstream reservoirs (Su_i^t)

Step 3—all admissible actions (releases), $A^i(s_i^t)$, should be properly defined (based on the optimistic or the pessimistic procedure)

Step 4—Initialize the values of Q-factor for each agent i in all possible state-action pairs

Step 5—Start with initial beginning storages for all reservoirs (s_i^1) and set $t = 1$

Step 6—For all agents,

- Calculate the beginning storages of upstream (Su_i^{t-1}) and non-upstream (Snu_i^{t-1}) reservoirs.
- Take an action (release) using one of the action-taking policies such as Softmax, ε-greedy, greedy, or random. For instance, if using ε-greedy policy, the probability of release (R_i^t) from reservoir i in period t in state s_i^t is calculated as follows:

$$P^{s_i^t}\left(R_i^t\right) = \begin{cases} 1 - \varepsilon + \dfrac{\varepsilon}{(|A^i(s_i^t)|)} & R_i^t = Rg_i^t \\[2mm] \dfrac{\varepsilon}{(|A^i(s_i^t)|)} & R_i^t \neq Rg_i^t \end{cases} \tag{8}$$

where Rg_i^t is the best (greedy) release (action) for a given state up to the current period for agent i. Note that Rg_i^t might be a vector, including multiple actions (releases) because there could be more than one outlet (decision variable) for one focus (actual) reservoir in a sup-problem.

Step 7—The next state of the actual reservoir in every sub-problem is calculated for all agents using the respective balance equation as the dynamic of the system.

$$s_i^{t+1} = s_i^t + I_i^t + Ru_i^t - R_i^t \tag{9}$$

where Ru_i^t is the total releases to focus (actual) reservoir i from its upstream reservoirs. The decision taken by an agent (R_i^t) in a sub-problem might not be feasible because the storage bounds are not satisfied. In this situation, the next state (the end-of-period storages) are replaced with the respective maximum or the minimum storages, and the releases are revised using the balance equation. Note that the final releases after revision process are used for computing the immediate rewards while the actions (releases before the revision process) are used for training (i.e., updating Q-factors).

Step 8—update the Q-factors for all agents as follows:

$$Q_i^t\left(Su_i^t, s_i^t, Snu_i^t, R_i^t\right) \leftarrow Q_i^t\left(Su_i^t, s_i^t, Snu_i^t, R_i^t\right) +$$
$$\alpha \times \left[r_i^t + \gamma \max_{b \in A^i(s_i^{t+1})} Q_i^{t+1}\left(Su_i^{t+1}, s_i^{t+1}, Snu_i^{t+1}, b\right) - Q_i^t\left(Su_i^t, s_i^t, Snu_i^t, R_i^t\right) \right] \tag{10}$$

where r_i^t is the reward used to update Q-factors for each action-state pair. There are two types of immediate reward for each agent in every sub-problem: actual and virtual. In the first type, the immediate reward of each agent is considered. Whereas the release from one reservoir might be used multiple times in the downstream, in the virtual type of immediate reward for each agent the benefits of this release should be calculated multiple times in those reservoirs which directly or indirectly receive this flow. For example, where all reservoirs generate power, the release from the most upstream reservoir can contribute the power generation multiple times with different benefit functions in all downstream reservoirs. The total immediate reward for every agent in the respective sub-problem is the summation of actual and virtual immediate rewards.

Step 9—If the stopping criterion is not satisfied, increment t and repeat steps 6–8; otherwise, go to the next step.

Step 10—Find the decision policy using the following equation:

$$Po^{ti}\left(Su_i^t, s_i^t, Snu_i^t\right) = arg \max_{b \in A^i(s_i^t)} Q\left(Su_i^t, s_i^t, Snu_i^t, b\right) \tag{11}$$

3. Problem Settings and Results

3.1. Case Study: Parambikulam–Aliyar Project (PAP)

We consider the Parambikulam–Aliyar Project (PAP) from India as this multi-reservoir system has been studied using AD-DP and FP methods in [28,49,53], respectively. We provide only the minimum details that are interesting to know here about the system and to correspond with the numbering of reservoirs different here than in [53]; detailed explanations of these reservoirs, their data and corresponding benefits and policies can be obtained from the above works. For the purpose of comparing results with those of four other methods reported in the literature, we use the same problem and objective functions, although that is not a restriction of the AD-RL method. In other words, any other highly non-linear or even discontinuous objective functions can easily be handled by the proposed model as RL is basically a simulation-based technique.

PAP, as studied, is presented in Figure 3. The PAP system comprises of two series of reservoirs in which the left-side reservoirs are more important in terms of the volume of inflow and demands (i.e., the demands and inflows are remarkably high compared to other side). The number inside the triangle depicting each reservoir in Figure 3 represent the live capacities, which can be considered as the maximum storage volume and the minimum (live) storage is zero. The subscript of inflow also represents the index of the reservoir and is used later for explanations.

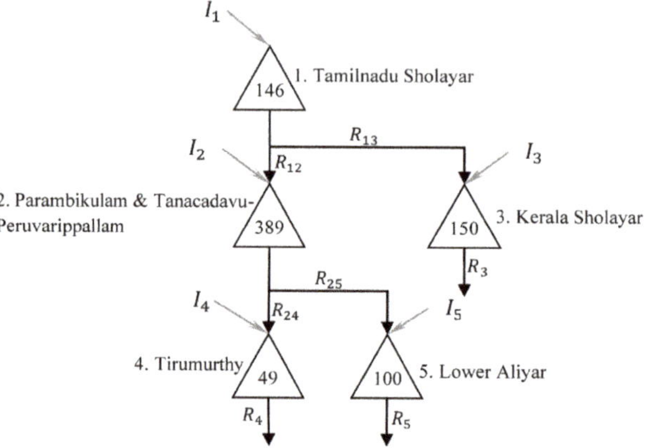

Figure 3. Parambikulam–Aliyar Project (PAP) system as studied.

The main purpose of this project is to conduct water from western slopes of Anamalai Mountains to irrigate the eastern arid area in two states (Tamilnadu and Kerala), and hydropower and fishing benefits are the secondary incomes of the project. Many operating constraints exist agreed upon in the inter-stage agreement between these two different states that should be taken into consideration for constructing any optimization model; the details are not provided here and are available in the original papers. The objective function is defined as;

$$Max\ Z(R_t) = Max \sum_{i=1}^{N=5} \sum_{j=1}^{N=5} \sum_{t=1}^{T=12} b_i^t \times R_{ij}^t \tag{12}$$

where b_i^t is the benefit per unit release which is given in Table 1 and is useful later to understand MAM-DP's objective function. Note that the above simple linear objective function has been chosen to be exactly the same as the objective function used in the literature for other methods to which we compare our proposed model, and it is not a restriction.

Table 1. The benefit per unit release.

Reservoir	Periods											
	1	2	3	4	5	6	7	8	9	10	11	12
1	0.6	1	1	0.05	0.2	0.2	0	0	0	0	0	0
2	0.8	0.9	1	0.4	0.4	0.7	0.8	0.8	0.8	0	0	0
3	0.1	0.25	0.3	0.22	0.25	0.4	0.5	0.4	0.4	0.3	0.2	0.2
4	0.55	0.9	1	0.22	0.28	0.42	0.58	0.62	0.44	0	0	0
5	0.25	0.35	0.35	0.25	0.35	0.3	0.3	0.3	0	0	0.2	0.25

It is worth noting that 20% loss should be considered for all releases flowing from the third reservoir (Parambikulam reservoir) to the fourth reservoir (Tirumurthy) because of the long tunnel used for water flow between these reservoirs.

The lower release bounds are set to zero for all periods. Table 2 also illustrates the upper bounds for all different connections in the PAP case study (i.e., the release from reservoir *i* to reservoir *j*). It is assumed that these bounds are the same for 12 months. The diagonal elements in this table indicate the total maximum releases for each of the five reservoirs.

Table 2. The upper bounds of release.

	Res. 1	**Res. 2**	**Res. 3**	**Res. 4**	**Res. 5**
Res. 1	173.62	123.9	49.72	-	-
Res. 2	-	115.4	-	57.7	57.7
Res. 3	-	-	66.67	-	-
Res. 4	-	-	-	105.12	-
Res. 5	-	-	-	-	49.23

The hydrological data available on a monthly basis for the five reservoirs is in Figure 4. The average monthly inflows to the first, second, third and fifth reservoirs are almost the same in which the rainy season starts from December to May. However, the main proportion of rainfall in the fourth reservoir occurs during the monsoon period (from May to September).

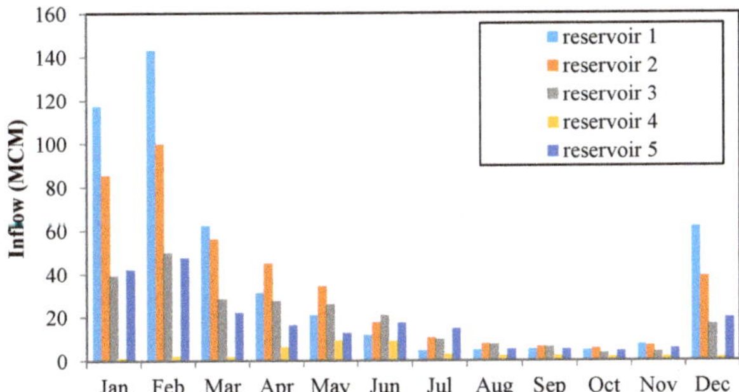

Figure 4. The average monthly inflows in million cubic meter (MCM).

Given the available inflows for each month and the non-normal highly skewed nature of the inflows, the Kumaraswamy distribution was found to be the best fit and a few examples are shown in Figure 5.

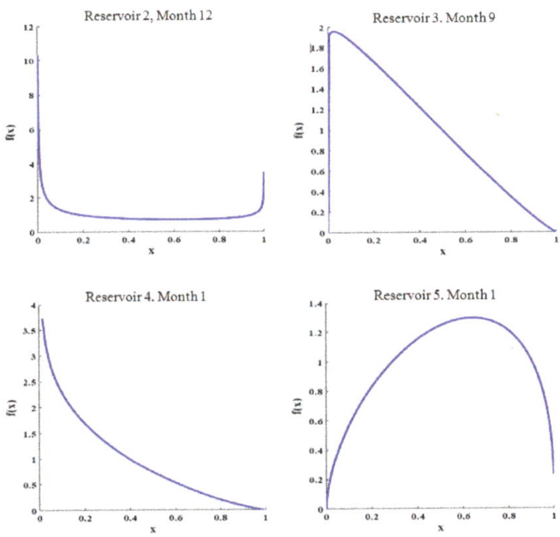

Figure 5. Examples for the probability density functions of inflows to reservoirs [19].

Furthermore, based on the nature of PAP case study, there are high correlations between the first reservoir (Tamilnadu Sholayar), the second reservoirs (Parambikalami and Tunacadavu-Peruvarippallam), the third reservoir (Lower Aliyar) and the fifth reservoir (Kerala Sholayar) in terms of natural inflows (Table 3). The inflow to reservoir 4 is independent of other reservoirs' inflows.

Table 3. The cross-correlation between inflows to reservoirs.

	Res. 1	Res. 2	Res. 3	Res. 4	Res. 5
Res. 1	1.0	0.8	0.8	−0.1	0.9
Res. 2	0.8	1.0	0.8	0.1	0.8
Res. 3	0.8	0.8	1.0	0.3	0.7
Res. 4	−0.1	0.1	0.3	1.0	0.0
Res. 5	0.9	0.8	0.7	0.0	1.0

The PAP project is solved using three other different optimization methods, including the MAM-DP, AD-DP and FP techniques in order to verify the performance of the proposed AD-RL. Two different methods of aggregation/decomposition are used in MAM-DP and AD-DP which are explained in the following sub-sections.

3.2. MAM-DP Method Applied to Parambikulam–Aliyar Project (PAP)

As observed in Figure 6, the PAP problem can be decomposed into four two-reservoir sub-problems for using the 2-level MAM-DP [28]. The values $\tilde{R}_{12}$, $\tilde{R}_{13}$ and $R_{24} + R_{25}$ are the releases whose conditional probabilities should be calculated from the respective sub-problems. Having solved all sub-problems, the optimal policy for the whole PAP case study can be obtained using the results of all sub-problems' policies (R_{13} and R_3 from sub-problem 1; R_{12} from sub-problem 2; R_{25} from sub-problem 3; R_{24}, R_4, R_5 and from sub-problem 4).

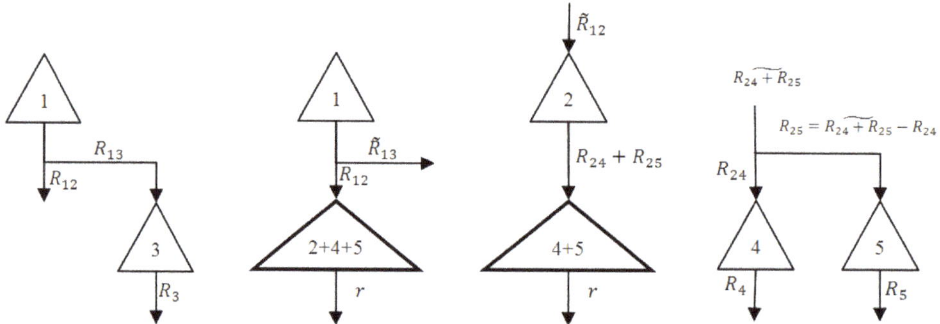

Figure 6. The sub-problems of PAP using the multilevel approximation-dynamic programming (MAM-DP) method [28].

Furthermore, defining an objective function for each sub-problem is a challenging issue. Here, it is assumed that the maximum possible benefit for each sup-problem is achieved from the releases in the downstream reservoirs. For instance, in the first period the benefit per unit release for R_{12} in the first sub-problem is 1.95 (the sum of the benefit per unit release from reservoirs 1, 2 and 4).

3.3. AD-DP Method Applied to PAP Project

All sub-problems of the PAP project using the AD-DP [21] aggregation method are demonstrated in Figure 7. The reservoir with a thick borderline in each sub-problem is a virtual reservoir whose release is not what actually occurs in reality. For instance, the release out of virtual reservoirs 1 and 3 in the second sub-problem does not really flow into reservoir 2, and it is divided between reservoirs 2 and 3.

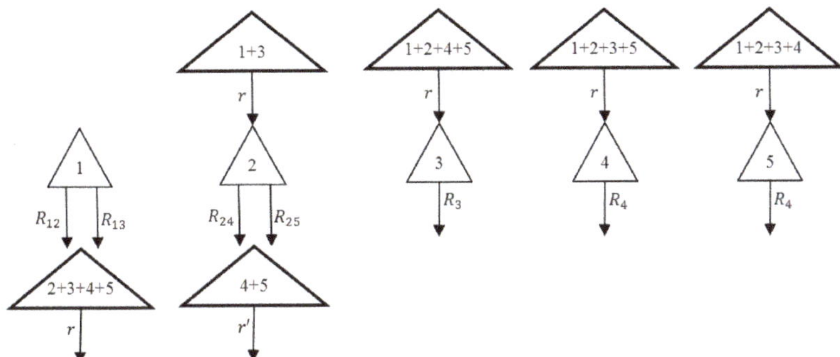

Figure 7. The sub-problems of PAP using the aggregation–decomposition dynamic programming (AD-DP) method [28].

As illustrated in Figure 7, the first sub-problem consists of two state variables (the storage of reservoir 1 and the storage of virtual reservoir which is made up from four reservoirs 2, 3, 4 and 5) and three decision variables (the next storage of the virtual reservoir, release from reservoir 1 to reservoir 2 and release from reservoir 1 to reservoir 3). The second sub-problem comprises three state variables and four decision variables. Other sub-problems have two state variables and two decision variables. Furthermore, for these sub-problems, one more assumption (in addition to the one explained in the section on AD-DP regarding the division of storages of the virtual reservoirs) should be taken into consideration in which the release from each virtual reservoir with more than one outlet downstream (e.g., reservoirs 1 and 2) should be proportionally divided based on the respective capacity of the outlets.

3.4. Fletcher–Ponnambalam (FP) Method Applied to PAP

In the FP [17] method, having assumed a linear decision rule, the first and the second moments of storages are available in analytical forms. Substituting these new moments in the objective function produces an analytical expression for the first and second moments of releases, spills, and deficits of each reservoir in each period. All these expressions are dependent on a single parameter for each reservoir in each period. A non-linear optimization is solved directly for the objective function while the statistical moments of storages, releases and probabilities of containments, spills and deficits are also available for all reservoirs. More details can be seen in [53].

3.5. AD-RL Method

To find a near-optimal policy by the Q-learning algorithm, a proper action-taking policy should be chosen. The discount factor (γ) for the learning step is 0.9 while the respective parameters including the learning factor (α) and ε (in ε-greedy policy) or τ (in Softmax policy) should be accurately set.

As previously mentioned, the main advantage of the Softmax policy is to explore more at the beginning of learning while increasing exploitation as learning continues. This behavior is controlled using the values of Q-factors while converging to the steady-state situation, that is, the action with a greater Q-factor should have a higher chance to be chosen in every interaction. However, while applying this policy, it is observed that the values of all Q-factors become almost close to each other as they converge to the steady-state situation. In other words, Softmax might end up with almost identical probabilities for admissible actions. Therefore, this action-taking policy may lead to a poor performance through converging to a far-optimal operating policy, which our numerical experiments accomplished herein confirmed this result too.

Despite the Softmax policy, in ε-greedy policy, the rate of exploration is constant over the learning process that may not guarantee reaching a near-optimal policy. To tackle this issue, the whole learning process can be implemented as episodic starting, with a big ε in the first episode and decrease the rate in the next episode. An episode comprises a predefined number of years that the learning (simulation) should be implemented. The initial state at the beginning of the learning in every episode can be randomly selected. It is worth mentioning that the value of ε in every episode is constant. To implement this way of learning in the PAP case study, we considered two different parameters: $\varepsilon 1$ and $\varepsilon 2$. The value of $\varepsilon 1$ equals one in the first episode, so there will be no difference between all admissible actions in terms of probability of being selected. Parameter $\varepsilon 2$ is the rate of exploration in the last episode, then the rate of exploration for other episodes is determined using a linear function of these two parameters.

The learning factor (α) is also an important parameter that should be precisely specified. There are different methods to set this parameter [54]. It has been specified based on the number of updating for each action-state pair using the following equation:

$$\alpha = \frac{B}{NOV\,(i,a)} \tag{13}$$

where B is a smaller-than-one predefined parameter. The role of this parameter is to cope with the asynchronous error in the learning process [54].

We use a robust ANN-based approach to tune the above-mentioned parameters. The respective input data for training the corresponding multilayer perceptron ANN is obtained using a combination of different values chosen for parameters $\varepsilon 2$ and B. To have a space-filling strategy, the values considered for each parameter should be uniformly distributed over its domain (for example B takes values of 0.1, 0.3, 0.5, 0.7 and 0.9). Given the sample values chosen for these parameters presented in Table 4, there will be 25 different input data representing all possible combinations for these parameter values. AD-RL will be implemented then for each individual set of these input combinations 10 times. Given the operating policy for each implementation of Q-learning, one can run the respective simulation to obtain the expected value of the benefit. The expected value ($\overline{\mu}$) and the semi variance (σ_{semi}) of all 10 expected values of benefit value obtained from the simulations are calculated using the following equations:

$$\overline{\mu} = \frac{\sum_i^{N_R} \mu_i}{N_R} \tag{14}$$

$$\sigma_{semi} = \frac{\sum_i^{N_R} \left\{ (\mu_i - \overline{\mu})^2 \middle| \mu_i < \overline{\mu} \right\}}{N_R - 1} \tag{15}$$

where N_R is the number of runs (10 in our experiments), and μ_i and σ_{semi} are respectively the mean and semi variance values of objective function obtained by ith run.

Table 4. Values of B, ε_1, ε_2 for making input data.

Parameters	Values				
B	0.1	0.3	0.5	0.7	0.9
Initial exploration factor (ε_1)	1	1	1	1	1
Final exploration factor (ε_2)	0	0.1	0.3	0.5	0.7

Recall that Q-learning is a simulation-based technique. Therefore, it could end up with different operating policies at the end of the learning process because of different sequences of synthetic data being generated over the learning phase. Obtaining more robust results for different runs of Q-learning (less variation of expected values at the end of simulation for different Q-learning implementations) is a good sign for the respective fine-tuned parameter values. To take the robustness of the results into account in ANN training task, one can use a function of the performance criterion ($\bar{\mu} - \sigma_{semi}$ in our experiments) as a desirable output.

To demonstrate how efficient the proposed tuning procedure is, one of the five reservoirs in the PAP case study (Tamilnadu Sholayar reservoir) is selected which can be considered a one-reservoir problem. Similar to what has been undertaken for training data, Q-learning is implemented for the test data set consisting of 100 data in our experiments, and the respective performance criterion is obtained for each test data. The outputs of the networks for these test data can be found using the trained network. The best outputs obtained from Q-learning and the trained networks are compared then to each other. If all or the most of these two different outputs are the same, we conclude that the training phase for tuning the parameters has been performed appropriately.

Table 5 demonstrates top- 3, 4, 5, 10, and 15 best sets of parameters in terms of the defined performance criterion ($\bar{\mu} - \sigma_{semi}$). For instance, 3 out of 5 best sets of parameters based on the output of the trained network are among the 5 best sets of parameters based on the Q-learning approach. Figure 8 illustrates the comparisons between the performance obtained from the neural network for the top 20 sets of parameters and the respective ones based on Q-learning. It shows that the function trained maps the input data to desirable data appropriately. Such a training procedure can, therefore, be used in cases with a larger number of reservoirs.

Table 5. The number of the same sets of parameters according to estimated-by-artificial neural network (ANN) and actual (obtained-by-Q learning) performances for top-N set.

	N				
N	3	4	5	10	15
The number of the same sets in top n sets	1	3	3	9	14

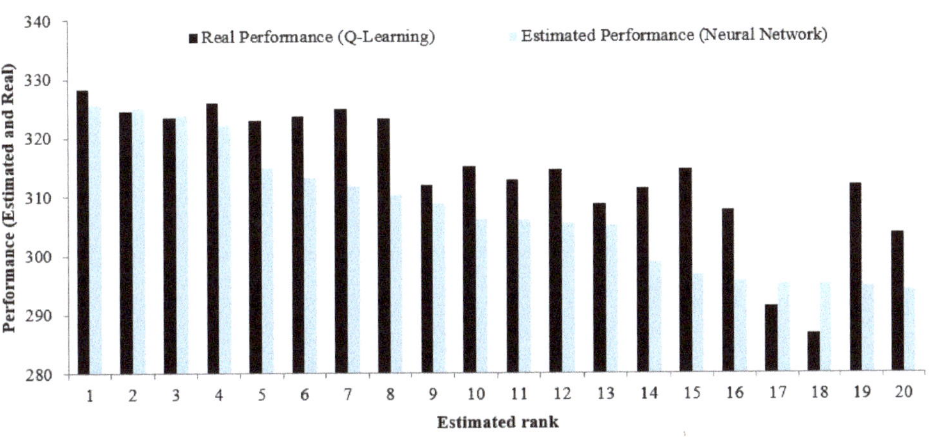

Figure 8. The estimated and real performance of top 20 sets based on estimated performance.

Having the trained network using 25 examples given in Table 5 for the five-reservoir case study (PAP project), one can select and sort a number of the best parameter sets using 100 test data (Table 6). We used the first set of parameters to implement AD-RL.

Table 6. Top three parameter sets obtained by the ANN-based parameter tuning approach for AD-RL algorithm applied to PAP.

Rank	Performance Criteria	ε_2	B.
1	1347.447	0.3	0.9
2	1346.325	0.4	0.9
3	1345.93	0.2	0.9

The AD-RL algorithm was implemented 10 times, each with one hundred episodes for the learning process. Each episode comprises 1000 years. Table 7 reports the average and the standard deviation of the annual benefit for all mentioned techniques applied in the PAP case study. It is worth noting that the average and standard deviation reported for the AD-RL method are their mean values obtained from running ten simulations. The AD-DP, MAM-DP, and AD-RL results are also determined by applying them to the same problem (PAP). Finally, the FP1 and FP2 results are from Mahootchi et al. [53] where FP1 and FP2 correspond to different approximations for estimating releases from upstream reservoirs for the Kumaraswamy distributed inflows. See [53] for details.

Table 7. The average expected value and the standard deviation of annual benefit.

Methods	Ave.	Std.
MAM-DP *	1432.2	249.6
AD-RL *	1353.9	224.1
FP2	1289.5	230.4
FP1	1262.5	211
AD-DP *	1268.0	231.6

* Results from the methods implemented in this paper.

Figure 9 compares the average benefit for all 10 runs of Q-learning. This is a good verification showing how robust the Q-learning algorithm is. It also verifies that the derived-by-Q learning policy for all different 10 runs outperforms the policies derived by FP and AD-DP.

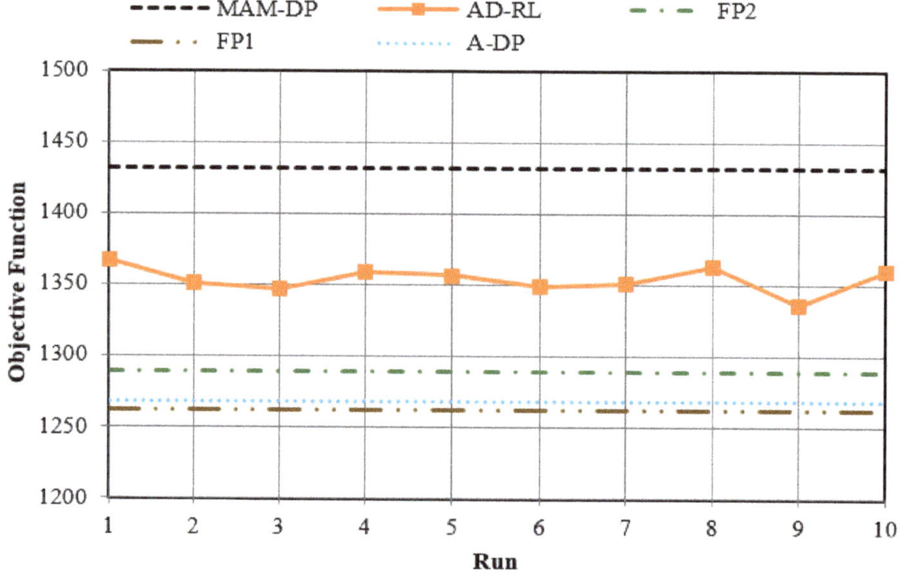

Figure 9. The average expected value of the objective function for 10 different runs.

4. Discussion

In above section, we presented the results of the proposed aggregation decomposition-reinforcement learning (AD-RL) approach for optimizing the PAP multi-reservoir system operations and compared them with those of three other stochastic optimization methods including MAM-DP, FP, and AD-DP. In terms of the average optimality criterion (objective function value), the AD-RL's solution was the second best result after that of MAM-DP. Because both mean and standard deviations are important, from Table 7 it is clear that the solutions of MAM-DP, AD-RL and FP1 are non-inferior and dominate the solutions of FP2 and AD-DP. Although computational run times are not always the best way to compare complexity in these methods where the number of simulations can be easily changed without much loss in optimality, the AD-RL method was three times faster than MAM-DP for the results presented above. This is important as MAM-DP and AD-RL were the top two ranked methods using mean objective function values. Moreover, as AD-RL was performed using the parameters tuned by a trained ANN, the solutions found by the proposed method in the PAP case study were reasonably robust. In other words, performing the AD-RL technique with different synthetic sequences of data leads to almost similar results. In this regard, assessing the performances of the investigated methods against different reservoir inflow scenarios associated with different runs, the standard deviation of the objective function for the AD-RL method was reasonably good compared to those of other stochastic methods investigated. This risk measure for the AD-RL approach was considerably lower than that resulting from the MAM-DP approach.

Overall, compared to other stochastic optimization methods, the results obtained by the proposed AD-RL approach were promising in terms of the optimality and robustness of the solutions found and the required computational burden.

5. Conclusions

In this paper, an aggregation/decomposition model combined with RL was developed for optimizing multi-reservoir systems operations, in which the whole system is controlled using a number of multiple co-operating agents. This method addresses the so-called dual curses of modeling and dimensionality in SDP. Each agent (operator) finds the best decision (release) for its own reservoir based on its current state and the feedback it receives from other agents represented by two virtual reservoirs, including one each for upstream and non-upstream reservoirs, respectively. An efficient approach based on neural networks was also proposed for tuning the model parameters in order to achieve a robust solution methodology. The developed methodology and original methods that did not use the RL method were applied to the Parambikulam–Aliyar Project (PAP) for which results from other methods were available for comparison.

For the PAP case studied herein, the average performance of the model (based on multiple runs) was compared with the performances of other stochastic optimization techniques, including multilevel approximation dynamic programming (MAM-DP), aggregate dynamic programming (AD-DP), and that of the Fletcher–Ponnambalam (FP) method (reported from the literature). The policies obtained by the AD-RL revealed that the proposed AD-RL approach outperformed FP and AD-DP methods in terms of the objective function criterion while having a comparable performance with MAM-DP but with a less computational time, which can be promising for large-scale problems.

As the proposed method is based on simulation, any other objective functions including non-linear/discontinuous ones can be considered which is a challenging issue for most other stochastic methods and is left for future studies.

Author Contributions: The work presented is mainly based on author M.H.'s MSc. thesis, supervised by authors S.J.M. and M.M. and the following contributions: conceptualization, K.P., M.M. and S.J.M.; methodology, S.J.M. and M.M.; software, M.H., M.M. and K.P.; validation, M.M., S.J.M., M.H. and K.P.; formal analysis, M.H., S.J.M. and M.M.; investigation, M.H., S.J.M. and M.M.; resources, K.P., M.M. and S.J.M.; data curation, K.P. and M.M.; writing—original draft preparation, M.M., S.J.M. and M.H.; writing—review and editing, K.P. and S.J.M.; visualization, M.H.; supervision, S.J.M., M.M. and K.P. All authors have read and agreed to the published version of the manuscript.

Funding: This research received no external funding.

Conflicts of Interest: The authors declare no conflict of interest.

References

1. Hiew, K.L.; Labadie, J.W.; Scott, J.F. Optimal operational analysis of the Colorado-Big Thompson project. In *Proceedings of the Computerized Decision Support Systems for Water Managers*; Labadie, J., Ed.; ASCE: Reston, VA, USA, 1989; pp. 632–646.
2. Willis, R.; Finney, B.A.; Chu, W.S. Monte Carlo Optimization for Reservoir Operation. *Water Resour. Res.* **1984**, *20*, 1177–1182. [CrossRef]
3. Hiew, K.L. *Optimization Algorithms for Large-Scale Multireservoir Hydropower Systems*; Colorado State University: Fort Collins, CO, USA, 1987.
4. Tejada-Guibert, J.A.; Stedinger, J.R.; Staschus, K. Optimization of Value of CVP's Hydropower Production. *J. Water Resour. Plan. Manag.* **1990**, *116*, 52–70. [CrossRef]
5. Arnold, E.; Tatjewski, P.; Wołochowicz, P. Two Methods for Large-Scale Nonlinear Optimization and Their Comparison on a Case Study of Hydropower Optimization. *J. Optim. Theory Appl.* **1994**, *81*, 221–248. [CrossRef]
6. Lee, J.H.; Labadie, J.W. Stochastic Optimization of Multireservoir Systems via Reinforcement Learning. *Water Resour. Res.* **2007**. [CrossRef]
7. Aiken, L.S.; West, S.G.; Pitts, S.C. *Multiple Linear Regression; Handbook of Psychology*; John Wiley & Sons: Hoboken, NJ, USA, 2003.
8. Hornik, K.; Stinchcombe, M.; White, H. Multilayer Feedforward Networks are Universal Approximators. *Neural Netw.* **1989**, *2*, 359–366. [CrossRef]
9. Wang, L.-X.; Mendel, J.M. Generating Fuzzy Rules by Learning from Examples. *IEEE Trans. Syst. Man. Cybern.* **1992**, *22*, 1414–1427. [CrossRef]
10. Mousavi, S.; Ponnambalam, K.; Karray, F. Reservoir Operation Using a Dynamic Programming Fuzzy Rule–Based Approach. *Water Resour. Manag.* **2005**, *19*, 655–672. [CrossRef]
11. Mousavi, S.J.; Ponnambalam, K.; Karray, F. Inferring Operating Rules for Reservoir Operations Using Fuzzy Regression and ANFIS. *Fuzzy Sets Syst.* **2007**, *158*, 1064–1082. [CrossRef]
12. Loucks, D.P.; Dorfman, P.J. An Evaluation of Some Linear Decision Rules in Chance-Constrained Models for Reservoir Planning and Operation. *Water Resour. Res.* **1975**, *11*, 777–782. [CrossRef]
13. Alizadeh, H.; Mousavi, S.J.; Ponnambalam, K. Copula-Based Chance-Constrained Hydro-Economic Optimization Model for Optimal Design of Reservoir-Irrigation District Systems under Multiple Interdependent Sources of Uncertainty. *Water Resour. Res.* **2018**, *54*, 5763–5784. [CrossRef]
14. Simonovic, S.P.; Marino, M.A. Reliability Programing in Reservoir Management: 1. Single Multipurpose Reservoir. *Water Resour. Res.* **1980**, *16*, 844–848. [CrossRef]
15. Simonovic, S.P.; Marino, M.A. Reliability Programing in Reservoir Management: 2. Risk-Loss Functions. *Water Resour. Res.* **1981**, *17*, 822–826. [CrossRef]
16. Simonovic, S.P.; Marino, M.A. Reliability Programing in Reservoir Management: 3. System of Multipurpose Reservoirs. *Water Resour. Res.* **1982**, *18*, 735–743. [CrossRef]
17. Fletcher, S.; Ponnambalam, K. Constrained State Formulation for the Stochastic Control of Multireservoir Systems. *Water Resour. Res.* **1998**, *34*, 257–270. [CrossRef]
18. Fletcher, S.; Ponnambalam, K. Stochastic Control of Reservoir Systems Using Indicator Functions: New Enhancements. *Water Resour. Res.* **2008**, *44*, 44. [CrossRef]
19. Mahootchi, M. Storage System Management Using Reinforcement Learning Techniques and Nonlinear Models. Ph.D. Thesis, University of Waterloo, Waterloo, ON, Canada, January 2009.
20. Thomas, H.; Watermeyer, P. *Mathematical Models: A Stochastic Sequential Approach*; Harvard University Press: Cambridge, MA, USA, 1962.
21. Archibald, T.; McKinnon, K.; Thomas, L. An Aggregate Stochastic Dynamic Programming Model of Multireservoir Systems. *Water Resour. Res.* **1997**, *33*, 333–340. [CrossRef]
22. Cervellera, C.; Chen, V.C.; Wen, A. Optimization of a Large-Scale Water Reservoir Network by Stochastic Dynamic Programming with Efficient State Space Discretization. *Eur. J. Oper. Res.* **2006**, *171*, 1139–1151. [CrossRef]

23. Foufoula-Georgiou, E. Convex Interpolation for Gradient Dynamic Programming. *Water Resour. Res.* **1991**, *27*, 31–36. [CrossRef]

24. Johnson, S.A.; Stedinger, J.R.; Shoemaker, C.A.; Li, Y.; Tejada-Guibert, J.A. Numerical Solution of Continuous-State Dynamic Programs Using Linear and Spline Interpolation. *Oper. Res.* **1993**, *41*, 484–500. [CrossRef]

25. Karamouz, M.; Mousavi, S.J. Uncertainty Based Operation of Large Scale Reservoir Systems: Dez and Karoon Experience. *J. Am. Water Resour. Assoc.* **2003**, *39*, 961–975. [CrossRef]

26. Mousavi, S.J.; Karamouz, M. Computational Improvement for Dynamic Programming Models by Diagnosing Infeasible Storage Combinations. *Adv. Water Resour.* **2003**, *26*, 851–859. [CrossRef]

27. Philbrick, C.R., Jr.; Kitanidis, P.K. Improved Dynamic Programming Methods for Optimal Control of Lumped-Parameter Stochastic Systems. *Oper. Res.* **2001**, *49*, 398–412. [CrossRef]

28. Ponnambalam, K.; Adams, B.J. Stochastic Optimization of Multi Reservoir Systems Using a Heuristic Algorithm: Case Study from India. *Water Resour. Res.* **1996**, *32*, 733–741. [CrossRef]

29. Saad, M.; Turgeon, A. Application of Principal Component Analysis to Long-Term Reservoir Management. *Water Resour. Res.* **1988**, *24*, 907–912. [CrossRef]

30. Saad, M.; Turgeon, A.; Bigras, P.; Duquette, R. Learning Disaggregation Technique for the Operation of Long-Term Hydroelectric Power Systems. *Water Resour. Res.* **1994**, *30*, 3195–3202. [CrossRef]

31. Saad, M.; Turgeon, A.; Stedinger, J.R. Censored-Data Correlation and Principal Component Dynamic Programming. *Water Resour. Res.* **1992**, *28*, 2135–2140. [CrossRef]

32. Stedinger, J.R.; Faber, B.A.; Lamontagne, J.R. Developments in Stochastic Dynamic Programming for Reservoir Operation Optimization. In Proceedings of the World Environmental and Water Resources Congress, Cincinnati, OH, USA, 19–23 May 2013.

33. Tejada-Guibert, J.A.; Johnson, S.A.; Stedinger, J.R. The Value of Hydrologic Information in Stochastic Dynamic Programming Models of a Multireservoir System. *Water Resour. Res.* **1995**, *31*, 2571–2579. [CrossRef]

34. Turgeon, A. Optimal Operation of Multireservoir Power Systems with Stochastic Inflows. *Water Resour. Res.* **1980**, *16*, 275–283. [CrossRef]

35. Turgeon, A. A Decomposition Method for the Long-Term Scheduling of Reservoirs in Series. *Water Resour. Res.* **1981**, *17*, 1565–1570. [CrossRef]

36. Pereira, M.V.F. Optimal Stochastic Operations Scheduling of Large Hydroelectric Systems. *Int. J. Electr. Power Energy Syst.* **1989**, *11*, 161–169. [CrossRef]

37. Pereira, M.V.F.; Pinto, L.M.V.G. Multi-Stage Stochastic Optimization Applied to Energy Planning. *Math. Program.* **1991**, *52*, 359–375. [CrossRef]

38. Poorsepahy-Samian, H.; Espanmanesh, V.; Zahraie, B. Management. Improved Inflow Modeling in Stochastic Dual Dynamic Programming. *J. Water Resour. Plan. Manag.* **2016**, *142*, 04016065. [CrossRef]

39. Rougé, C.; Tilmant, A. Using Stochastic Dual Dynamic Programming in Problems with Multiple Near-Optimal Solutions. *Water Resour. Res.* **2016**, *52*, 4151–4163. [CrossRef]

40. Zhang, J.L.; Ponnambalam, K. Stochastic Control for Risk Under Deregulated Electricity Market—A Case Study Using a New Formulation. *Can. J. Civ. Eng.* **2005**, *32*, 719–725. [CrossRef]

41. Kelman, J.; Stedinger, J.R.; Cooper, L.A.; Hsu, E.; Yuan, S.Q. Sampling Stochastic Dynamic Programming Applied to Reservoir Operation. *Water Resour. Res.* **1990**, *26*, 447–454. [CrossRef]

42. Mahootchi, M.; Tizhoosh, H.; Ponnambalam, K. Opposition-based reinforcement learning in the management of water resources. In Proceedings of the IEEE International Symposium on Approximate Dynamic Programming and Reinforcement Learning, Honolulu, HI, USA, 1–5 April 2007; pp. 217–224.

43. Castelletti, A.; Pianosi, F.; Restelli, M. A Multiobjective Reinforcement Learning Approach to Water Resources Systems Operation: Pareto Frontier Approximation in a Single Run. *Water Resour. Res.* **2013**, *49*, 3476–3486. [CrossRef]

44. Bhattacharya, B.; Lobbrecht, A.; Solomatine, D. Neural Networks and Reinforcement Learning in Control of Water System. *J. Water Resour. Plan. Manag.* **2003**, *129*, 458–465. [CrossRef]

45. Castelletti, A.; Galelli, S.; Restelli, M.; Soncini-Sessa, R. Tree-Based Reinforcement Learning for Optimal Water Reservoir Operation. *Water Resour. Res.* **2010**, *46*, W09507. [CrossRef]

46. Pianosi, F.; Castelletti, A.; Restelli, M. Tree-Based Fitted Q-Iteration for Multi-Objective Markov Decision Processes in Water Resource Management. *J. Hydroinform.* **2013**, *15*, 258–270. [CrossRef]

47. Bertoni, F.; Giuliani, M.; Castelletti, A. Integrated Design of Dam Size and Operations via Reinforcement Learning. *J. Water Resour. Plan. Manag.* **2020**, *146*, 04020010. [CrossRef]
48. Bellman, R. Dynamic Programming and Lagrange Multipliers. *Proc. Natl. Acad. Sci. USA* **1956**, *42*, 767. [CrossRef] [PubMed]
49. Ponnambalam, K.; Adams, B.J. Experiences with integrated irrigation system optimization analysis. In *Proceedings of the Irrigation and Water Allocation (Proc., Vancouver Symposium)*; IAHS: Wallingford, UK, 1987; pp. 229–245.
50. Sutton, R.S.; Barto, A.G. *Reinforcement Learning: An Introduction*; MIT Press: Cambridge, MA, USA, 1998; Volume 1.
51. Watkins, C.J.C.H.; Dayan, P. Q-Learning. *Mach. Learn.* **1992**, *8*, 279–292. [CrossRef]
52. Robbins, H.; Monro, S. A Stochastic Approximation Method. *Ann. Math. Stat.* **1951**, *22*, 400–407. [CrossRef]
53. Mahootchi, M.; Ponnambalam, K.; Tizhoosh, H. Operations Optimization of Multireservoir Systems Using Storage Moments Equations. *Adv. Water Resour.* **2010**, *33*, 1150–1163. [CrossRef]
54. Gosavi, A. *Simulation-Based Optimization: Parametric Optimization Techniques and Reinforcement Learning*; Springer: Berlin/Heidelberg, Germany, 2014; Volume 55.

 *water*

Review

A Water Footprint Review of Italian Wine: Drivers, Barriers, and Practices for Sustainable Stewardship

Eirini Aivazidou [1,2,*] **and Naoum Tsolakis** [1,3,4]

[1] Laboratory of Statistics and Quantitative Analysis Methods (LASCM), Department of Industrial Management, School of Mechanical Engineering, Aristotle University of Thessaloniki, 54124 Thessaloniki, Greece; nt377@cam.ac.uk

[2] Department of Computer Science and Engineering (DISI), School of Engineering, Alma Mater Studiorum–University of Bologna, 40126 Bologna, Italy

[3] Centre for International Manufacturing, Institute for Manufacturing (IfM), Department of Engineering, School of Technology, University of Cambridge, Cambridge CB3 0FS, UK

[4] Institute for Bio-Economy and Agri-Technology (iBO), Centre for Research and Technology–Hellas (CERTH), 57001 Thessaloniki, Greece

* Correspondence: aveirini@auth.gr

Received: 21 December 2019; Accepted: 25 January 2020; Published: 29 January 2020

Abstract: Wine constitutes the dominant Italian agricultural product with respect to both production quantity and economic value. Italy is the top wine producer worldwide in terms of volume and the second one below France in terms of national income. As the Italian agricultural production accounts for 85% of the national freshwater appropriation, the country's agricultural sector strains freshwater resources, especially in the central and southern regions, which constitute important winemaking areas in terms of quantity and quality. To this end, we first perform a review of the existing research efforts on wine water footprint assessment to investigate the water dynamics of wine production in Italy compared to the rest of the world. The results indicate a prevalence of studies on the water footprint of Italian wine, emphasising the need for deeper research on the sector's water efficiency. Then, we aim at exploring the major drivers, barriers, and good practises for systematic water stewardship in the Italian winemaking industry, considering the product and territorial characteristics. This research is anticipated to contribute towards providing insights for practitioners in the Italian wine sector to develop water-friendly corporate schemes for enhancing the added value of their products.

Keywords: freshwater resources; water footprint; water management; wine production; winemaking sector; Italy

---

1. Introduction

The winemaking industry plays a critical role in the economy of the primary sector of the Southern European and Mediterranean regions [1]. Thus, there is an increased pressure towards minimising the environmental impacts of wine production [2] to improve the sustainability of the sector in terms of climate change and natural resources [3]. More specifically, consumers' environmental expectations further motivate winemakers to adopt green technological interventions for efficient water use during irrigation or wastewater reuse [4]. In addition, given that a considerable number of consumers, especially young ones [5], express willingness to pay a premium for a sustainable wine label [6], the production of water-friendly wine could be an ambitious strategy for increasing profitability through quality improvement [7].

The winemaking efficiency in terms of freshwater use can be expressed through the water footprint (WF) concept, which refers to the total volume of freshwater consumed and polluted at national, corporate, or product levels [8]. Specifically, WF is a multidimensional indicator that consists of three components: (i) green water addresses the absorption of rainwater by plants (i.e., the proportion of precipitation that infiltrates into the unsaturated soil zone and is temporarily stored in soil and vegetation canopy [9]), (ii) blue water refers to the consumption of surface or groundwater during irrigation and processing activities, and (iii) grey water constitutes the freshwater quantity used for assimilating pollutants during farming and manufacturing given specific water quality standards [8].

In Italy, agricultural production is responsible for 85% of the country's freshwater appropriation (This percentage is calculated as the ratio of the WF of crop production, grazing, and animal water supply to the total WF of national production in Italy, which are both provided in annual average values during the reference period 1996–2005), of which 81% refers to green WF, 8% refers to blue WF, and 11% refers to grey WF [10]. In the case of the winemaking sector, the average WF of Italian grapes equals to 488 L per kg of fresh fruit, of which 76% corresponds to green water, 7% corresponds to blue water, and 17% corresponds to grey water [11]. In terms of wine, the average WF of a glass (0.125 L) of Italian wine is 88 L [12]. Notably, a considerable number of research papers further quantify the WF of different wine varieties across the regions of the country. Apart from water use, emphasis is further placed on water scarcity issues of the Italian territory. In fact, the national agricultural sector poses considerably high stress on freshwater resources [13], particularly in Southern Italy [14], which constitutes an important winemaking area in terms of wine quantity and quality [15].

Although the water impact of wine is relatively low compared to other agricultural commodities [11], its high production volume and its economic value in Italy render research on the WF of Italian wine essential. In fact, wine constitutes the top national agricultural product in terms both of quantity and value [16]. Compared to the rest of the world in 2017, Italy constituted the first producer concerning wine volume (4.25 billion L, excluding juice and must) [1], and the second one below France regarding economic value (12.1 billion Euro) [17]. In addition, the country came third following the United States and France (2.26 billion L) in respect to wine consumption, while it was second both below Spain in terms of export volume (2.14 billion L) and below France concerning export value (5.87 billion Euro) [1].

Notably, scientific research on the WF assessment of wine is growing rapidly [18]. As water management across supply chains is considered as vital for the long-term sustainability of the winemaking industry [19], this work aims at (i) reviewing the existing WF assessment efforts during wine production to explore how the wine WF research is diffused worldwide (Section 2), (ii) investigating the drivers and barriers of water stewardship in the Italian winemaking sector as an identified global leader in the field of study (Section 3), and (iii) discussing water stewardship policies applicable to the Italian wine production (Section 4). Overall, this paper aims at highlighting the need for water management in the wine industry, especially in water-scarce countries where it constitutes a major economic activity. To this end, we anticipate that this research will contribute towards supporting winemaking practitioners in identifying good practices in water management and launching efficient water-related corporate schemes through overcoming barriers motivated by impelling drivers.

2. Water Footprint of Wine: Literature Background

In this section, we perform a review of the global wine WF literature to identify the position of the Italian case studies in this research field. Within the extant literature, we have identified 20 articles in total that include the terms "water footprint" (or "water management" or "freshwater resources") and "wine" (or "winemaking") in the Scopus and Web of Science databases. Then, we present the major descriptive statistics of the review findings along with a brief discussion. Finally, the taxonomy of the Italian literature provides a detailed analysis of the papers under study in a structured manner.

The WF of a product is defined as the total volume of freshwater used directly or indirectly across its end-to-end supply chain [20]. Figure 1 illustrates the different stages of a wine supply chain, highlighting the viticulture and the vinification phases as the prevalent WF contributors. To quantify the WF of wine, several methodological approaches exist; the WF assessment manual that focusses on the volumetric measurement of water consumption and pollution [8] and the life cycle assessment (LCA) techniques (e.g., the ISO 14046 [21]), further including the impact of water scarcity (which varies spatially and temporally) on the WF indicator [14], constitute the most common ones. Although a comparison among the different WF assessment approaches is considered out of the scope of this research, a more detailed analysis of their unique charasterics is provided by Chenoweth et al. [22].

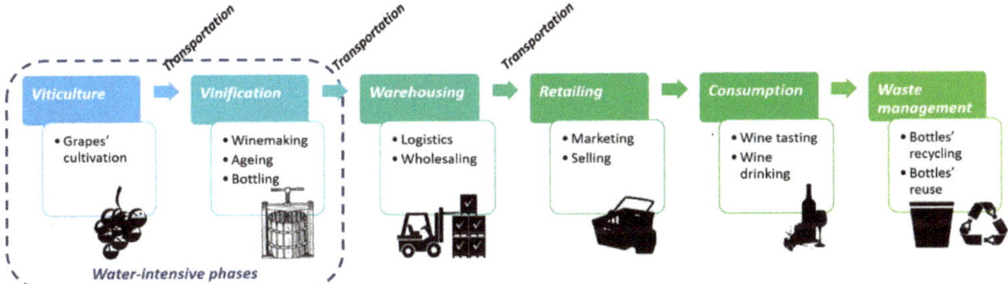

Figure 1. Typical wine supply chain.

2.1. Research Efforts Worldwide

Several case studies on wine WF assessment have been identified in the literature, indicating the increasing academic interest in evaluating the use of freshwater resources in the winemaking sector worldwide. Outside Europe, efforts have been made to quantify the WF of New Zealand's wines. First, a combination of an LCA-based approach and a hydrological water-balance technique was implemented to quantify the volume of water consumed and polluted during wine production in two New Zealand regions [23]. In a later study, the authors extended the WF research of New Zealand's wine by comparing the results obtained using diverse methodologies, further including the traditional WF assessment method, to investigate freshwater utilisation from different perspectives [24]. In North America, a US study assessed the greenhouse gas emissions, the energy use, and the freshwater use across the life cycle of wine produced in California, beginning from the cultivation of grapes up to their delivery at the winery gate, to provide a holistic evaluation of the wine's environmental impact [25]. Moreover, a preliminary research effort was made for assessing the grey WF associated with wastewater produced during the winemaking process in a Canadian winery and co-treated by municipal wastewater treatment plants [26]. In Latin America, a recent study quantified the consumptive blue and green WF of several varieties of grapes for wine production in five Argentinian regions, using different irrigation systems [27].

Within Europe, although the majority of research on wine WF assessment has been documented across southern countries, two publications refer to Northern Europe. In Romania, the WF of a bottled wine produced in a medium-sized winemaking plant was quantified in the stages of viticulture and vinification, further evaluating the socio-economic potential of winemaking and the related water-related schemes within the country [28]. In Hungary, a recent study developed a framework for WF assessment during grapes' cultivation and processing to optimise the consumption of both rainwater and freshwater consumed [29]. Moving to the south, several researchers evaluated the WF of Iberian wines. More specifically, an evaluation of both direct and indirect freshwater use was performed for a Portuguese white wine variety during the viticulture and the winemaking stages, further analysing the related environmental impacts of water use [30]. In addition, a more comprehensive analysis included the LCA of the carbon, water, and energy footprints, as well as the material intensity and solid and water wastes, of a bottle wine during the phases of grapes' cultivation, wine production, bottling and packaging in Portugal [31]. More recently, the water-related ISO 14046 was used to analyse the WF profile of a Spanish grape variety for vinification and to address the impacts due to water scarcity and degradation from a life cycle perspective [32]. Moreover, an indicator of water depletion, mainly due to irrigation during the viticulture stage, was evaluated in the context of a complete LCA of an aged red wine produced in Catalonia, Spain [33].

2.2. Italian Case Studies

An increased number of case studies on wine WF has been mapped across the Italian territory. Lamastra et al. [34] proposed a new WF quantification approach (Valutazione Impatto Viticoltura sull'Ambiente – V.I.V.A. tool) to improve the WF assessment manual technique [8], emphasising in detail the calculation of the grey WF of six different wine varieties of a Sicilian winery. Bonamente et al. [35] quantified the direct green, blue, and grey WFs of a typical red wine produced from a blend of grape varieties by a medium-sized winery in Umbria, based on the V.I.V.A. tool [34] and following the ISO 14046 principles [21]. In a later study, the authors performed a combined carbon and WF assessment in the life cycle of the Italian red wine using the same dataset and methodological approach [36]. In the same vein, Rinaldi et al. [37] performed a cradle-to-grave analysis for juxtaposing the carbon and WF indicators of a red and a white wine of an Umbrian producer, using the same system boundaries, functional unit, and input data, based on the relevant ISO guidelines [21]. In Umbria again, Bartocci et al. [38] calculated the carbon, ecological, and WF, along with several LCA-related environmental impacts, for two different varieties of grapes during cultivation, wine production, vinegar ageing, and bottling, following the ISO approach [21]. Recently, Borsato et al. [39] compared the WF outcomes of a volumetric (i.e., the V.I.V.A. tool [34]) and two LCA-based approaches (i.e., Available WAter REmaing – AWARE [40] and Water Scarcity Index [14]) during the production of a white wine variety in Northeast Italy to improve water management. Miglietta et al. [41] investigated the WF of two types of wines indicated with designation of origin whose vineyards are situated in Northern (Piedmont) and Southern (Sicily) Italy to compare the geographical impact of grapes' cultivation on freshwater consumption and pollution. More recently, Miglietta et al. [42] quantified the water efficiency (i.e., the ratio of total wine WF to total wine production) and the economic water productivity (i.e., the ratio of wine price to wine WF) of all Italian wines indicated with appellation of origin. In addition, Miglietta and Morrone [43] studied the virtual water flows and economic water productivity of wine trade between Italy and Balkan countries. The latter three research efforts were conducted based on the WF assessment manual estimates [8].

Figure 2 illustrates the distribution of the case studies on wine WF assessment by country. Notably, Italy dominates the wine WF research (i.e., nine out of 20 studies), confirming (i) the leading role of the Italian winemaking industry both within the country [16] and abroad [1], and (ii) the increased water scarcity concerns in the region [14], followed by New Zealand, Portugal, and Spain (i.e., two out of 20 studies each). Notably, there is an apparent absence of WF studies for French wines, despite the major economic impact of the country's winemaking sector worldwide [17], which is potentially due to lower water scarcity indices compared to Italy [14]. In addition, Figure 3 depicts the distribution of the studies by year of publication. In fact, the research on wine WF assessment has received a rather constant interest during the last 6.5 years (i.e., the first paper was identified in 2013), exhibiting an average of 2.9 studies per year worldwide and 1.2 studies annually in Italy.

Finally, Table 1 provides a taxonomy of the literature in the field of wine WF assessment in Italy. More specifically, the type of the study, the period in which the data were collected, the location of the study, the wine variety examined, the winemaking phase considered, the WF assessment method used, as well as the type and volume of the WF quantified, are documented to provide detailed information in the field of wine WF assessment in a supplementary manner. Notably, a comparative analysis of the studies could be challenging due to significant differences concerning the (i) methodological approaches implemented, (ii) databases utilised, (iii) assumptions articulated, and (iv) temporal or spatial characteristics considered. However, as the diverse WF assessment approaches exhibited vary with respect to the manner that they quantify water use [22], it is not infeasible to compare WF results derived from different methods, even though the calculations are performed using the same dataset [44]. In fact, Bonamente et al. [36] confirm this statement through providing different results compared to Bonamente et al. [35], although they use the same input data. Even by applying the same methodology, the different wine variety types [34], as well as the diverse climatic and geographical conditions of the Italian regions from North to South [41], influence the wine WF assessment findings. Nevertheless, in general, green water emerges as the typical source of water for wine production, even in semi-arid environments such as Central Italy.

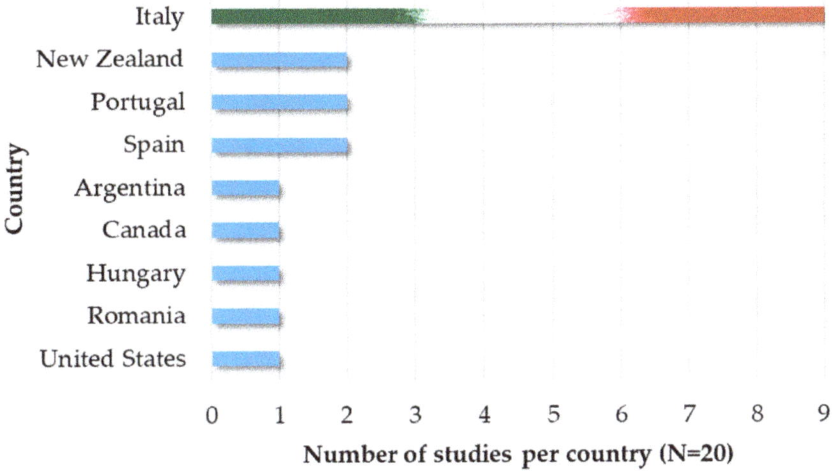

Figure 2. Distribution of studies on wine WF assessment by country.

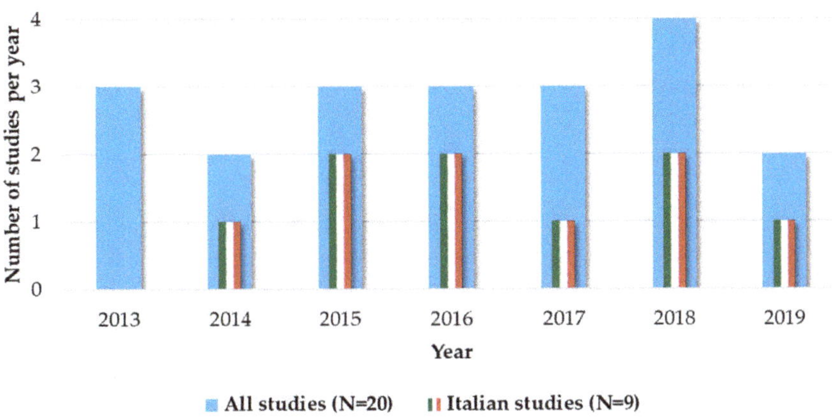

Figure 3. Distribution of studies on wine WF assessment by year.

Table 1. Taxonomy of wine WF research in Italy.

Reference	Study Type	Study Period	Location	Wine Variety	Winemaking Phase	WF Assessment Method	WF Type and Volume		
							Green	Blue	Grey
Lamastra et al. [34]	Real case study	Not specified	Province of Palermo, Region of Sicily (13.49° N, 13.51° E)	Cabernet Sauvignon; Chardonnay; Nero d'Avola; White Pinot; Grecanico	Viticulture; Vinification	WF assessment manual [9]; V.I.V.A. tool [34]	694.5–902.9 (WF manual); 689.5–915.9 (V.I.VA.) L/L of wine	2.6–42.5 L/L of wine (Same for both methods)	0–228.6 (WF manual); 0–389.8 (V.I.V.A) L/L of wine
Bonamente et al. [35]	Real case study	2012	Region of Umbria	Sangiovese with small percentages of Merlot and Cabernet Sauvignon	Viticulture; Vinification	V.I.V.A. tool [34]; ISO 14046 [21] (only as a framework)	621.4 L/bottle of 0.75 L	3.4 L/bottle of 0.75 L	7.4 L/bottle of 0.75 L
Bonamente et al. [36]	Real case study	2012	Region of Umbria	Sangiovese with small percentages of Merlot and Cabernet Sauvignon	Viticulture; Vinification	ISO 14046 [21]	450.6 L/bottle of 0.75 L	7.1 L/bottle of 0.75 L	120.4 L/bottle of 0.75 L
Rinaldi et al. [37]	Real case study	2012	Region of Umbria	Red wine; white wine (specific variety not specified)	Viticulture; Vinification	ISO 14046 [21]	450.6 (red); 496.6 (white) L/bottle of 0.75 L	10 (red); 9.8 (white) L/bottle of 0.75 L	43.5 (red); 44.6 (white) L/bottle of 0.75 L
Bartocci et al. [38]	Real case study	2012	Province of Perugia, Region of Umbria	Grechetto; Sarantino	Viticulture; Vinification	ISO 14046 [21]	830 (Grechetto); 592 (Sarantino) L/L of vinegar	446 (Grechetto); 301 (Sarantino) L/L of vinegar	616 (Grechetto); 439 (Sarantino) L/L of vinegar
Borsato et al. [39]	Real case study	2017	Northeast Italy (45.87° N, 12.70° E)	White wine (specific variety not specified)	Viticulture; Vinification	V.I.V.A. tool [34]; AWARE [40]; Water Scarcity Index [14]	0.988 m^3/ bottle of 0.75 L (V.I.V.A.)	0.181 m^3/ bottle of 0.75 L (V.I.V.A.)	0.024 m^3/ bottle of 0.75 L (V.I.V.A.)
							1.44 (AWARE); 0.01 (Water scarcity index) m^3/ bottle of 0.75 L (No type categorisation)		
Miglietta et al. [41]	Secondary data analysis	Not specified	Region of Piedmont; Region of Sicily	Barolo; Moscato di Pantelleria	Viticulture	WF assessment manual [8]	487–548 L/L of wine (Sum of all types)		
Miglietta et al. [42]	Secondary data analysis	2011–2015	Whole Italian territory	65 varieties with appellation of origin	Viticulture	WF assessment manual [8]	3.03–6.68 m^3/ha of vineyard (sum of all types)		
Miglietta and Morrone [43]	Secondary data analysis	2007–2016	Whole Italian territory (Average)	All varieties (average)	Viticulture	WF assessment manual [8]	460 m^3/ton of wine	40 m^3/ton of wine	101 m^3/ton of wine

3. Water Stewardship in the Italian Wine Industry: Drivers and Barriers

The food and beverage industry's contribution to global freshwater withdrawal is well documented in the extant scientific literature, while freshwater resources are dwindling at an alarming rate [18]. As satisfying the supply of food products requires a sufficient and consistent availability of freshwater resources, collaborative and harmonised interventions across supply chains are essential in order to ensure sustainable and efficient water use [45]. Considering that the Italian wines exhibit substantial WFs especially in water-stressed areas, we discuss the main drivers and barriers of water management in the wine sector based on the existing Italian research efforts. Table 2 summarises the identified drivers and barriers along with a taxonomy of the citing articles.

Table 2. Drivers and barriers of water stewardship in Italy.

Type	Description	References
Drivers	Linkage between water-related environmental aspects to space–temporal pressures	Lamastra et al. [34]; Bonamente et al. [35]; Miglietta et al. [42]; Miglietta and Morrone [43]
	Global trade and makers' attentiveness to sustainable wine supplies and sustainable marketing	Bonamente et al. [35]; Bartocci et al. [38]; Borsato et al. [39]; Miglietta et al. [42]; Miglietta and Morrone [43]
	Consumers' profitable purchasing behaviours towards sustainable wine supplies, particularly when linked to particular territorial culture and history	Bartocci et al. [38]; Miglietta et al. [42]
	Correlation between freshwater quantity/quality and wine quality	Lamastra et al. [34]; Miglietta et al. [41]
	Proliferation of the literature with studies and methodologies on water management allowing for benchmarking	Bonamente et al. [36]; Rinaldi et al. [37]
	Institutional policies and funding schemes supporting water management initiatives	Borsato et al. [39]; Miglietta et al. [41]
	Production effectiveness deriving from water stewardship, particularly from an end-to-end supply chain perspective	Bonamente et al. [36]; Bartocci et al. [38]; Miglietta et al. [42]; Miglietta and Morrone [43]
Barriers	Lack of standardisation of system boundaries to apply and assess the impact of water management policies and practises	Bonamente et al. [36]; Rinaldi et al. [37]; Borsato et al. [39]
	Limited contextualisation of water management operations, particularly with reference to the economic water productivities	Lamastra et al. [34]; Miglietta et al. [42]
	Structural and computational diversification of methodologies assessing the impact of water management policies and practises	Bonamente et al. [35]; Bonamente et al. [36]; Borsato et al. [39]; Miglietta et al. [41]
	Variations in functional characteristics of wine production setting (e.g., local climatic conditions, production processes, etc.)	Bonamente et al. [36]; Borsato et al. [39]; Miglietta and Morrone [43]
	Proliferation of eco-labelling options limiting business differentiation possibilities	Miglietta and Morrone [43]

3.1. Drivers

From an environmental point of view, the elevated global water stress levels foster the adoption and application of water management policies and practices in the wine industry [19]. More specifically, wine quality is correlated to grapes' quality, thus motivating the wine industry to investigate irrigation practises [34]. In particular, a wine's identify is defined by grape maturation, aroma, and coloration [46], which are attributes that are amenable to the vine's geographical location and climate conditions that determine the chemical composition and sensory characteristics of grapes [47].

As freshwater appropriation is characterised by space–temporal dimensions, the adoption of advocated practises in the winemaking industry is eminent to mitigate water stress phenomena at both local and global freshwater bodies [33,35,43]. This need is even more pronounced in regions where the nexus of water scarcity, vineyards, production seasonality, and climatic conditions' severity aggravate water consumption. In this regard, targeted institutional and state-specific policies and directives (e.g., European Program of Sustainability, New Zealand Winegrowing Program, Italian initiative on Valutazione dell' Impatto in Vitivinicoltura sull' Ambiente) motivate circular economy and water-use minimisation in wine [39], while they further support investments in related infrastructure to protect water quality and quantity [41]. Furthermore, the plethora of research studies and corporate reports pertinent to water consumption across the wine supply chains operations allows benchmarking [36,37], thus further enabling the continuous improvement and proliferation of water management policies and practises among industry stakeholders.

From a socio-economic angle, water security has a prominent role in the United Nations Sustainable Development Goals, as it is recognised as a key determinant to the delivery of a viable ecosystem to future generations and a critical factor towards ensuring continuity to food manufacturing operations [48]. In particular, securing freshwater resources' sustainability allows winegrowers to improve economic water productivity of their wine supply chains (i.e., monetary value attained per cubic meter of water used), hence ensuring high-quality winery products at a low level of water use [42]. In addition, the implementation of water management policies and practises (e.g., water reuse) allows grape growers and wine manufacturers to reduce the resources' scarcity burden linked with their production [35,36]. To a greater extent, reducing the utilisation of freshwater as a production material results in operational cost savings for companies [43].

From a market perspective, consumers' awareness and attentiveness over the sustainability impact of wine products drive demand growth in the sector, especially in case water-related eco-certification is provided [39]. Indicatively, Rugani et al. [49] critically analysed LCA and carbon footprint-based studies on the wine-making industry and stressed that carbon footprint labelling in wines provides a market differentiation element that could influence consumers' purchasing behaviour. Moreover, wine produced with sustainable techniques has a greater export potential [42], while customers have a willingness to pay a premium price for environmentally friendly wine products [38]. Specifically, given that around 80% of wine sales occur in-store, clear communication of sustainably produced wine is deemed critical for increasing sales [50]. Therefore, as sustainable marketing has nowadays a dominant role in consumer purchasing behaviour and market sales, the communication of the water-related identity of wines could be an additional driver for approaching water-sensitive market segments. To some extent, consumers and policy-makers should also become aware of the virtual water flows, particularly blue water, embedded in international wine trade, considering that agri-food trade greatly influences water appropriation in a country [43]. To this end, national initiatives and businesses in the wine sector actively engage multispectral WF mitigation initiatives to reduce operating costs and communicate the water stewardship of their products to increase consumer value. In particular, better communication to consumers could be achieved via calculating a single-score indicator for labelling purposes [5,51].

In the case of the Italian wines, which are traded under the "controlled designation of origin" and "controlled and guaranteed designation of origin" labels, the adoption of water management policies can deliver a compelling marketing narrative linked to the territorial culture and history of each specific production wine site, promote the valorisation of local freshwater resources, and ultimately drive rural development [41]. Notably, on average, young wine consumers in Italy value water saving labelled wines and demonstrate a willingness to pay a premium price for such product offerings; determinats include wine consumption frequency, environmental-friendly attitude, label use, and label trust [5]. Therefore, as young consumers represent the most common market segment regarding wine consumption, policy-makers could act as a driving force for supporting the winemaking industry to adopt more environmental-friendly production methods (e.g., the Common Agricultural Policy of the EU [52]).

3.2. Barriers

Notwithstanding the pronounced need to apply water management policies and practises in the winemaking industry [46], dominant barriers hinder their adoption and maturity. The greatest peril in this process regards the poor alignment between water and agricultural policies [43], which is supported by the existence of different views among scientists regarding the system boundaries to apply water management policies (e.g., indirect WF from raw materials, transportation, end-of-life processes, etc.) [36,37,39]. Indicatively, Italian wines are associated with a lower WF exclusively due to the particular production specifications to guarantee designation of origin (i.e., irrigation and fertilisation are prohibited) [41]. To that effect, WF assessment methodologies generate different results even in the case that the same water management techniques are considered [35].

To a greater extent, established methodologies used for the ex ante evaluation of manufacturing operations' water impact (e.g., LCA) myopically leverage secondary data sources and neglect geographically related characteristics, such as diverse climatic conditions and applied production techniques [36,39]; thus, inconsistencies and discrepancies in the derived results are possible, but they have a detrimental effect on specific regions considering the localised supply of the embedded production inputs. At a more granular level, an evident gap in existing databases regarding indirect water consumption (e.g., green water) further raises evaluation challenges [36]. The lack of detailed data input further inhibits the contextualisation of the results in real-world operations [34], subsequently affecting the decision-making over the investments in related practises.

Furthermore, most water management-related studies focus on the academic merit of the applied methodological approaches in the pursuit of accuracy and precision of calculations. However, business stakeholders, who in principal operationalise water mitigation policies, cannot make inferences about the associated economic water productivities [42]. Moreover, the diverse alternative eco-labels for certifying the adoption of good practices for freshwater utilisation does not always provide businesses with an opportunity to differentiate from the competition [43].

4. Discussion

Food production and consumption are considered to have a rather detrimental impact on the environment [39,53]. Particularly, in the winemaking industry, sustainability is a key driver for competitiveness, market differentiation, and process innovation [4]. WF could become a meaningful indicator in sustainability initiatives for wines (as lower water consumption is also connected with a better wine quality and taste [41,42]); thus, the winemaking industry is exploring practises to improve the related environmental impact.

At a national level (e.g., Chile, Australia, New Zealand), frameworks to inform sustainability in the winemaking industry exist. Flores [3] reviewed the process-based winemaking frameworks in six countries and reported three categories where common water management practises are recognised: (i) soil management—protection of water resources from pollution, (ii) water management—registration of water use, selection of irrigation system, and control of water quality, and (iii) wastewater—monitoring of effluents and treatment of winery wastewater. Focussing on wastewater treatment, in countries such as France, Italy, and Spain, where the wine cellars are generally located close to urban areas, the use of advanced biological processes is crucial [54].

At a more granular level of operations and to cultivate grapes that result in high-quality wines, agricultural practices are required to enable the control of particular properties of grapes, such as the concentration of phenols, which determine the taste, color, and mouthfeel of wine. From a WF point of view, regulated deficit irrigation at the phenological stage is applied to increase the phenols' content [27]. Alongside the different quality of the cultivated grapes for winemaking, irrigation systems and practises are also dictated by the edaphoclimatic and related infrastructure conditions at each region. Indicatively, in the province of Mendoza, Argentina, 88% of the vineyards are irrigated through surface irrigation (i.e., gravity-based systems), whereas the remaining vine-growing area is irrigated through pressurised systems (e.g., dripping) to grow wine grapes of different qualities [27]. Moreover, the selection of good practices in water management for the winemaking industry could be influenced by the assessment methodology applied [39]. Overall, this selection depends on a range of decision-making constituents. More specifically, alternative irrigation options (e.g., drip, deficit) [55,56] and wastewater treatment techniques (i.e., aerobic, anaerobic, or their combination) [54] result in different levels of water savings. In addition, the implementation of digital technologies (e.g., sensors used during viticulture) [57,58] or holistic approaches for water-friendly activities across the whole wine supply chain [59] can further support an advanced and complete water stewardship plan. An indicative list of WF mitigation practices in the winemaking industry is tabulated in Table 3.

Table 3. Indicative good practices for water stewardship in the winemaking industry.

Good Practice	Description	Aims	References
Application of drip irrigation	Drip water slowly to the roots of plants, either above the soil surface (via micro-spray heads) or below the soil surface (via buried dripperline or drip tape)	■ Minimise evaporation ■ Improve water-use efficiency ■ Save nutrients	Borsato et al. [39]; Christ and Burritt [55]
Application of deficit irrigation	Irrigate during drought-sensitive growth stages of a crop and leverage available rainfall in other crop cycles	■ Improve water-use efficiency ■ Control vegetative vigour and production quality of grapevines	Civit et al. [27]; Chaves et al. [56]
Application of partial root-zone drying techniques	Irrigate about half of the root system of a crop and leave the other half to dry	■ Improve water-use efficiency	Christ and Burritt [55]
Digitalisation of irrigation system	Monitor water requirements and use via sensors	■ Monitor evapotranspiration, precipitation, soil/leaf water content ■ Improve water-use efficiency	Aiello et al. [57]; Tsolakis et al. [58]
Treatment of winery wastewater	Use aerobic or anaerobic techniques to biodiograde organic compounds, remove nitrogen, phosphorous, heavy metals, and pathogens	■ Monitor effluents ■ Purify industrial water ■ Promote water reuse (e.g., for irrigation purposes)	Bolzonella et al. [54]
Training of employees and application of water-friendly processes along the production line	Apply process changes and reuse water during wine processing	■ Reduce industrial water consumption in cleaning, disinfecting, cooling, and heating operations ■ Monitor/mitigate effluents	Oliver et al. [59]

5. Conclusions

The Italian wine industry constitutes a global leader in terms of production quantity and quality. Within the country, the economic scale of winemaking renders the industry as the key production component of the Italian agrifood sector related to respective freshwater appropriation. To that end, the investigation of water use needs during viticulture and vinification is imperative to support water stewardship within the country's wine sector. According to the scientific literature, Italian wine dominates the research efforts with respect to WF assessment, thus validating the important role of economic and environmental sustainability in the national winemaking industry. To a greater extent, given the consumers' awareness and positive purchasing behaviour towards water-efficient wines, the sustainability profile of the Italian wines could further link to the designation of origin and receive international market's appreciation.

As research regarding water management in the winemaking industry is limited, this paper acts as an initial mapping that captures the major drivers and barriers of water stewardship considering the unique geographic and socio-economic characteristics of the Italian landscape. Our research findings indicate that the environmental, socio-economic, and market drivers outperform the existing, mainly technical and methodological, barriers. This proliferation of drivers and the identification of good practises in the industry motivate the development of operationalisable water stewardship frameworks for the Italian wine sector.

5.1. Practical Implications

Based on the literature evidence, it is critical that the practitioners of the Italian wine industry should act towards the direction of water management to support the preservation of freshwater resources and enhance the economic water productivity of their products. This research validates that water management policies and practices should be systematically applied in the Italian winemaking industry, from an end-to-end supply chain perspective, to enhance the sustainable brand image of the national production and foster its trade potential and market appreciation.

At an operational level, this research suggests that vine growers and winemaking practitioners should focus on water management interventions at three levels, including (i) soil management, (ii) freshwater management, and (iii) wastewater treatment [3]. In particular, we propose that the type of irrigation systems and practises to be applied should also consider the edaphoclimatic and related infrastructure conditions at each winemaking region to increase the efficiency of water resources appropriation [27]. Regarding wastewater treatment, aerobic processes (e.g., membrane bioreactors [60]) could offer an efficient and easy-to-use solution compared to anaerobic ones that constitute a more economic option [54].

To a greater extent, we propose the digitalisaiton of the wine supply chain, particularly at the farming echelon. In this regard, sensor technologies are reported to support the decision-making process concerning the water stewardship of agrifood commodities [57], which could also be pertinent to the case of the wine supply chain. The use of sensors in grapes' farming for monitor freshwater use and other related parameters (e.g., soil moisture) is an indicative digital intervention that relates to the quality of the wine production. The introduction of advanced technologies can assist in (i) mitigating methodological errors in water-use estimations, (ii) gathering field-level data, (iii) calculating water consumption in viticulture in a more accurate way, (iv) extrapolating information with regard to the WF of their supply chain, and (v) devising sound marketing strategies to engage with consumers [61].

5.2. Future Research

With regard to future research directions, both researchers and practitioners of the winemaking production field may focus on developing analytical and computer-based tools for multi-objective analysis and simulation to solve freshwater resource planning and operational problems. To that end, Aivazidou et al. [7] suggest a framework that guides the ex ante evaluation of applied water management policies through developing a pertinent simulation model that enbables the assessment of water utilisation on the supply chain financial performance. Notably, the modelling effort captures the concept of consumers' environmental sensitivity with regard to blue WF efficiency as a supply chain profitability factor. In addition, the economic evaluation of the green WF to the overall production value of wine, as inspired by the study of Grammatikopoulou et al. [9] for the case of cereals, is highly recommended considering that the majority of wine grapes across the Mediterranean are grown under rainfed conditions. To a greater extent, the economic water productivity could be combined with water scarcity indicators to account for the inter-annual variability of the green and blue WFs at a regional level to improve the management of grapes' production, supply, and wine trade in the winemaking sector [62].

To wrap up, based on the environmental, economic, and technical managerial insights obtained by the analysis of the major drivers, barriers, and good practises, it is crucial that industry stakeholders should systematically focus towards developing a concrete water management scheme in the Italian winemaking sector for fostering its sustainability.

Author Contributions: Both authors have read and agreed to the published version of the manuscript. Conceptualisation, E.A.; writing—original draft preparation, review, and editing, E.A. and N.T.; supervision, E.A. All authors have read and agreed to the published version of the manuscript.

Funding: This research received no external funding.

Conflicts of Interest: The authors declare no conflict of interest.

References

1. State of the Vitiviniculture World Market—April 2018. Available online: http://www.oiv.int/public/medias/5958/oiv-state-of-the-vitiviniculture-world-market-april-2018.pdf (accessed on 15 July 2019).
2. Christ, K.L.; Burritt, R.L. Water management accounting: A framework for corporate practice. *J. Clean. Prod.* **2017**, *152*, 379–386. [CrossRef]
3. Flores, S.S. What is sustainability in the wine world? A cross-country analysis of wine sustainability frameworks. *J. Clean. Prod.* **2018**, *172*, 2301–2312. [CrossRef]
4. Fiore, M.; Silvestri, R.; Contò, F.; Pellegrini, G. Understanding the relationship between green approach and marketing innovations tools in the wine sector. *J. Clean. Prod.* **2017**, *142*, 4085–4091. [CrossRef]
5. Pomarici, E.; Asioli, D.; Vecchio, R.; Næs, T. Young consumers' preferences for water-saving wines: An experimental study. *Wine Econ. Pol.* **2018**, *7*, 65–76. [CrossRef]
6. Schäufele, I.; Hamm, U. Consumers' perceptions, preferences and willingness-to-pay for wine with sustainability characteristics: A review. *J. Clean. Prod.* **2017**, *147*, 379–394. [CrossRef]
7. Aivazidou, E.; Tsolakis, N.; Vlachos, D.; Iakovou, E. A water footprint management framework for supply chains under green market behaviour. *J. Clean. Prod.* **2018**, *197*, 592–606. [CrossRef]
8. Hoekstra, A.Y.; Chapagain, A.K.; Aldaya, M.M.; Mekonnen, M.M. *The Water Footprint Assessment Manual*, 1st ed.; Earthscan: London, UK, 2011; ISBN 978-1-84971-279-8.
9. Grammatikopoulou, I.; Sylla, M.; Zoumides, C. Economic evaluation of green water in cereal crop production: A production function approach. *Water Resour. Econ.* **2019**, 10048. [CrossRef]
10. Hoekstra, A.Y.; Mekonnen, M.M. The water footprint of humanity. *Proc. Natl. Acad. Sci. USA* **2012**, *109*, 3232–3237. [CrossRef]
11. Mekonnen, M.M.; Hoekstra, A.Y. The green, blue and grey water footprint of crops and derived crop products. *Hydrol. Earth Syst. Sci.* **2011**, *15*, 1577–1600. [CrossRef]
12. Lamastra, L. The Virtual Water in A Bottle of Wine. In *The Water We Eat: Combining Virtual Water and Water Footprints*, 1st ed.; Antonelli, M., Greco, F., Eds.; Springer International Publishing: Cham, Switzerland, 2015; pp. 209–228. ISBN 978-3-319-16392-5.
13. Gassert, F.; Reig, P.; Luo, T.; Maddocks, A. *Aqueduct Country and River Basin Rankings: A Weighted Aggregation of Spatially Distinct Hydrological Indicators*, 1st ed.; Working Paper; World Resources Institute: Washington, DC, USA, 2013.
14. Pfister, S.; Koehler, A.; Hellweg, S. Assessing the Environmental Impacts of Freshwater Consumption in LCA. *Environ. Sci. Technol.* **2009**, *43*, 4098–4104. [CrossRef]

15. La Produzione di vino in Italia nel 2017—Dati Finali ISTAT. Available online: http://www.inumeridelvino.it/2018/05/la-produzione-di-vino-in-italia-nel-2017-dati-finali-istat.html (accessed on 15 July 2019).

16. L'agricoltura Italiana Conta 2017. Available online: http://www.condifesafoggia.it/wp-content/uploads/2018/01/Itaconta-2017.pdf (accessed on 15 July 2019).

17. Il Valore Della Produzione di vino nel Mondo—Stima INDV 2017. Available online: http://www.inumeridelvino.it/tag/valore-mercato-del-vino (accessed on 15 July 2019).

18. Aivazidou, E.; Tsolakis, N.; Iakovou, E.; Vlachos, D. The emerging role of water footprint in supply chain management: A critical literature synthesis and a hierarchical decision-making framework. *J. Clean. Prod.* **2016**, *137*, 1018–1037. [CrossRef]

19. Christ, K.L. Water management accounting and the wine supply chain: Empirical evidence from Australia. *Br. Account. Rev.* **2014**, *46*, 379–396. [CrossRef]

20. Hoekstra, A.Y. *Water Neutral: Reducing and Offsetting the Impact of Water Footprints*, 1st ed.; Value of Water Research Report Series No. 28; UNESCO-IHE Institute for Water Education: Delft, The Netherlands, 2008; ISSN 1812-2108.

21. ISO. *ISO 14046:2014—Environmental Management—Water Footprint—Principles, Requirements and Guidelines*; International Organization for Standardization: Geneva, Switzerland, 2014.

22. Chenoweth, J.; Hdjikakou, M.; Zoumides, C. Quantifying the human impact on water resources: A critical review of the water footprint concept. *Hydrol. Earth Syst. Sci.* **2014**, *18*. [CrossRef]

23. Herath, I.; Green, S.; Singh, R.; Horne, D.; van der Zijpp, S.; Clothier, B. Water footprinting of agricultural products: A hydrological assessment for the water footprint of New Zealand's wines. *J. Clean. Prod.* **2013**, *41*, 232–243. [CrossRef]

24. Herath, I.; Green, S.; Horne, D.; Singh, R.; McLaren, S.; Clothier, B. Water footprinting of agricultural products: Evaluation of different protocols using a case study of New Zealand wine. *J. Clean. Prod.* **2013**, *44*, 159–167. [CrossRef]

25. Steenwerth, K.L.; Strong, E.B.; Greenhut, R.F.; Williams, L.; Kendall, A. Life cycle greenhouse gas, energy, and water assessment of wine grape production in California. *Int. J. Life Cycle Assess.* **2015**, *20*, 1243–1253. [CrossRef]

26. Johnson, M.-B.; Mehvar, M. An assessment of the grey water footprint of winery wastewater in the Niagara Region of Ontario, Canada. *J. Clean. Prod.* **2019**, *214*, 623–632. [CrossRef]

27. Civit, B.; Piastrellini, R.; Curadelli, S.; Arena, A.-P. The water consumed in the production of grapes for vinification (Vitis vinifera). Mapping the blue and green water footprint. *Ecol. Indic.* **2018**, *85*, 236–243. [CrossRef]

28. Ene, S.-A.; Teodosiu, C.; Robu, B.; Volf, I. Water footprint assessment in the winemaking industry: A case study for a Romanian medium size production plant. *J. Clean. Prod.* **2013**, *43*, 122–135. [CrossRef]

29. Bujdosó, B.; Waltner, I. Water Footprint Assessment of a Winery and its Vineyard. *Hung. Agr. Res.* **2017**, *1*, 10–13.

30. Quinteiro, P.; Dias, A.-C.; Pina, L.; Neto, B.; Ridoutt, B.G.; Arroja, L. Addressing the freshwater use of a Portuguese wine ('vinho verde') using different LCA methods. *J. Clean. Prod.* **2014**, *68*, 46–55. [CrossRef]

31. Martins, A.A.; Araújo, A.R.; Morgado, A.; Graça, A.; Caetano, N.S.; Mata, T.M. Sustainability Evaluation of a Portuguese "Terroir" Wine. *Chem. Eng. Trans.* **2017**, *57*, 1945–1950. [CrossRef]

32. Villanueva-Rey, P.; Quinteiro, P.; Vánquez-Rowe, I.; Rafael, S.; Arroja, L.; Moreira, M.T.; Feijoo, G.; Dias, A.C. Assessing water footprint in a wine appellation: A case study for Ribeiro in Galicia, Spain. *J. Clean. Prod.* **2018**, *172*, 2097–2107. [CrossRef]

33. Meneses, M.; Torres, C.M.; Castells, F. Sensitivity analysis in a life cycle assessment of an aged red wine production from Catalonia, Spain. *Sci. Total Environ.* **2016**, *562*, 571–579. [CrossRef] [PubMed]

34. Lamastra, L.; Suciu, N.-A.; Novelli, E.; Trevisan, M. A new approach to assessing the water footprint of wine: An Italian case study. *Sci. Total Environ.* **2014**, *490*, 748–756. [CrossRef]

35. Bonamente, E.; Scrucca, F.; Asdrubali, F.; Cotana, F.; Presciutti, A. The Water Footprint of the Wine Industry: Implementation of an Assessment Methodology and Application to a Case Study. *Sustainability* **2015**, *7*, 12190–12208. [CrossRef]

36. Bonamente, E.; Scrucca, F.; Rinaldi, S.; Merico, M.-C.; Asdrubali, F.; Lamastra, L. Environmental impact of an Italian wine bottle: Carbon and water footprint assessment. *Sci. Total Environ.* **2016**, *560–561*, 274–283. [CrossRef]

37. Rinaldi, S.; Bonamente, E.; Scrucca, F.; Merico, M.-C.; Asdrubali, F.; Cotana, F. Water and Carbon Footprint of Wine: Methodology Review and Application to a Case Study. *Sustainability* **2016**, *8*, 621. [CrossRef]

38. Bartocci, P.; Fantozzi, P.; Fantozzi, F. Environmental impact of Sagrantino and Grechetto grapes cultivation for wine and vinegar production in central Italy. *J. Clean. Prod.* **2017**, *140*, 569–580. [CrossRef]

39. Borsato, E.; Giubilato, E.; Zabeo, A.; Lamastra, L.; Criscione, P.; Tarolli, P.; Marinello, F.; Pizzol, L. Comparison of Water-focused Life Cycle Assessment and Water Footprint Assessment: The case of an Italian wine. *Sci. Total Environ.* **2019**, *666*, 1220–1231. [CrossRef]

40. Boulay, A.-M.; Bare, J.; Benini, L.; Berger, M.; Lathuillière, M.J.; Manardo, A.; Margni, M.; Motoshita, M.; Núñez, M.; Pastor, A.-V.; et al. The WULCA consensus characterization model for water scarcity footprints: Assessing impacts of water consumption based on available water remaining (AWARE). *Int. J. Life Cycle Assess.* **2018**, *23*, 368–378. [CrossRef]

41. Miglietta, P.-P.; De Leo, F.; Massari, S. Water footprint assessment of some Italian wines: A territorial perspective. *Int. J. Environ. Policy Decis. Making* **2015**, *1*, 320–331. [CrossRef]

42. Miglietta, P.-P.; Morrone, D.; Lamastra, L. Water footprint and economic water productivity of Italian wines with appellation of origin: Managing sustainability through an integrated approach. *Sci. Total Environ.* **2018**, *633*, 1280–1286. [CrossRef] [PubMed]

43. Miglietta, P.-P.; Morrone, D. Managing Water Sustainability: Virtual Water Flows and Economic Water Productivity Assessment of the Wine Trade between Italy and the Balkans. *Sustainability* **2018**, *10*, 543. [CrossRef]

44. EcoWater Report—Comparing Water Footprint Methods: The Importance of a Life Cycle Approach in Assessing Water Footprint. Available online: https://www.ivl.se/download/18.343dc99d14e8bb0f58b7739/1445517875373/C98.pdf (accessed on 13 December 2019).

45. Hoekstra, A.Y.; Chapagain, A.K.; van Oel, P.R. Progress in Water Footprint Assessment: Towards Collective Action in Water Governance. *Water* **2019**, *11*, 1070. [CrossRef]

46. Fayolle, E.; Follain, S.; Marchal, P.; Chéry, P.; Colin, F. Identification of environmental factors controlling wine quality: A case study in Saint-Emilion Grand Cru appellation, France. *Sci. Total Environ.* **2019**, *69*, 133718. [CrossRef]

47. Vaudour, E. The quality of grapes and wine in relation to geography: Notions of terroir at various scales. *J. Wine Res.* **2002**, *13*, 117–141. [CrossRef]

48. Special Edition: Progress towards the Sustainable Development Goals. Report of the Secretary-General. Available online: https://undocs.org/E/2019/68 (accessed on 11 October 2019).

49. Rugani, B.; Vázquez-Rowe, I.; Benedetto, G.; Benetto, E. A comprehensive review of carbon footprint analysis as an extended environmental indicator in the wine sector. *J. Clean. Prod.* **2013**, *54*, 61–77. [CrossRef]

50. Lockshin, L.; Corsi, A.M. Consumer behaviour for wine 2.0: A review since 2003 and future directions. *Wine Econ. Policy* **2012**, *1*, 2–23. [CrossRef]

51. Ridoutt, B.G.; Pfister, S. A new water footprint calculation method integrating consumptive and degradative water use into a single stand-alone weighted indicator. *Int. J. Life Cycle Assess.* **2013**, *18*, 204–207. [CrossRef]

52. The Common Agricultural Policy at a Glance. Available online: https://ec.europa.eu/info/food-farming-fisheries/key-policies/common-agricultural-policy/cap-glance_en (accessed on 13 December 2019).

53. Notarnicola, B.; Tassielli, G.; Renzulli, P.A.; Castellani, V.; Sala, S. Environmental impacts of food consumption in Europe. *J. Clean. Prod.* **2017**, *140*, 753–765. [CrossRef]

54. Bolzonella, D.; Papa, M.; Da Ros, C.; Muthukumar, L.-A.; Rosso, D. Winery wastewater treatment: A critical overview of advanced biological processes. *Crit. Rev. Biotechnol.* **2019**, *39*, 489–507. [CrossRef]

55. Christ, K.L.; Burritt, R.L. Critical environmental concerns in wine production: An integrative review. *J. Clean. Prod.* **2013**, *53*, 232–242. [CrossRef]

56. Chaves, M.M.; Santos, T.P.; Souza, C.R.; Ortuño, M.F.; Rodrigues, M.L.; Lopes, C.M.; Maroco, J.P.; Pereira, J.S. Deficit irrigation in grapevine improves water-use efficiency while controlling vigour and production quality. *Ann. Appl. Biol.* **2007**, *150*, 237–252. [CrossRef]

57. Aiello, G.; Cannizzaro, L.; La Scalia, G.; Muriana, C. An expert system for vineyard management based upon ubiquitous network technologies. *Int. J. Services Oper. Inform.* **2011**, *6*, 230–247. [CrossRef]

58. Tsolakis, N.; Aivazidou, E.; Srai, J.S. Sensor applications in agrifood systems: Current trends and opportunities for water stewardship. *Climate* **2019**, *7*, 44. [CrossRef]

59. Oliver, P.; Rodríguez, R.; Udaquiola, S. Water use optimization in batch process industries. Part 1: Design of the water network. *J. Clean. Prod.* **2008**, *16*, 1275–1286. [CrossRef]

60. Bolzonella, D.; Fatone, F.; Pavan, P.; Cecchi, F. Application of a membrane bioreactor for winery wastewater treatment. *Water Sci. Technol.* **2010**, *62*, 2754–2759. [CrossRef]

61. Van der Laan, M.; Jarmain, C.; Bastidas-Obando, E.; Annandale, J.G.; Fessehazion, M.; Haarhoff, D. Are water footprints accurate enough to be useful? A case study for maize (*Zea mays* L.). *Agr. Water Manage.* **2019**, *213*, 512–520. [CrossRef]

62. Zoumides, C.; Bruggeman, A.; Hadjikakou, M.; Zachariadis, T. Policy-relevant indicators for semi-arid nations: The water footprint of crop production and supply utilization of Cyprus. *Ecol. Indic.* **2014**, *43*, 205–214. [CrossRef]

Article

Socio-Hydrology: A New Understanding to Unite or a New Science to Divide?

Kaveh Madani [1,2,*] **and Majid Shafiee-Jood** [3]

[1] Department of Political Science and the MacMillan Center for International and Area Studies, Yale University, New Haven, CT 06511, USA
[2] Department of Physical Geography, Stockholm University, SE-106 91 Stockholm, Sweden
[3] Department of Civil and Environmental Engineering, University of Illinois at Urbana-Champaign, Urbana, IL 61801, USA; shafiee2@illinois.edu
* Correspondence: kaveh.madani@yale.edu; Tel.: +1-(203)-436-8913

Received: 3 June 2020; Accepted: 3 July 2020; Published: 8 July 2020

Abstract: The socio-hydrology community has been very successful in promoting the need for taking the human factor into account in the mainstream hydrology literature since 2012. However, the interest in studying and modeling human-water systems is not new and pre-existed the post-2012 socio-hydrology. So, it is critical to ask what socio-hydrology has been able to offer that would have been unachievable using the existing methods, tools, and analysis frameworks. Thus far, the socio-hydrology studies show a strong overlap with what has already been in the literature, especially in the water resources systems and coupled human and natural systems (CHANS) areas. Nevertheless, the work in these areas has been generally dismissed by the socio-hydrology literature. This paper overviews some of the general concerns about originality, practicality, and contributions of socio-hydrology. It is argued that while in theory, a common sense about the need for considering humans as an integral component of water resources systems models can strengthen our coupled human-water systems research, the current approaches and trends in socio-hydrology can make this interest area less inclusive and interdisciplinary.

Keywords: socio-hydrology; hydro-sociology; human-water systems; human-nature systems; water resources systems; social-ecological systems; CHANS; SES; socio-hydrologic modeling; integrated water resources management; IWRM; water resources management; hydrology

---

1. Introduction

The increasing interest in more explicit representation of human behavior and decisions in hydrologic models is undeniably a positive change that must be welcomed and promoted. The water resources community must appreciate the courage of those hydrologists who have been questioning the reliability and practical relevance of our sophisticated mathematical models in which the human dimension of water resource systems is overlooked. However, the interest in coupled human-water systems is certainly not new. For decades, people in natural/social science and engineering have been exploring human-water systems.

The need to push the envelope and expand the boundaries of our models has resulted in the emergence of interdisciplinary methods, interest areas, and even fields of study. The recent decade might be a turning point in the history of human-water systems studies as we have observed a tremendous increase in the interest of researchers and funding agencies in incorporating complexity and the human dimension into our water resources models.

In pursuit of their interest in better understanding human-water systems, Murugesu Sivapalan, Hubert Savenije, and Günter Blöschl "welcomed" their peers in "traditional hydrology" to "a new science" called socio-hydrology in an invited commentary in 2012 [1]. Blaming "traditional hydrology" for ignoring the human factor for too long, the authors proposed socio-hydrology as "a new science that is aimed at understanding the dynamics and co-evolution of coupled human-water systems". Demetris Koutsoyiannis [2], a reviewer of this invited commentary, who published his review comments online, criticized the authors for discounting the attention to the human factor in classical hydrology, downgrading the significance of Integrated Water Resources Management (IWRM), and dismissing the human-water systems analysis studies. Koutsoyiannis was not convinced that proposing a "new science" was necessary and found the authors' claims "immodest". The supposedly novel idea of socio-hydrology was also harshly criticized by Sivakumar [3], who believed that socio-hydrology was "not a new science, but a recycled and re-worded hydro-sociology" that had been originally proposed by Falkenmark [4] to study human-water interactions. Since 2012, similar critiques have been also expressed in different publications (e.g., [5]) and at informal and formal water gatherings (e.g., the annual meetings of World Environmental and Water Resources Congress) about the approach, novelty, claims, and contributions of socio-hydrology.

Despite the cold welcome, proposers of socio-hydrology have been certainly successful in creating a new space of interest and engaging an international group of researchers. To date, the invited commentary of Sivapalan et al. [1] has been cited more than 400 times (Web of Science (WoS); more than 600 times on Google Scholar). Although some of the significant critiques of socio-hydrology have not been directly addressed by its leaders and followers, so far, about 180 socio-hydrology papers have been published that have been cited nearly 4000 times in total according to the WoS (Figure 1). A considerable number of early career researchers have joined the community of researchers that identify themselves as socio-hydrologists. The socio-hydrologists have a working group with the Penta Rhei initiative of the International Association for Hydrological Science (IAHS), run summer schools and training workshops, publish special issues in different journals, and have been successful in receiving funding from major research agencies in Europe and North America for doing socio-hydrology research. The proposers of socio-hydrology have also received major international awards and recognitions for their "new science".

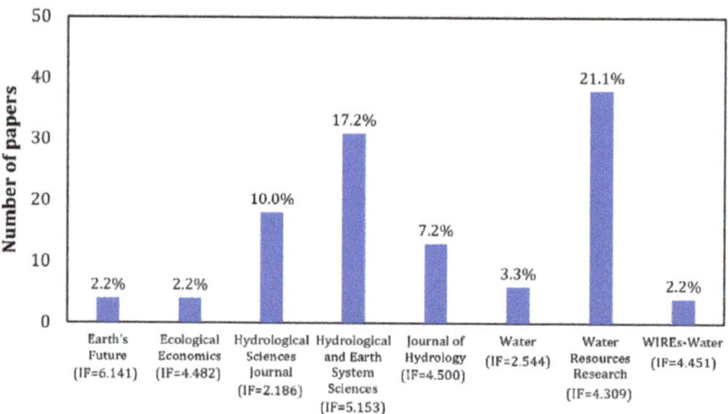

Figure 1. Number and percentage of socio-hydrology papers published in different journals. IF values show the impact factors of the journals in 2019 according to the Clarivate Analytics' Journal Citation Report. Only journals that have published at least four papers are shown. In our literature search, we first searched for the keywords "socio-hydrology", "socio-hydrology", "socio hydrology", "socio-hydrologic", "socio hydrologic", "socio-hydrologic", "socio-hydrological", "socio-hydrological", and "socio hydrological" in the WoS database, resulting in 278 peer-reviewed journal papers. After reviewing these papers, we discarded those which did not focus on topics that are directly related to socio-hydrology. Our final database includes 180 papers, cited 3756 times overall based on WoS citation report, with 593 contributing authors.

As researchers who are interested in understanding and managing coupled, dynamic, complex human-water systems problems, we celebrate this growing interest in studying human-water systems problems. As colleagues, who have had the chance of interacting with socio-hydrologists in various meetings and discussions, we also acknowledge that the socio-hydrology community leaders have been effective in creating a useful momentum and an invaluable opportunity to expand the traditional research and education horizon in hydrology. Yet, as the outsiders of the group, who have been enthusiastically following their products, we struggle to comprehend what this "new science" is, what it is trying to do that is different from previous works in this space, how it is going to do it, and what it has accomplished to date.

Given the norms and cultures of our field, we have been hesitant to write this paper in anticipation of any misinterpretation or unintentional offense. Nevertheless, we feel that the ongoing development trends in socio-hydrology as well as dismissing the outstanding concerns and critiques of human-water systems experts can create a counterproductive divide between the relevant research communities, wasting valuable economic resources and human talents. Thus, we feel ethically and scientifically responsible to ask some questions about socio-hydrology at this stage of the coupled human-water systems field's history when many young students and researchers are joining the socio-hydrology club in their pursuit of new methods, insights, and scientific inquiries.

Our intention is not to write a critique or start a debate. Instead, in this paper, we simply ask some questions that over the years have kept us wondering why socio-hydrology must be regarded as a "new science" or even a new field. As outsiders, some of our questions and comments are perhaps rooted in our ignorance and poor understanding of socio-hydrology. Nevertheless, we think that answering these questions in future socio hydrology publications can address some of the main concerns of the many researchers who have been working or going to work on coupled human-water systems problems. We believe that the increasing interest in the human dimension of water resources problems has the potential to unite and strengthen our large, but currently fragmented communities, once we better understand each other, improve our communication skills, and sharpen our messages. So, we hope that our colleagues find these sincere concerns and questions constructive in shaping the future of socio-hydrology and training new generations of socio-hydrologists.

2. Socio-Hydrology: Originality, Practicality, and Contributions

2.1. Is Socio-Hydrology a New Science?

Socio-hydrology was introduced by its proposers as a "new science" and "a discovery-based fundamental science, whose practice is informed through observing, understanding, and predicting socio-hydrologic phenomena" [1]. The authors insisted that their proposed "interdisciplinary" science of people and water "must" strive to be quantitative "with ambition to make predictions of water cycle dynamics". They compared socio-hydrology with eco-hydrology, arguing that eco-hydrology is similar to socio-hydrology in coupling water with another system, while the former studies "the co-evolution and self-organization of vegetation in the landscape in relation to water availability" and the latter would focus on "the co-evolution and self-organization of people in the landscape with respect to water availability". In this comparison, however, they referred to eco-hydrology as a "field" and to socio-hydrology as a "science".

Before getting into what socio-hydrology has achieved, let us focus on the ambitious claim of creating a "new science". The socio-hydrology proposers' argument that many hydrologists overlook the interdependence and interrelated dynamics of water and humans in their models is valid and reaffirms what many experts in the water resources field have been talking about for decades. But does the lack of attention to the human dimension by a group of hydrologists justify the need for creating a new science? If many hydrologists are missing a significant component in their models, do we need to revisit hydrology or create a new science?

Like any other field, the water resources field is full of limitations, and over time, we detect new restrictions and develop new interests, expectations, and questions. Accordingly, we revise our approaches, methods, models, and even buzzwords. The review of hyphenated hydrology by McCurley and Jawitz [6] is a good proof that our new questions can even form sub-disciplines and evolve our understanding and expectations from our disciplines. As a result, hydrology in the 21st century is very different from hydrology in the 12th century. If recognizing each new need justifies creating a new science, hydrologists must have created numerous branches of "science" by now.

The precedent of inventing a new science based on the recognition of existing limitations can be counterproductive and weaken the hydrology community by generating tendency and competition for creating new spaces (for example, the suggestions to create socio-meteorology, socio-climatology [7] and socio-hydrogeology [8] have been motivated by the suggestion to create the "new science" of socio-hydrology). Instead, our discipline and community must be open and prepared to evolve in response to our new needs. Even when the new need is felt, the process of proposing a new interest area, developing new research questions, and identifying a research gap must be scientific. How can we suggest a new science without scientifically proving that what we offer is different from what is available?

When proposing socio-hydrology, Sivapalan et al. [1] did not make any reference to: (1) hydro-sociology that had been around since the 1970's [3,6]; (2) coupled human-water systems studies by the system dynamics and water resources systems communities in the last five decades; (3) the popular coupled human-environment systems (also known as coupled human and natural systems (CHANS)) and socio-ecological system (SES) research areas; (4) the major human-water studies by social scientists and economists; as well as (5) other studies that had specifically referred to "socio-hydrology" earlier. For example, Smakhtin et al. [9] had suggested that considering riparian communities as an integral part of the riverine ecosystem could lead to new fields of work such as "socio-hydrology" or "socio-ecology". Kock [10] had developed "agent-based models of socio-hydrological systems for exploring the institutional dynamics of water resources conflict". He had based his "socio-hydrologic" approach on Mohorjy's [11] idea of integrating the hydrologic and socio-economic aspects of water resources planning and defined socio-hydrologic systems as systems in which "social, economic and hydrologic subsystems are causally linked".

Overlooking the existing literature might have been the result of the authors' unfamiliarity with the human-water systems space. Yet, the trend has continued in later publications. Although the later publications of the socio-hydrology community have cited a very limited number of water-human systems studies, the past and ongoing human-water systems research has been largely dismissed by the socio-hydrology literature. Apparently, the peer review system has failed to provide constructive feedback to socio-hydrologists, refer them to the similar work done by others, ask for better explanation of why and how socio-hydrology is different, and remind the socio-hydrologists that what's new to them, might not be new in science, and if they insist that there is a need for a new science, this need must be justified through scientific evidence and gap analysis that is based on a comprehensive review of the existing literature.

Socio-hydrology can be a new interest area or sub-field or even a new field once it clarifies what it is searching for, can clearly communicate how it is different from the existing fields, and through a number of solid analyses, proves the practicality of its goals and the value of the insights it can offer. However, creating a "new science" based on simple inquiries in an invited commentary seems extremely ambitious and perhaps very unscientific!

2.2. What Is New about Socio-Hydrology?

A review of 180 socio-hydrology papers suggests that these papers have been mainly written by three networks (Figure 2), led by three hydrologists/civil (water/environmental) engineers, i.e., Sivapalan, Blöschl, and Di Baldassarre. While the proposers of socio-hydrology had insisted that socio-hydrology must be quantitative [1], a significant portion of socio-hydrology papers is dedicated to opinion papers and commentaries that provide verbal and often ambitious discussions on why socio-hydrology is essential, what it is going offer, or what it must do without solid proof. Other non-quantitative papers include reviews of the socio-hydrology literature and their accomplishments. Most of these papers have been written by hydrologists with an occasional leadership or co-authorship of social scientists. Quantitative socio-hydrology papers have grown in number over the years and include system dynamics modeling and analyzing time-series and survey data.

As a quantitative science, socio-hydrology intended to make advancements in three areas [1]: (1) historical socio-hydrology: learning from the past; (2) comparative socio-hydrology: compare and contrast different human-water systems; and (3) process socio-hydrology: understanding existing human-water systems to "predict possible trajectories in the future". So far, most of the socio-hydrology publications with quantitative elements that deal with real cases belong to the first two areas. In these studies, historical correlations and/or survey data are used to derive a hypothesis that can explain the past dynamics of the studied human-water systems. These studies are valuable but the different insights they offer by being branded as socio-hydrology studies remain unclear. In the reviews of their past work, socio-hydrologists (e.g., [12,13]) refer to their developed mathematical models for human-water systems as system dynamics models. Yet, their modeling papers do not make a proper connection to the water resources system dynamics literature. Higher mathematical sophistication does not necessarily provide additional insights. Nonetheless, given the interest of socio-hydrologists in quantitative science, one must note that most of the system dynamics models developed by the non-socio-hydrologists are mathematically much more sophisticated than the socio-hydrology system dynamics models. Assuming that the past studies have been ignored simply because of unfamiliarity with the human-water resources system dynamics literature, one still can ask: what has socio-hydrology added to these modeling studies that system dynamics could have not offered?

The same question applies to the papers written by social scientists (e.g., [14–16]) that do not have a strong hydrologic component but have some quantitative elements. In these studies, water has simply been the study domain for the social scientists. So, it is not clear what makes these socio-hydrology papers any different from the many studies of humans in water systems in the past. Even for developing her conceptual human-water system model, Leong [14] borrows the conceptual system dynamics model of Newell and Wasson [17], which preceded the 2012 socio-hydrology paper. What the science of socio-hydrology has added to these social science studies or what insight these studies have offered that could have not been gained by the existing methods that are not branded as socio-hydrologic tools remains unclear.

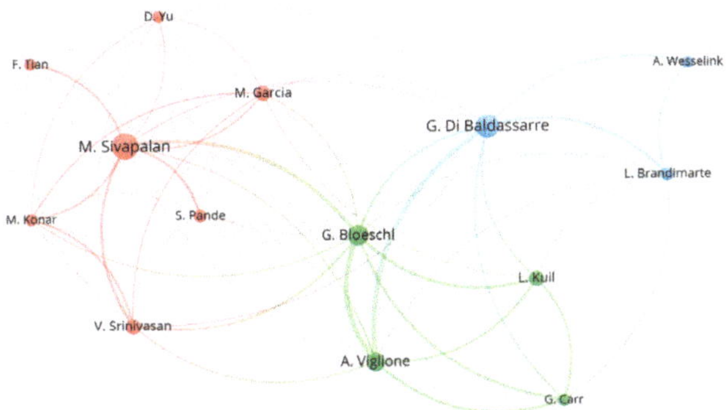

Figure 2. Co-authorship network based on the selected 180 papers. Only authors with at least 6 socio-hydrology papers are shown. Size of nodes corresponds to the number of papers written by an author. Thickness of a link between two authors corresponds to the number of papers they have co-authored. The network was plotted with VOSviewer [18,19] using bibliographic records extracted from WoS.

The idea of studying the evolution of complex, coupled human-water systems is not new. Over the last two decades, many scholars have used system dynamics as a framework to study and explain the evolution of coupled human-water systems, causal relationships, non-linear feedbacks, paradoxical behavior, counter-intuitive dynamics, and the unintended consequences of structural or non-structural interventions (e.g., see the review by Mirchi et al. [20]). The idea of internalizing the human component (not treating it as an exogenous element), suggested by Sivapalan et al. [1] is not new either. Before their paper, many people had developed coupled human-water system dynamics models for explaining the underlying mechanisms of these systems and making future projections to guide policy (e.g., [21–38]).

The idea of predicting the future through socio-hydrology as was originally suggested by Sivapalan et al. [1] seems to be in contrast to the socio-hydrologists' claim that socio-hydrology is not interested in scenario-based approaches as well as their expressed interest in advising policy. Complex human-water systems have some essential characteristics, including uncertainty, bounded rationality, indeterminate causality, limited predictability, evolutionary change, and non-stationarity [39]. So, predicting or forecasting the long-term evolution of coupled human-water systems is impossible. Nevertheless, projecting the future of systems involving human and nature is a common approach (note the difference between *prediction* and *projection* [40]). Any projection of the future evolution of human-nature systems, involves scenarios or if-then statements. Even when different scenarios are not evaluated, the inherent assumption is that business-as-usual is the underlying scenario. In addition, the essential characteristics of these systems are what cause surprises, unintended consequences, and "black swan events" [41]. Expanding the boundaries of our models can help us better capture feedbacks and detect undesirable patterns, especially when studying the past. Nonetheless, the future evolution of human-water systems will involve many surprises, unintended consequences, and black swans that cannot be detected and avoided with the help of socio-hydrology.

In response to the critiques on the capacity of socio-hydrology to predict social behavior and politics (e.g., [42]), socio-hydrologists claimed that "predictions in socio-hydrology do not aim at predicting time series" [43]. Instead, they aim at predicting phenomena emerging in human-water systems "in a quantitative and generalizable way". This could be in contrast with the earlier emphasis on branding socio-hydrology as a "quantitative" science.

In Srinivasan et al. [44], the socio-hydrologists criticize the existing "prediction" paradigms and call for a "fundamental change" in our understanding of "prediction". In this opinion paper, the authors "argued" that socio-hydrology can replace the traditional "predictions" that are "mere sets of scenarios that present snapshots of the world at some future date" with "projection of alternative, plausible and co-evolving trajectories of the socio-hydrological system" that can provide insights into causal relationships and help identify a desirable operating space. This authors' argument and suggestion, in this case, are not new either. As proof, let us look at a relevant section of an example paper in the coupled human-water systems literature in 2009 that has used system dynamics as the analysis framework [24]:

> "System dynamics which provides a unique framework for integrating the disparate physical and social systems important to water resource management is formulated on the premise that the structure of a system, the network of cause and effect relations between system elements, governs the overall system's behavior. [45]

> The systems approach is a discipline for seeing the structures that underlie complex domains. System dynamics is a framework for seeing interrelationships rather than things, for seeing patterns of change rather than static snapshots, and for seeing processes rather than objects [21]. The major concept of the system dynamics simulation approach is feedback which is used as the basis for structuring description of complex systems and their economic, social, political, and environmental implications. [46]

> The typical purpose of a system dynamics study is to realize how and why the dynamics of concern are generated and to look for managerial policies that can improve the situation". [47]

In another section of the paper, the authors write that:

"In system dynamics studies the emphasis is on understanding trends and behaviors rather than values and numbers".

This suggests that neither the idea of coupled human-water systems modeling, nor the idea of projecting trends and behavior rather than snapshot prediction is novel. They existed in the water resources systems literature before the 2012 proposal of socio-hydrology. So, it is justified to ask what different questions socio-hydrology is asking, what alternative study approaches it is proposing, and why the vast literature on this subject was dismissed before proposing a "new science" in 2012 [1] or calling for a "fundamental change" in 2017 [44].

The desire to develop generic models of human-water systems to "predict" future patterns has shaped some of the endeavors in socio-hydrology in which the researchers have tried to identify and introduce what they refer to as "classes of emergent phenomena" [13]. These "phenomena" have been defined as "actual outcomes, paradoxical dynamics, or unintended consequences that arise from water management" in the analyzed human-water systems. While the effort is valuable, one might wonder why the socio-hydrologists have dismissed the efforts of other water resource researchers in studying, developing, and explaining generic structures, causal dynamics, and evolution, and even projecting and explaining the evolving structures of human-water systems based on real-world systems or hypothetical examples using methods such as system dynamics and game theory (e.g., [39,48–60]).

Many, if not all, of the syndromes, prototypes, phenomena, or sub-phenomena that the socio-hydrologists have proposed, detected, modeled, or expressed interest in, are already in the literature under similar or other names. For example, the safe development/government paradox [61–63], "levee effect" and "reservoir effect" [13], all result from the "shifting the burden" archetype in complex systems [51,64] (Figure 3). This archetype explains the unintended consequence of rectifying the obvious problem symptoms by simple solutions (e.g., raising a levee or building a reservoir) while overlooking the primary causes with the potential of causing addiction to the symptomatic remedies as the problem worsens (raising levees further or building more reservoirs).

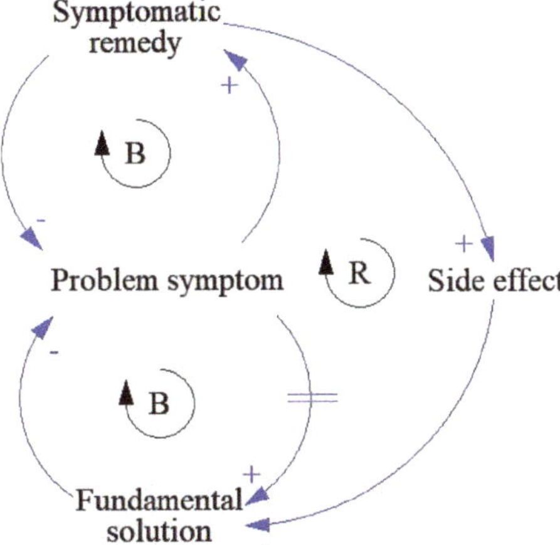

Figure 3. The Causal Loop Diagram (CLD) of the "shifting the burden" systems archetype. In this particular setting, addressing the problem symptom using a quick fix rather than a fundamental measure will cause increasing dependence on symptomatic fixes. "B" refers to a balancing loop and R refers to a reinforcing loop. Double bars reflect lag time.

Rebound effect (Jevons' paradox [65]), vicious supply-demand cycles [66], and irrigation efficiency paradox [67,68], are all produced by the "fix that backfires" archetype in complex systems [20,51,56,64,69]

(Figure 4). This archetype explains how quick-fix and short-sighted solutions (increasing irrigation efficiency, inventing a stronger pump, building a desalination plant or water supply reservoir, digging or deepening wells, and inter-basin water transfer) that cure the problem symptoms (e.g., water shortage) can worsen the situation in the long-run through unintended/side effects (e.g., increased water consumption or demand).

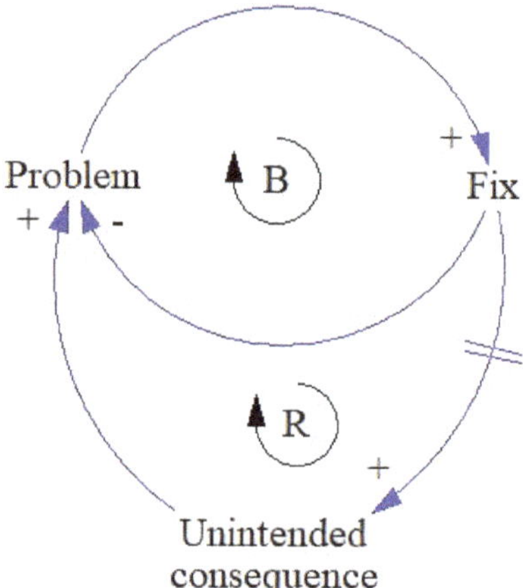

Figure 4. The Causal Loop Diagram (CLD) of the "fixes that fail" or "fixes that backfire" system archetype. In this particular setting, the implemented fix alleviates the problem in the short-run but results in unintended consequences in the long-run that can make the situation worse and necessitate additional fixes. "B" refers to a balancing loop and R refers to a reinforcing loop. Double bars reflect lag time.

Reintroducing what exists in the literature without referring to it is not supposed to be a scientific contribution. Finding that overall water consumption/demand can increase as the result of increasing water supply might be new to some hydrologists, who are used to modeling water demand as an exogenous variable, but this finding is definitely not new in the water resources literature.

A range of other concepts that have been mentioned or used in the socio-hydrology like risk attitude, loss aversion, memory, trust, heuristics and cognitive biases to model and explain trends in human-water systems are not new to the water resources literature either and have been used by people in water resources systems, economics, and social sciences in the past (e.g., [70–82]). Concepts such as path dependency, adaptation pathways, and facilitated stakeholder participation that were discussed in Srinivasan et al. [44] are also among the well-known concepts in the water resources systems literature (e.g., [83–92]).

By comparing the selected 180 socio-hydrology papers to what has been already in the scientific literature, it is hard to find what new insights and approaches have been offered by the "new science" of socio-hydrology. Seidl and Barthel's [5] concerns about socio-hydrology's lack of protocols for interdisciplinary research and failure to acknowledge "much previous work on integrated models undertaken by hydrologists and social scientists" in addition to comments of other scholars on the dismissed works in the water-human systems space remain outstanding.

We emphasize that many of the publications that have identified themselves as socio-hydrologic research are significant and provide very useful insights. The scientific findings and suggestions of papers such as [14,15,93–109] are significant and potentially helpful for policymaking. Though, given that their questions, methods of study, and analysis frameworks are similar to what already existed in the literature, it is not clear why it was necessary to identify them as socio-hydrologic research. It is important to note that while socio-hydrology as a "science" might not have methodologically helped the research of these authors in a unique way, their work in this space might have been inspired by the activities of the socio-hydrology community. Indeed, the socio-hydrology community has effectively increased the number of hydrologists who care about the human dimension in their research, especially in the case of early career scientists. This is a positive development and must be appreciated.

2.3. Where Are the Boundaries of Socio-Hydrology?

The proposers of socio-hydrology have had the ambition of proposing a science that is unique and can do novel things. Though, it seems that over time, perhaps after receiving feedback from their peers and becoming familiar with part of the existing human-water systems literature, their expectations and goals have evolved. For example, projecting qualitative behavior is now being promoted while admitting that modeling behavior is not a trivial task [43]. In 2012, however, the goal was to use quantitative science to predict the evolution of human-water systems. This change in communicating the objectives and expectations is, of course, a good sign, and in agreement with what many water resources scholars have said in the past, i.e., trends, patterns, governing problem structures, and pathways to get to the ideal space generally matter more than numbers and snapshots in complex human-water systems studies. Yet, the coordinates of socio-hydrology in the scientific space remain hard to map as they seem to have been defined by the socio-hydrology leaders in reference to imaginary or perceived limits and capacities of other fields or interest areas.

In their original paper, Sivapalan et al. [1], refer to integrated water resources management (IWRM) as a "science". They subsequently define the socio-hydrology's point of departure from IWRM by arguing that IWRM is "unrealistic, especially for long-term predictions, as it does not account for the dynamics of the interactions between water and people". The authors introduce IWRM as a method which "often uses scenario-based approaches" to explore the human-water interactions. They end up promoting socio-hydrology as a "fundamental science" that can underpin IWRM. But, how accurate are these statements?

The "new science" of socio-hydrology was given an identity by being compared with the "science of IWRM" while IWRM is not a science! IWRM is simply an ambitious "process" (not product) and a recommended cross-sectoral policy-making approach [110] for holistic management of water instead of fragmented and sectoral water management. Similar to other ambitious targets and processes that are set through international negotiations by high-level politicians (e.g., Agenda 21, Sustainable Development Goals (SDGs), and Paris Agreement) IWRM has not fully implemented for many practical reasons [111–114]. The understanding of policymakers, researchers, stakeholders, and the general public of the IWRM concept has been evolving and will continue to evolve as we detect more problems and face new practical challenges. Just like sustainability, IWRM is a moving and unachievable, but useful target that we chase [115]. Enforcing IWRM principles requires insights from natural/social scientists, engineers, stakeholders, policymakers, practitioners, and the general public. Socio-hydrologists can also provide useful insights but socio-hydrology on its own will not make IWRM practical.

Ceola et al. [43] recognize the need for a systems approach to study human-water systems. Nevertheless, socio-hydrologists seem to stay reluctant to acknowledge the contributions of the water resources systems community in the coupled human-water systems study space as discussed earlier. Instead, they try to draw an arbitrary line between socio-hydrology and water resources systems. Di Baldassarre et al. [13] claim that water resources systems is focused on "optimization" with the goal of combining "hydrology and economics" to "design and operate optimal infrastructure projects". The authors put socio-hydrology in contrast with their understanding of water resources systems and state that unlike water resources systems, socio-hydrology is focused on "understanding why certain water management outcomes arise rather than proposing actual management solutions".

Classification of water resources systems as a normative approach that is only limited to optimization and combines hydrology and economics must be due to unfamiliarity of the socio-hydrologists as well as the reviewers of their papers with the water resources systems research space. Water resources systems analysis involves normative and positive approaches, develops simulation and optimization models, and uses quantitative and qualitative approaches [116–119]. Water resources systems studies do not always have economics as a component (e.g., [85,120–136]) and are not always focused on designing and running water infrastructure (e.g., [31,82,130,137–148]). Additionally, the water resources systems studies do not always limit themselves to proposing solutions as Di Baldassare et al. [13] argue. There are many studies in the water resources systems literature that look at "why" certain behavior or evolution path has emerged or might emerge in coupled human-water systems (e.g., [49,52,53,57,59,76,145,149–158]).

Di Baldassarre et al. [13] also distinguish socio-hydrology from CHANS and complex systems science by stating that the former has an explicit focus on water and the hydrologic cycle. If this is the case, one might wonder if socio-hydrology is just a sub-area of CHANS. If so, it is not clear why the existence of CHANS was never recognized by the original proposers of socio-hydrology. What can socio-hydrology offer that CHANS cannot? In an era of increasing interest in the nexus of water with food, energy, environment, etc. what does justify limiting the dynamics of the natural systems to water and the hydrologic cycle?

We certainly know that the natural systems around us are not limited to water and we, as humans, do not only interact with the hydrologic system [159]. So, limiting the scope of the natural component of the "new science" to the hydrologic cycle and not acknowledging that socio-hydrology is a limited subset of CHANS needs a strong justification if the socio-hydrology community insists on keeping the boundaries of socio-hydrology as stated in Di Baldassarre et al. [13]. Even CHANS, with a much larger scope and age, does not recognize itself as a science. Hence, referring to socio-hydrology with a much smaller focus than CHANS as a "new science" might not be fair. The ideas that socio-hydrology is promoting are not new in the CHANS literature either and existed before 2012. As a proof, let us have a look at an abstract of a 2007 review paper in the CHANS literature [160] that suggests that the socio-hydrologists' idea of coupling social and natural systems to reveal complex patterns and new insights into nonlinear dynamics with thresholds, reciprocal feedback loops, time delays, surprises, etc. are not novel:

> "Integrated studies of coupled human and natural systems reveal new and complex patterns and processes not evident when studied by social or natural scientists separately. Synthesis of six case studies from around the world shows that couplings between human and natural systems vary across space, time, and organizational units. They also exhibit nonlinear dynamics with thresholds, reciprocal feedback loops, time lags, resilience, heterogeneity, and surprises. Furthermore, past couplings have legacy effects on present conditions and future possibilities".

While the socio-hydrologists have tried to claim a new territory, the available evidence so far does not suggest that their approach, inquiries, and goals are unique and different. Consequently, mapping the boundaries of the socio-hydrology "science" and finding its overlaps with and points of departure from the existing fields, disciplines and interest areas are nearly impossible.

2.4. Is Socio-Hydrology Practical?

Based on what was discussed so far, there is no strong evidence that socio-hydrology has been successful as a "discovery-based fundamental science" in accommodating the "dynamics we never had to deal with" as claimed by Sivapalan et al. [1]. Deriving causal relationships based on statistical correlations is not necessarily novel and has its own limitations [161]. Improving the mathematical sophistication of socio-hydrology models might help capture some significant dynamics in both human and water components that are currently missing from socio-hydrologic analyses [161]. Improving the behavior resolution of the socio-hydrologic models through replacing differential equations with game theory and agent-based models to better capture the heterogeneity across humans and their interactions networks with careful consideration trade-offs of disaggregation [162] is another possible are of improvement in socio-hydrology. Yet, the unique role of socio-hydrology in providing insights into water-human systems problems remains unclear. Some of the socio-hydrologic modeling papers cited in Section 2.2 have relatively sophisticated mathematical components. However, the developed systems models are comparable to the models that had existed in the water resources systems, water resources economics, and the CHANS literature. So, it is not quite clear how these studies have benefitted from socio-hydrology.

Developing models and generic mechanisms to replicate the historical observations is possible with proxy variables (e.g., memory, culture, emotion, trust, satisfaction, and utility). Nevertheless, extrapolation with the models that replicate historical trends in complex systems has major caveats. Such models and mechanisms are mostly incapable of predicting the future evolution of human-water systems which are associated with uncertainty, bounded rationality, indeterminate causality, limited predictability, evolutionary change, and non-stationarity. What has happened in the past in complex human-nature systems will not necessarily repeat in the future. So even if socio-hydrology is only interested in future trajectories and trends, its ability to project the future is very limited.

Given that water is only one, and from the policy-making standpoint often a small, component of complex human-nature systems [159], limiting the study boundaries to water can create additional sources of unreliability for socio-hydrology models in future projections. When the other components of natural systems are not included in the model but exist in reality, the modeler must inevitably rely on exogenous variables or scenarios. Nonetheless, models that exclude other natural resource components cannot make reliable projections about the future patterns and evolutions, increasing our inability to detect unintended consequences and the actual interrelated dynamics of the human-nature systems around us. To address this issue, other components of the natural systems can be added to human-water systems models. But in that case, CHANS becomes the analysis domain and socio-hydrology gets redundant.

Fully internalizing the exogenous human and water components is also impractical, even when the goal is modeling the past events and evolutions. For example, modeling the Syrian conflict solely based on the interactions of water and people (only endogenous variables) might suggest that the Syrian crisis was caused by a drought, which is a misleading conclusion. While drought has certainly been effective as a trigger or catalyst of the crisis, one must not overlook major variables such as the Syrian politics, ideological conflicts, political economy, foreign interventions, and the accumulation of problems in multiple sectors over the years as the result of bad governance, poor economy, unemployment, etc. Developing a reliable socio-hydrologic model that can meaningfully replicate the Syrian crisis is either impossible or requires considering a significant number exogenous socio-economic and climatic variables, making the idea of internalizing all exogenous variables impractical.

Socio-hydrologists have shown a strong desire in advising policy [13,43] but no interest in scenario analysis and "proposing management solutions". They have also indicated that their projections of the future will not include time-series [43]. At the end of the day, policymakers are in desperate need of solutions and need to make decisions based on scenarios in uncertain environments. Also, in most cases, they require quantitative data, especially if they are operating in the water sector. All decision analysis studies and future projections inherently include exogenous variables as internalizing all exogenous variables is just impossible. No matter how complex, the modeler must choose some arbitrary boundaries for the complex systems models and make some variables exogenous. So, it is not clear how much socio-hydrologists can support the decision-making process if they are not interested in scenario-analysis, do not like to propose solutions, and insist on avoiding exogenous variables. Nevertheless, if they like using scenarios to project the future trajectories and "stress test" the human-water systems as suggested by Srinivasan et al. [44], their literature must avoid actively implying that exogenous scenario-analysis is an improper practice that is popular among the non-socio-hydrologists.

So far, the socio-hydrology literature has set a lot of ambitious targets. Yet, there is little evidence that what socio-hydrology has offered is original. Without a clear understanding of the boundaries of socio-hydrology and its unique analysis methods/tools, judging about the practicality of the declared goals is not easy. Nevertheless, socio-hydrology seems to be more practical in studying the past rather than predicting the future, unless it evolves its targets and waives some of the unnecessary restrictions it has imposed on itself such as remaining a quantitative, advising policymakers without proposing solutions, avoiding scenarios or exogenous variables, and not going beyond water systems and hydrologic cycles.

2.5. Is Socio-Hydrology Converging to Water Resources Systems or Coupled Human and Natural Systems?

Assuming that the socio-hydrologists have been fully unaware of the existing literature, one can conclude that their efforts have resulted in reproducing and recognizing what existed. Unless clear new paths are defined and alternative methods and approaches are developed by them in the near future, the convergence of socio-hydrology to the existing domains such as water resources systems and coupled human-environment systems might be inevitable.

The socio-hydrology literature includes numerous papers that focus on what must be ideally done. Speaking of what is missing in hydrology and what needs to be done is good but we often get too busy talking about our ideals that "we forget that we also need means and realistic pathways to the end goal" [163]. To reach their targets, the socio-hydrologists have recognized the need for a systems approach [43] and have developed mathematical methods using this approach. Nevertheless, the tools offered by the quantitative socio-hydrology literature already exist in the water resources systems and CHANS literature. The insights gained by the socio-hydrologists through different exercises were already available in the literature (as discussed above) or/and could have been achieved for the case studies of interest using the methods that existed in the water resources systems and CHANS literature.

Some of the papers that have been written and branded as socio-hydrology papers could have also been published without making a reference to socio-hydrology. An example of this case is Ishtiaque et al. [164]. This paper which focuses on flood management in coastal Bangladesh and has been published in the Ecology and Society journal (one of the reputable journals in CHANS/SES) shares two authors with Yu et al. [95] and Sung et al. [98] which have been published in the special issue of "Socio-hydrology: Spatial and Temporal Dynamics of Coupled Human-Water Systems" in Water Resources Research with 32 publications. The Ishtiaque et al.'s paper identifies itself as an SES study and although it uses socio-hydrology as a keyword, it does not use this term even once in the whole paper. Two alternative conclusions can be made here. First, that the SES and CHANS study frameworks are more comprehensive and mature than socio-hydrology frameworks. So, making a reference to socio-hydrology or limiting the scope of the work to socio-hydrology was unnecessary. Second, although flood management was at the center of the studied case, limiting the natural system to water would have made the analysis unreliable and the authors had to use the SES study frameworks to be able to include other natural components in their SES problem. Both of these conclusions undermine the value of socio-hydrology when compared with CHANS.

It is noteworthy that just like hydrology or any other field, the water resources systems field has evolved over the last seven decades [117,118,165–170]. The increasing understanding of the limitations of the past models, improved computational power, and data availability, as well as the recognition of the real-world complexities have forced the water resources systems community to revisit and broaden their scope and improve their modeling approaches. Among such improvements, one must note the increasing interest in better representation of humans and stakeholders' opinion in the models using complex systems theories, system dynamics modeling, game theory, multi-criteria assessment, agent-based modeling, and different methods of human behavior analysis in social sciences, operations research, and economics. Indeed, the water resources systems community has also shown interest in and has been working for some time on coupled human-water systems as evidenced by the many references cited earlier. So, it is natural for socio-hydrology and water resources systems to converge under the umbrella of coupled human-water systems.

Over time, the water resources systems modelers have also broadened the scope of their models beyond water systems and added other components such as food, energy, climate, and ecology (e.g., [144,171–182]). So, their research is organically transitioning into the CHANS space. Socio-hydrology will fail in understanding and explaining human-water systems if it limits its scope to the hydrologic cycle. Getting into food, energy, and natural systems is inevitable for the socio-hydrologists, in an increasingly complex world, especially if they want to help decision makers solve and navigate through contemporary problems involving humans and water. Expanding the boundaries of the natural side of socio-hydrology beyond water implies converging into CHANS as it has occurred in the case of water resources systems analysis.

2.6. Is Socio-Hydrology Reinventing the Wheel?

Sivakumar [3] labeled socio-hydrology as a "recycled and re-worded hydro-sociology". McCurley and Jawitz [6] also argued that socio-hydrology was prominent at the beginning of "modern hydrology" (1989) but later decayed and remerged. McCurley and Jawitz believed that the studies of coevolution of water and humans are not novel and existed for decades but the interest in socio-hydrology has considerably increased in the last decade.

Dismissing the existing literature by the socio-hydrologists has led to the reintroduction of existing concepts under new terms, sometimes as new discoveries and socio-hydrology-specific jargons. As explained in Section 2.2, for example, a concept such as "shifting the burden" (Figure 3) already exists in the system dynamics, complex systems, and water resources literature that can explain the generic structure and evolution of coupled human-water systems with certain characteristics. This general structure can illustrate the unintended consequences of implementing certain policy or engineering solutions (e.g., building a reservoir or raising a levee) in a certain setting. While the general structure is the same, the components of the problem can change (e.g., from a levee to a reservoir) without an impact on the overall behavior of the coupled systems. So, it is unclear why new jargons such as "levee effect", "reservoir effect", etc. must be developed and promoted. In the same setting (Figure 3), implementing a certain intervention such as offering an insurance policy, providing water subsidies, granting loans or cheap farmlands, inter-basin water transfers, digging deeper wells, and building a desalination plant can have the same impact. Is socio-hydrology going to develop new jargons like "insurance effect", "subsidies effect", "loan effect", "cheap land effect", "water transfer effect", etc., for each specific example?

These terms can encourage attention to a specific component of the water system (e.g., levee and reservoir) rather than the generic structure of the coupled human-water system. This is in contrast with the declared interest in identifying generalizable human-water systems evolution patterns, something that has been also done by other researchers in the past (see Section 2.2). Inventing new terms and concepts that already exist under different names makes communication with researchers of other fields harder. This is not an effective strategy for marriage between disciplines. As an example, economists, social/natural scientists, engineers, and water resources systems scholars have frequently used the famous prisoner's dilemma setting (game structure) to explain the tragedy of the commons. A shared aquifer is perhaps the most popular example in the water resources literature to explain the tragedy of commons. Rather than creating a new game (e.g., "farmers' dilemma"), prisoner's dilemma has been used in the water resources literature to explain this setting that has a specific dominant structure (see [53,54], for example). The strategy, in this case, has been to use groundwater sharing as an accessible example for the water resources community to seek attention to the institutional setting which drives a certain trajectory in evolution of the coupled human-water system, rather than giving an unnecessary weight to the type of water resources component (i.e., groundwater). Once the structure is introduced, people can be directed to similar settings in which groundwater is replaced with a shared wetland [59], an irrigation system [52,183], a water transfer system [57,184], a trans-boundary river [185], a water quality control system [186], or the atmosphere [187–189] while the general insights and trends will remain unchanged. This approach could facilitate communication with people in other disciplines which might not be familiar with specific water resources settings, our disciplinary jargons, or the hydrologic cycle but understand prisoner's dilemma.

Similarly, socio-hydrology can benefit from using the well-known generic structures in complex systems to explain trajectories in water-human systems problems without a need to invent new terms and concepts. Levees, reservoirs, irrigation efficiency, etc. can be used as examples in our classrooms and conferences to make the complex systems concepts more accessible for the people in hydrology without compromising the capacity of socio-hydrologists to communicate with people in other disciplines, if developing an interdisciplinary domain is a goal. Socio-hydrologists are encouraged to take Seidl and Barthel's [5] feedback on their approach to interdisciplinarity more seriously:

> *"Socio-hydrology is still dominated by hydrologists, who have adopted a perceived hegemonic attitude toward inter-disciplinary collaboration".*

The socio-hydrology community is dominated by hydrologists. If the socio-hydrology's objective is to break the existing boundaries through a systems approach that leads to a holistic understanding of complex human-water systems, hydrologists must be open to actively learning from and working with other disciplines and do not need to pull every exciting concept toward their own domain. Rephrasing the concepts from other fields might not the most constructive step toward scientific interdisciplinarity.

The socio-hydrology literature has shown a strong tendency to borrow fashionable phrases and exciting concepts from other fields (e.g., co-evolution, tipping points, self-organization, Anthropocene, black swan, ludic fallacy, unintended consequences). Using "popular terms and fashionable phraseology" socio-hydrology [2] and the fame of the socio-hydrology proposers have made the socio-hydrology papers highly cited and popular. Invention of socio-hydrology has also sparked the idea of developing socio-hydrogeology, socio-climatology, socio-meteorology. Yet, the popularity of socio-hydrology has mainly emerged in networks that are less familiar with the water resources systems and CHANS literature. Socio-hydrology remains to be dominated by hydrologists who are less "traditional" than the mainstream hydrologists but more "traditional" than the water experts who have been working in the human-water systems space for decades. We noted many papers in the literature that use socio-hydrology as a keyword or in their title but do not make a proper connection to socio-hydrology "concepts" or literature throughout their articles. Seidl and Barthel [5] also noted that 26% of the socio-hydrology papers they reviewed mention the term socio-hydrology in their title, abstracts, and keywords but "do not refer to the concept of socio-hydrology presented by Sivapalan et al. [1]". Seidl and Barthel expressed their "surprise", claiming that when they contacted the authors of several of those papers, they confirmed that they were not aware of the socio-hydrology concepts.

While the desire for adopting fashionable concepts from other disciplines is high among the socio-hydrologists, the inappropriate use of such concepts significantly hurt the scientific rigor of some socio-hydrology publications. The interest of the original proposers of socio-hydrology [1] in predicting black swan events (also questioned by Koutsoyiannis [2] in his review report), reflects their misunderstanding of the black swan theory [41] which asserts that black swans might be predictable retrospectively, not prospectively. As another example, Di Baldassarre et al. [104] explained the known mechanism of increase in water demand as the result of increasing water supply (e.g., [24,35,38,69]) as the Jevons' paradox (rebound effect). However, Jevons' paradox in economics is about the increased efficiency in use of a resource (an intervention on the demand side), not an intervention on the supply side. In other words, the total resource volume remains constant and the rebound is caused due to increase in the rate of resource consumption. But, in the featured examples of Di Baldassarre et al. [104], changes in the supply side, i.e., increasing the volume of resource (through building a new reservoir) had led to an increase in demand (by enabling "agricultural, industrial or urban expansion") that ended up offsetting the increase in supply. In their examples, the intervention is on the supply side (reservoirs) as opposed to the Jevons' paradox which involves an intervention on the demand side (e.g., increasing resource use efficiency through technology advancements). While the governing structure of the two problems and the overall behavior in the variables of interest (e.g., supply-demand gap) might be the same, the interventions that cause this behavior can be different. That is why using generic coupled human-water systems archetypes for this type of problems has been proposed [20,51,56]. The "fixes that fail" archetype (Figure 4) can explain the behaviors of both problems (Jevons' paradox and viscous supply-demand cycles) as discussed earlier (see Section 2.2).

Borrowing concepts and terms from other disciplines can be illuminating and valuable as long as they are used correctly and are not rephrased unnecessarily. Relabeling and rewording the syntaxes and terminologies of other fields might create short-term popularity. Though, in the long run, it can increase our distance from scholars in other fields rather than breaking the boundaries and barriers to facilitate interdisciplinary and trans-disciplinary interactions. We already have too many terminologies in different disciplines that are worded differently but mean the same thing.

3. Efficiency of Our Peer-Review Systems

We raised some concerns about the claims that have been made by socio-hydrology literature. These concerns are about the general interest area of socio-hydrology and not specific to certain publications. Some of these concerns are not new but have remained unaddressed since the day the "new science" was proposed in 2012 [1]. The socio-hydrology community could have avoided most of these concerns through essential scientific efforts on a proactive basis (e.g., by a comprehensive review of the existing literature). However, these basic concerns could have also been addressed reactively and in response to peer review comments. If we attribute the systematic overlooking of the water resources systems, system dynamics, and CHANS research in the socio-hydrology publications to the unfamiliarity of socio-hydrologists with the existing literature, then we must be seriously concerned about the major flaws and deficiencies of our peer review system. Evidently, the peer review system has systematically failed in providing constructive feedback to our colleagues in socio-hydrology.

As discussed in Section 1, the questions asked in this paper are not supposed to undermine socio-hydrology. We consider the increasing interest in socio-hydrology as a positive development, whilst remaining surprised that these issues have not been raised in the peer review of 180 socio-hydrology publications by some of the best water resources journals that are supposed to have a fundamental role in setting the science agenda and recognizing the value of novel insights [190]. This calls for revisiting our peer review system and asking how an effective review system can be set up when a "new science" is proposed. Who is qualified to review papers that belong to a "new" space? How must we set up a reliable review system that promotes innovation and interdisciplinarity but does not compromise our scientific research procedures and standards?

Another issue that has contributed to the existing deficiencies in the socio-hydrology literature is our culture in academia. We do not want to challenge, shame, and disrespect our colleagues. So, we have strong reservations about putting our thoughts, comments, and feedback in writing in fear of offending our colleagues. When our colleagues are more senior and famous, our concerns grow further. The two of us also had serious reservations about writing this article. However, this culture must change. We have to be more helpful to each other and not be afraid of providing constructive feedback to our colleagues if we care about them and really believe in the power of science and interdisciplinary approaches. The concerns we listed here are not new and we have heard our colleagues talking about them for years but did not share them formally. When our formal peer review systems fail, we have a stronger responsibility to protect the integrity of science and help our colleagues flourish and succeed.

While the flaws of our peer review system require great attention, we should not forget that as academics, we still have an ethical responsibility to ensure that to the best of our knowledge what we propose as a "new" contribution does not already exist. It is very hard to imagine that the socio-hydrology community has remained unaware of the ongoing and past research in other areas such as system dynamics with a strong overlap with their work, especially when they refer to their early products as system dynamics models, use CLDs, and frequently use the popular terms such as unintended consequences, complexity, feedbacks, evolution, etc.

4. Conclusions and Final Remarks

Undoubtedly, the seminal paper of Sivapalan et al. [1] and their subsequent leadership has created an increasing interest in social systems in the hydrologic science community. The socio-hydrologists' success is evidenced by the number of published socio-hydrology papers since 2012, as well as the number of researchers who identify themselves as socio-hydrologists or use the socio-hydrology term in their publications. Yet, socio-hydrology seems to have become mostly popular among the mainstream hydrologists with limited familiarity with the past work in the human-water systems space, not to those who have worked in this space for decades. The socio-hydrologists' interest in holistic understanding of human-water systems, which necessitates systems approaches, together with their desire to advise policy creates a natural overlap between socio-hydrology and research in water resources systems, SES and CHANS areas. Nevertheless, socio-hydrologists insist that their work is different from the works of other groups without providing sufficient scientific evidence.

Socio-hydrology was originally proposed as a "new science" but so far it has not been more than an interest area or a sub-field of hydrology. While the socio-hydrology literature shows a great tendency to borrow fashionable syntaxes and popular concepts from the literature of other fields, socio-hydrologists have systematically overlooked the past and ongoing work in the coupled human-water systems space and this has led to reproduction of some existing concepts under new names.

Some concerns about the practicality of socio-hydrology goals, the types of unique tools/methods it uses, the new questions it asks, and its boundaries and points of departure from existing domains have been around and remained unaddressed since 2012. Thus, it is not clear to us why socio-hydrology insists on creating arbitrary boundaries with water resources systems and CHANS. The overviewed deficiencies in the socio-hydrology literature reflect the existing shortfalls of our peer review system that require serious attention, especially in the era of increasing interest in developing new interdisciplinary fields in response to our research needs.

Generally, the ex-ante creation of a new science or even a new field is not scientific and sets the counterproductive precedent of creating new sciences based on personal judgement rather than scientific proof. Establishing a new field must be done through a scientific procedure that recognizes the new needs, carefully examines what is available, identifies the gap, proposes meaningful and new questions, and suggests solid evidence for the possibility of answering such questions using new, old, or revised approaches and methods. Certainly, this procedure was not followed by socio-hydrology and as a result, after eight years, we still have a hard time figuring out what socio-hydrology means and what it is trying to do that is novel. This can be attributed to our ignorance, but we believe that this confusion has been contagious and common for a lot of non-socio-hydrologists.

In our opinion, the valuable contribution of the socio-hydrologists is not their "new science", models, ambitious statements, and exciting agenda, but their perseverance and dedication to reminding the mainstream hydrologists about the need for taking the human factor into account. For the reasons discussed in the paper, a good number of socio-hydrology papers would have been nearly unpublishable in their current forms in traditional water resources management or water resources systems journals. The same is true if these papers had been reviewed by those familiar with the larger water resources systems, system dynamics, and CHANS literature. Nevertheless, the appearance of these papers in the most reputable hydrology/water resources journals and the dedicated efforts of the socio-hydrology leaders to promote their "new science" at major gatherings of the field have resulted in: (1) an increasing recognition of the need for incorporating the human element into traditional hydrologic models; (2) an increasing interest, especially among the early-career researchers to study and model coupled human-water systems; and (3) a big surge in the production of coupled human-water systems literature (mainly based on systems dynamics methods) that provide potentially useful insights for policymaking. Thanks to the efforts of the socio-hydrology leaders, the coupled human-water systems community now has a bigger and more enthusiastic task force. This success must be celebrated and promoted as it can strengthen the current efforts of those who have been working on coupled human-water systems for a long time.

Meanwhile, to avoid wasting economic resources and human talents, the socio-hydrology leaders have a great responsibility to ensure that their community is aware of and recognizes the major contributions of the water resources systems and CHANS communities to studying coupled human-water systems. The presented boundaries and coordinates of the socio-hydrology "science" seem to be suffering from a misunderstanding of what other communities (e.g., water resources systems) are doing. Socio-hydrologists have an ethical responsibility to review what other groups have done and clearly indicate their point of departure, if such departure is necessary at all.

Hjorth and Madani [191] warned that within the water resources profession, our mental frames [192], beliefs, expectations, and judgements could converge over time as we continue to read the same journals and go to the same conferences. Repetitions make our frames stronger and empower them in our brains. To conserve our mental frames, we have a tendency to preoccupy ourselves with issues that are closer to our comfort zones. We disregard the observations that do not fit into our frames as we know well that "a frame modification would imply adjustment, insecurity, and even confusion, possibly not just for the individual but for an entire community" [191]. This issue could be among the reasons that the socio-hydrology, water resources systems, and CHANS communities have not successfully interacted with each other in a complementary fashion. Instead, each group has remained defensive of its own framing of problems and its own unique solutions to it. As proponents of the systems approach in decision making in the real world, we have failed to adopt such an approach in our own world, i.e., academia, where most of the real-world complexities do not exist and the stakes are supposedly much lower. Instead, as academics with strong interest overlaps, we have competed to create and lead our own territories, dismissing what others have done. It is true and very unfortunate that "we work hard, but separately, to solve interlinked problems" [191].

New challenges require changed priorities and new thinking. We need to update our common sense and come to grips with our mistaken beliefs [115]. Common sense can unite us but creating new science might divide us. Thus, if we want a unified effort, we must function within a common-sense framework [191] instead of developing and protecting our own science. The lack of common understanding makes scientists compete to interpret problems [115], propose new terms, "correct" evaluation methods [193] and disciplinary jargons [194], and prescribe solutions based on their own preferences and knowledge of their own domain instead of focusing on problem solving. If the subject is more important, more people will try to compete and pull it toward their domain where they feel most empowered by their own capabilities and perspectives [195]. Creating a common sense for hydrologists that humans must be an integral element of water models can unite the water resources community. However, insisting on creating a "new science" of socio-hydrology while undermining the existing work in the human-water systems space might be a frustrating precedent that can lead to further fragmentation of the already siloed scientists.

Lastly, despite the possible misinterpretations of our intention, we decided to write this paper as outsiders who remain interested but confused about the developments and contributions of socio-hydrology. The ability to give constructive and critical feedback, without causing resentment, is a superpower that we might not possess. Nevertheless, we remain hopeful that these comments encourage our colleagues in socio-hydrology to sharpen their messages, more comprehensively consider the existing literature, and, if appropriate, join their forces and merge their work with other scholars who are working on human-nature and human-water systems problems as unity can certainly make us more powerful.

Author Contributions: K.M. and M.S.-J. did the analysis and literature review through extensive discussions. M.S.-J. did the literature search in the WoS database and produced the figures. K.M. wrote the manuscript with input from M.S.-J. All authors have read and agreed to the published version of the manuscript.

Funding: This research received no external funding.

Acknowledgments: We thank the Guest Editor, Slobodan Simonovic, for the invitation to write this paper. We are also thankful to the anonymous reviewers for their constructive feedback.

Conflicts of Interest: The authors declare no conflict of interest.

References

1. Sivapalan, M.; Savenije, H.H.G.; Blöschl, G. Socio-hydrology: A new science of people and water. *Hydrol. Process.* **2012**, *26*, 1270–1276. [CrossRef]
2. Koutsoyiannis, D. *Review Report of "Socio-Hydrology: A New Science of People and Water"*; National Technical University of Athens: Athens, Greece, 2011. Available online: http://www.itia.ntua.gr/en/docinfo/1991/ (accessed on 20 May 2020).
3. Sivakumar, B. Socio-hydrology: Not a new science, but a recycled and re-worded hydrosociology. *Hydrol. Process.* **2012**, *26*, 3788–3790. [CrossRef]
4. Falkenmark, M. Main problems of water use and transfer of technology. *GeoJournal* **1979**, *3*, 435–443. [CrossRef]
5. Seidl, R.; Barthel, R. Linking scientific disciplines: Hydrology and social sciences. *J. Hydrol.* **2017**, *550*, 441–452. [CrossRef]
6. McCurley, K.L.; Jawitz, J.W. Hyphenated hydrology: Interdisciplinary evolution of water resource science. *Water Resour. Res.* **2017**, *53*, 2972–2982. [CrossRef]
7. Keys, P.W.; Wang-Erlandsson, L. On the social dynamics of moisture recycling. *Earth Syst. Dyn.* **2018**, *9*, 829–847. [CrossRef]
8. Hynds, P.; Regan, S.; Andrade, L.; Mooney, S.; O'Malley, K.; DiPelino, S.; O'Dwyer, J. Muddy Waters: Refining the Way Forward for the "Sustainability Science" of Socio-Hydrogeology. *Water* **2018**, *10*, 1111. [CrossRef]
9. Smakhtin, V.U.; Shilpakar, R.L.; Hughes, D.A. Hydrology-based assessment of environmental flows: An example from Nepal. *Hydrol. Sci. J.* **2006**, *51*, 207–222. [CrossRef]
10. Kock, B. Agent-Based Models of Socio-Hydrological Systems for Exploring the Institutional Dynamics of Water Resources Conflict. Ph.D. Thesis, Massachusetts Institute of Technology, Cambridge, MA, USA, 2008.
11. Mohorjy, A.M. Multidisciplinary planning and managing of water reuse. *JAWRA J. Am. Water Resour. Assoc.* **1989**, *25*, 433–442. [CrossRef]
12. Pande, S.; Sivapalan, M. Progress in socio-hydrology: A meta-analysis of challenges and opportunities. *WIREs Water* **2016**, *4*, e1193. [CrossRef]
13. Di Baldassarre, G.; Sivapalan, M.; Rusca, M.; Cudennec, C.; Garcia, M.; Kreibich, H.; Konar, M.; Mondino, E.; Mård, J.; Pande, S.; et al. Sociohydrology: Scientific challenges in addressing the sustainable development goals. *Water Resour. Res.* **2019**, *55*, 6327–6355. [CrossRef]
14. Leong, C. The role of narratives in sociohydrological models of flood behaviors. *Water Resour. Res.* **2018**, *54*, 3100–3121. [CrossRef]
15. Sanderson, M.R.; Bergtold, J.S.; Heier Stamm, J.L.; Caldas, M.M.; Ramsey, S.M. Bringing the "social" into sociohydrology: Conservation policy support in the Central Great Plains of Kansas, USA. *Water Resour. Res.* **2017**, *53*, 6725–6743. [CrossRef]
16. Haeffner, M.; Jackson-Smith, D.; Flint, C.G. Social position influencing the water perception gap between local leaders and constituents in a socio-hydrological system. *Water Resour. Res.* **2018**, *54*, 663–679. [CrossRef]
17. Newell, B.; Wasson, R. Social system vs solar system: Why policy makers need history. In *Conflict and Cooperation Related to International Water Resources: Historical Perspectives*; Castelein, S., Otte, A., Eds.; UNESCO: Paris, France, 2002; p. 22.
18. Van Eck, N.J.; Waltman, L. Software survey: VOSviewer, a computer program for bibliometric mapping. *Scientometrics* **2010**, *84*, 523–538. [CrossRef]
19. Van Eck, N.J.; Waltman, L. Visualizing Bibliometric networks. In *Measuring Scholarly Impact*; Springer International Publishing: Cham, Switzerland, 2014; pp. 285–320, ISBN 9783319103778.
20. Mirchi, A.; Madani, K.; Watkins, D.; Ahmad, S. Synthesis of system dynamics tools for holistic conceptualization of water resources problems. *Water Resour. Manag.* **2012**, *26*, 2421–2442. [CrossRef]
21. Simonovic, S.P.; Fahmy, H. A new modeling approach for water resources policy analysis. *Water Resour. Res.* **1999**, *35*, 295–304. [CrossRef]
22. Guo, H.C.; Liu, L.; Huang, G.H.; Fuller, G.A.; Zou, R.; Yin, Y.Y. A system dynamics approach for regional environmental planning and management: A study for the Lake Erhai Basin. *J. Environ. Manag.* **2001**, *61*, 93–111. [CrossRef]
23. Madani, K. Water Transfer and watershed development: A system dynamics approach. In Proceedings of the World Environmental and Water Resources Congress, Tampa, FL, USA, 15–19 May 2007; pp. 1–15.

24. Madani, K.; Mariño, M.A. System dynamics analysis for managing Iran's Zayandeh-Rud river basin. *Water Resour. Manag.* **2009**, *23*, 2163–2187. [CrossRef]
25. Ahmad, S.; Prashar, D. Evaluating municipal water conservation policies using a dynamic simulation model. *Water Resour. Manag.* **2010**, *24*, 3371–3395. [CrossRef]
26. Shahbazbegian, M.; Bagheri, A. Rethinking assessment of drought impacts: A systemic approach towards sustainability. *Sustain. Sci.* **2010**, *5*, 223–236. [CrossRef]
27. Bagheri, A.; Darijani, M.; Asgary, A.; Morid, S. Crisis in urban water systems during the reconstruction period: A system dynamics analysis of alternative policies after the 2003 earthquake in Bam-Iran. *Water Resour. Manag.* **2010**, *24*, 2567–2596. [CrossRef]
28. Madani, K. *Towards Sustainable Watershed Management: Using System Dynamics for Integrated Water Resources Planning*; VDM Verlag Dr. Müller: Saarbrücken, Germany, 2010.
29. Davies, E.G.R.; Simonovic, S.P. Global water resources modeling with an integrated model of the social-economic-environmental system. *Adv. Water Resour.* **2011**, *34*, 684–700. [CrossRef]
30. Zarghami, M.; Akbariyeh, S. System dynamics modeling for complex urban water systems: Application to the city of Tabriz, Iran. *Resour. Conserv. Recycl.* **2012**, *60*, 99–106. [CrossRef]
31. Simonovic, S.P. World water dynamics: Global modeling of water resources. *J. Environ. Manage.* **2002**, *66*, 249–267. [CrossRef]
32. Xu, Z.X.; Takeuchi, K.; Ishidaira, H.; Zhang, X.W. Sustainability analysis for Yellow River water resources using the system dynamics approach. *Water Resour. Manag.* **2002**, *16*, 239–261. [CrossRef]
33. Martínez Fernández, J.; Selma, M.A.E. The dynamics of water scarcity on irrigated landscapes: Mazarrón and Aguilas in south-eastern Spain. *Syst. Dyn. Rev.* **2004**, *20*, 117–137. [CrossRef]
34. Simonovic, S.P.; Rajasekaram, V. Integrated analyses of Canada's water resources: A system dynamics approach. *Can. Water Resour. J.* **2004**, *29*, 223–250. [CrossRef]
35. Madani Larijani, K. *Watershed Management and Sustainability—A System Dynamics Approach (Case Study: Zayandeh-Rud River Basin, Iran)*; Lund University: Lund, Sweden, 2005.
36. Bagheri, A. *Sustainable Development: Implementation in Urban Water Systems*; Lund University: Lund, Sweden, 2006.
37. Leal Neto, A.D.C.; Legey, L.F.L.; González-Araya, M.C.; Jablonski, S. A system dynamics model for the environmental management of the Sepetiba Bay Watershed, Brazil. *Environ. Manag.* **2006**, *38*, 879–888. [CrossRef]
38. Bagheri, A.; Hjorth, P. A framework for process indicators to monitor for sustainable development: Practice to an urban water system. *Environ. Dev. Sustain.* **2007**, *9*, 143–161. [CrossRef]
39. Hjorth, P.; Bagheri, A. Navigating towards sustainable development: A system dynamics approach. *Futures* **2006**, *38*, 74–92. [CrossRef]
40. MacCracken, M. Prediction versus projection—Forecast versus possibility. *WeatherZine* **2001**, *26*, 3–4.
41. Taleb, N.N. *The Black Swan: The Impact of Highly Improbable*, 1st ed.; Random House: New York, NY, USA, 2007.
42. Loucks, D.P. Debates-perspectives on socio-hydrology: Simulating hydrologic-human interactions. *Water Resour. Res.* **2015**, *51*, 4789–4794. [CrossRef]
43. Ceola, S.; Montanari, A.; Krueger, T.; Dyer, F.; Kreibich, H.; Westerberg, I.; Carr, G.; Cudennec, C.; Elshorbagy, A.; Savenije, H.; et al. Adaptation of water resources systems to changing society and environment: A statement by the International Association of Hydrological Sciences. *Hydrol. Sci. J.* **2016**, *61*, 2803–2817. [CrossRef]
44. Srinivasan, V.; Sanderson, M.; Garcia, M.; Konar, M.; Blöschl, G.; Sivapalan, M. Prediction in a socio-hydrological world. *Hydrol. Sci. J.* **2017**, *62*, 338–345. [CrossRef]
45. Sterman, J. *Systems Thinking and Modeling for a Complex. World*; McGraw Hill: Boston, MA, USA, 2000, ISBN 0-07-231135-5.
46. Bender, M.J.; Simonovic, S.P. A systems approach for collaborative decision support in water resources planning. In Proceedings of the 1996 International Symposium on Technology and Society Technical Expertise and Public Decisions, Princeton, NJ, USA, 21–22 June 1996; pp. 357–363. [CrossRef]
47. Saysel, A.K.; Barlas, Y.; Yenigün, O. Environmental sustainability in an agricultural development project: A system dynamics approach. *J. Environ. Manag.* **2002**, *64*, 247–260. [CrossRef]
48. Bagheri, A.; Hjorth, P. Planning for sustainable development: A paradigm shift towards a process-based approach. *Sustain. Dev.* **2007**, *15*, 83–96. [CrossRef]

49. Hui, R.; Lund, J.R.; Madani, K. Game theory and risk-based leveed river system planning with noncooperation. *Water Resour. Res.* **2016**, *52*, 119–134. [CrossRef]

50. Gohari, A.; Mirchi, A.; Madani, K. System dynamics evaluation of climate change adaptation strategies for water resources management in Central Iran. *Water Resour. Manag.* **2017**, *31*, 1413–1434. [CrossRef]

51. Bahaddin, B.; Mirchi, A.; Watkins, D.; Ahmad, S.; Rich, E.; Madani, K. System Archetypes in Water Resource Management. In Proceedings of the World Environmental and Water Resources Congress 2018, Minneapolis, MN, USA, 3–7 June 2018; pp. 130–140.

52. Ristić, B.; Madani, K. A game theory warning to blind drivers playing chicken with public goods. *Water Resour. Res.* **2019**, *55*, 2000–2013. [CrossRef]

53. Madani, K. Game theory and water resources. *J. Hydrol.* **2010**, *381*, 225–238. [CrossRef]

54. Madani, K.; Hipel, K.W. Non-cooperative stability definitions for strategic analysis of generic water resources conflicts. *Water Resour. Manag.* **2011**, *25*, 1949–1977. [CrossRef]

55. Mirchi, A. System Dynamics Modeling As a Quantitative- Qualitative Framework for Sustainable Water Resources Management: Insights for Water Quality Policy in the Great Lakes Region. Doctoral Thesis, Michigan Technological University, Houghton, MI, USA, 2013.

56. Mirchi, A.; Watkins, D.W.; Madani, K. Water resources system archetypes: Towards a holistic understanding of persistent water resources problems. In Proceedings of the AGU Fall Meeting, San Francisco, CA, USA, 5–9 December 2011; p. H11F-1142.

57. Madani, K.; Lund, J.R. California's Sacramento-San Joaquin delta conflict: From cooperation to chicken. *J. Water Resour. Plan. Manag.* **2012**, *138*, 90–99. [CrossRef]

58. Mirchi, A.; Watkins, D. A systems approach to holistic total maximum daily load policy: Case of Lake Allegan, Michigan. *J. Water Resour. Plan. Manag.* **2013**, *139*, 544–553. [CrossRef]

59. Madani, K.; Zarezadeh, M. The significance of game structure evolution for deriving game-theoretic policy insights. In Proceedings of the 2014 IEEE International Conference on Systems, Man, and Cybernetics (SMC), San Diego, CA, USA, 5–8 October 2014; pp. 2715–2720. [CrossRef]

60. Mirchi, A.; Watkins, D.W.; Huckins, C.J.; Madani, K.; Hjorth, P. Water resources management in a homogenizing world: Averting the Growth and Underinvestment trajectory. *Water Resour. Res.* **2014**, *50*, 7515–7526. [CrossRef]

61. White, G.F. *Human Adjustments to Floods: A Geographical Approach to the Flood Problem in the United States*; The University of Chicago–Department of Geography: Chicago, IL, USA, 1945.

62. Burton, I.; Kates, R.W.; White, G.F. *The Human Ecology of Extreme Geophysical Events*; FMHI Publications: Tampa, FL, USA, 1968. Available online: https://scholarcommons.usf.edu/fmhi_pub/78/ (accessed on 7 July 2020).

63. Burby, R.J. Hurricane Katrina and the paradoxes of government disaster policy: Bringing about wise governmental decisions for hazardous Areas. *Ann. Am. Acad. Pol. Soc. Sci.* **2006**, *604*, 171–191. [CrossRef]

64. Senge, P.M. *The Fifth Discipline: The Art and Practice of the Learning Organization*; Doubleday/Currency: New York, NY, USA, 1990.

65. Alcott, B. Jevons' paradox. *Ecol. Econ.* **2005**, *54*, 9–21. [CrossRef]

66. Kallis, G. Coevolution in water resource development. The vicious cycle of water supply and demand in Athens, Greece. *Ecol. Econ.* **2010**, *69*, 796–809. [CrossRef]

67. Ward, F.A.; Pulido-Velazquez, M. Water conservation in irrigation can increase water use. *Proc. Natl. Acad. Sci. USA* **2008**, *105*, 18215–18220. [CrossRef]

68. Grafton, R.Q.; Williams, J.; Perry, C.J.; Molle, F.; Ringler, C.; Steduto, P.; Udall, B.; Wheeler, S.A.; Wang, Y.; Garrick, D.; et al. The paradox of irrigation efficiency. *Science* **2018**, *361*, 748–750. [CrossRef]

69. Gohari, A.; Eslamian, S.; Mirchi, A.; Abedi-Koupaei, J.; Massah Bavani, A.; Madani, K. Water transfer as a solution to water shortage: A fix that can Backfire. *J. Hydrol.* **2013**, *491*, 23–39. [CrossRef]

70. Sage, A.; White, E. On the value dependent role of the identification processing and evaluation of information in risk/benefit analysis. In *Risk/Benefit Analysis in Water Resources Planning and Management*; Haimes, Y.Y., Ed.; Springer: Boston, MA, USA, 1981; pp. 245–262.

71. Howitt, R.E.; Msangi, S.; Reynaud, A.; Knapp, K.C. Estimating intertemporal preferences for natural resource allocation. *Am. J. Agric. Econ.* **2005**, *87*, 969–983. [CrossRef]

72. Li, M.; Xu, W.; Rosegrant, M.W. Irrigation, risk aversion, and water right priority under water supply uncertainty. *Water Resour. Res.* **2017**, *53*, 7885–7903. [CrossRef] [PubMed]

73. Whateley, S.; Palmer, R.N.; Brown, C. Seasonal Hydroclimatic Forecasts as Innovations and the Challenges of Adoption by Water Managers. *J. Water Resour. Plan. Manag.* **2015**, *141*, 04014071. [CrossRef]

74. OBeidi, A.; Hipel, K. Strategic and dilemma analyses of a water export conflict. *INFOR Inf. Syst. Oper. Res.* **2005**, *43*, 247–270. [CrossRef]

75. Kuruppu, N.; Liverman, D. Mental preparation for climate adaptation: The role of cognition and culture in enhancing adaptive capacity of water management in Kiribati. *Glob. Environ. Change* **2011**, *21*, 657–669. [CrossRef]

76. Madani, K.; Dinar, A. Non-cooperative institutions for sustainable common pool resource management: Application to groundwater. *Ecol. Econ.* **2012**, *74*, 34–45. [CrossRef]

77. Hall, J.; Borgomeo, E. Risk-based principles for defining and managing water security. *Philos. Trans. R. Soc. A Math. Phys. Eng. Sci.* **2013**, *371*. [CrossRef]

78. Madani, K.; Hooshyar, M. A game theory-reinforcement learning (GT-RL) method to develop optimal operation policies for multi-operator reservoir systems. *J. Hydrol.* **2014**, *519*, 732–742. [CrossRef]

79. Gallagher, J. Learning about an infrequent event: Evidence from flood insurance take-up in the United States. *Am. Econ. J. Appl. Econ.* **2014**, *6*, 206–233. [CrossRef]

80. Berglund, E.Z. Using agent-based modeling for water resources planning and management. *J. Water Resour. Plan. Manag.* **2015**, *141*, 1–17. [CrossRef]

81. DeCaro, D.A.; Arnold, C.A.; Frimpong Boamah, E.; Garmestani, A.S. Understanding and applying principles of social cognition and decision making in adaptive environmental governance. *Ecol. Soc.* **2017**, *22*, 33. [CrossRef] [PubMed]

82. Du, E.; Cai, X.; Brozović, N.; Minsker, B. Evaluating the impacts of farmers' behaviors on a hypothetical agricultural water market based on double auction. *Water Resour. Res.* **2017**, *53*, 4053–4072. [CrossRef]

83. Palmer, R.N.; Werick, W.J.; MacEwan, A.; Woods, A.W. Modeling water resources opportunities, challenges and trade-offs: The use of shared vision modeling for negotiation and conflict resolution. In Proceedings of the WRPMD'99, Tempe, AZ, USA, 6–9 June 1999; pp. 1–13. [CrossRef]

84. Cohen, S.; Neilsen, D.; Smith, S.; Neale, T.; Taylor, B.; Barton, M.; Merritt, W.; Alila, Y.; Shepherd, P.; Mcneill, R.; et al. Learning with local help: Expanding the dialogue on climate change and water management in the Okanagan Region, British Columbia, Canada. *Clim. Change* **2006**, *75*, 331–358. [CrossRef]

85. Langsdale, S.; Beall, A.; Carmichael, J.; Cohen, S.; Forster, C. An exploration of water resources futures under climate change using system dynamics modeling. *Integr. Assess.* **2007**, *7*, 51–79.

86. Heinmiller, T. Path dependency and collective action in common pool governance. *Int. J. Commons* **2009**, *3*, 131. [CrossRef]

87. Harris, E. The impact of institutional path dependence on water market efficiency in Victoria, Australia. *Water Resour. Manag.* **2011**, *25*, 4069–4080. [CrossRef]

88. Libecap, G.D. Institutional path dependence in climate adaptation: Coman's "some unsettled problems of irrigation". *Am. Econ. Rev.* **2011**, *101*, 64–80. [CrossRef]

89. Carr, G.; Loucks, D.; Blöschl, G. An analysis of public participation in the Lake Ontario—St. Lawrence river study. In *Water Co-Management*; Grover, V.I., Krantzberg, G., Eds.; CRC Press: Boca Raton, FL, USA, 2013; pp. 48–77.

90. Haasnoot, M.; Kwakkel, J.H.; Walker, W.E.; ter Maat, J. Dynamic adaptive policy pathways: A method for crafting robust decisions for a deeply uncertain world. *Glob. Environ. Change* **2013**, *23*, 485–498. [CrossRef]

91. Langsdale, S.; Beall, A.; Bourget, E.; Hagen, E.; Kudlas, S.; Palmer, R.; Tate, D.; Werick, W. Collaborative modeling for decision support in water resources: Principles and best practices. *J. Am. Water Resour. Assoc.* **2013**, *49*, 629–638. [CrossRef]

92. Marshall, G.R.; Alexandra, J. Institutional path dependence and environmental water recovery in Australia's Murray-Darling Basin. *Water Altern.* **2016**, *9*, 679–703.

93. Elshafei, Y.; Tonts, M.; Sivapalan, M.; Hipsey, M.R. Sensitivity of emergent sociohydrologic dynamics to internal system properties and external sociopolitical factors: Implications for water management. *Water Resour. Res.* **2016**, *52*, 4944–4966. [CrossRef]

94. Van Emmerik, T.H.M.; Li, Z.; Sivapalan, M.; Pande, S.; Kandasamy, J.; Savenije, H.H.G.; Chanan, A.; Vigneswaran, S. Socio-hydrologic modeling to understand and mediate the competition for water between agriculture development and environmental health: Murrumbidgee River Basin, Australia. *Hydrol. Earth Syst. Sci. Discuss.* **2014**, *11*, 3387–3435. [CrossRef]

95. Garcia, M.; Portney, K.; Islam, S. A question driven socio-hydrological modeling process. *Hydrol. Earth Syst. Sci.* **2016**, *20*, 73–92. [CrossRef]

96. Yu, D.J.; Sangwan, N.; Sung, K.; Chen, X.; Merwade, V. Incorporating institutions and collective action into a sociohydrological model of flood resilience. *Water Resour. Res.* **2017**, *53*, 1336–1353. [CrossRef]

97. Gunda, T.; Turner, B.L.; Tidwell, V.C. The influential role of sociocultural feedbacks on community-managed irrigation system behaviors during times of water stress. *Water Resour. Res.* **2018**, *54*, 2697–2714. [CrossRef]

98. Kuil, L.; Evans, T.; McCord, P.F.; Salinas, J.L.; Blöschl, G. Exploring the influence of smallholders' perceptions regarding water availability on crop choice and water allocation through socio-hydrological modeling. *Water Resour. Res.* **2018**, *54*, 2580–2604. [CrossRef]

99. Sung, K.; Jeong, H.; Sangwan, N.; Yu, D.J. Effects of flood control strategies on flood resilience under sociohydrological disturbances. *Water Resour. Res.* **2018**, *54*, 2661–2680. [CrossRef]

100. Xu, L.; Gober, P.; Wheater, H.S.; Kajikawa, Y. Reframing socio-hydrological research to include a social science perspective. *J. Hydrol.* **2018**, *563*, 76–83. [CrossRef]

101. O'Keeffe, J.; Moulds, S.; Bergin, E.; Brozović, N.; Mijic, A.; Buytaert, W. Including farmer irrigation behavior in a sociohydrological modeling framework with application in North India. *Water Resour. Res.* **2018**, *54*, 4849–4866. [CrossRef]

102. O'Keeffe, J.; Moulds, S.; Scheidegger, J.M.; Jackson, C.R.; Nair, T.; Mijic, A. Isolating the impacts of anthropogenic water use within the hydrological regime of north India. *Earth Surf. Process. Landf.* **2020**, *45*, 1217–1228. [CrossRef]

103. Srinivasan, V.; Lambin, E.F.; Gorelick, S.M.; Thompson, B.H.; Rozelle, S. The nature and causes of the global water crisis: Syndromes from a meta-analysis of coupled human-water studies. *Water Resour. Res.* **2012**, *48*, 1–16. [CrossRef]

104. Di Baldassarre, G.; Viglione, A.; Carr, G.; Kuil, L.; Salinas, J.L.; Blöschl, G. Socio-hydrology: Conceptualising human-flood interactions. *Hydrol. Earth Syst. Sci.* **2013**, *17*, 3295–3303. [CrossRef]

105. Di Baldassarre, G.; Wanders, N.; AghaKouchak, A.; Kuil, L.; Rangecroft, S.; Veldkamp, T.I.E.; Garcia, M.; van Oel, P.R.; Breinl, K.; Van Loon, A.F. Water shortages worsened by reservoir effects. *Nat. Sustain.* **2018**, *1*, 617–622. [CrossRef]

106. Di Baldassarre, G.; Martinez, F.; Kalantari, Z.; Viglione, A. Drought and flood in the Anthropocene: Feedback mechanisms in reservoir operation. *Earth Syst. Dyn.* **2017**, *8*, 225–233. [CrossRef]

107. Gober, P.; Wheater, H.S. Socio-hydrology and the science—Policy interface: A case study of the Saskatchewan river basin. *Hydrol. Earth Syst. Sci.* **2014**, *18*, 1413–1422. [CrossRef]

108. Elshafei, Y.; Sivapalan, M.; Tonts, M.; Hipsey, M.R. A prototype framework for models of socio-hydrology: Identification of key feedback loops and parameterisation approach. *Hydrol. Earth Syst. Sci.* **2014**, *18*, 2141–2166. [CrossRef]

109. Elshafei, Y.; Coletti, J.Z.; Sivapalan, M.; Hipsey, M.R. A model of the socio-hydrologic dynamics in a semiarid catchment: Isolating feedbacks in the coupled human-hydrology system. *Water Resour. Res.* **2015**, *51*, 6442–6471. [CrossRef]

110. Global Water Partnership. What Is IWRM? Available online: https://www.gwp.org/en/GWP-CEE/about/why/what-is-iwrm/ (accessed on 30 May 2020).

111. Biswas, A.K. Integrated water resources management: A reassessment. *Water Int.* **2004**, *29*, 248–256. [CrossRef]

112. Biswas, A.K. Integrated water resources management: Is it working? *Int. J. Water Resour. Dev.* **2008**, *24*, 5–22. [CrossRef]

113. Molle, F. Nirvana concepts, narratives and policy models: Insights from the water sector. *Water Altern.* **2008**, *1*, 131–156.

114. Medema, W.; McIntosh, B.S.; Jeffrey, P.J. From premise to practice: A critical assessment of integrated Water resources management and adaptive management approaches in the water sector. *Ecol. Soc.* **2008**, *13*, 29. [CrossRef]

115. Hjorth, P.; Madani, K. Sustainability monitoring and assessment: New challenges require new thinking. *J. Water Resour. Plan. Manag.* **2014**, *140*, 133–135. [CrossRef]

116. Simonovic, S.P. *Managing Water Resources: Methods and Tools for a Systems Approach*; Routledge: London, UK, 2009, ISBN 9781844075546.

117. Mirchi, A.; Watkins, D.J.; Madani, K. Modeling for watershed planning, management, and decision making. In *Watersheds: Management, Restoration and Environmental Impact*; Nova Science Pub Inc.: London, UK, 2010, ISBN 9781616686673.

118. Loucks, D.P.; van Beek, E. *Water Resources Systems Planning and Management: An. Introduction to Methods, Models and Applications*; UNESCO: Paris, France, 2005.

119. Loucks, D.P.; Stedinger, J.R.; Haith, D.A. *Water Resource Systems Planning and Analysis*, 1st ed.; Prentice Hall: Upper Saddle River, NJ, USA, 1981.

120. Nayak, M.A.; Herman, J.D.; Steinschneider, S. Balancing flood risk and water supply in California: Policy search integrating short-term forecast ensembles with conjunctive use. *Water Resour. Res.* **2018**, *54*, 7557–7576. [CrossRef]

121. Aljefri, Y.M.; Fang, L.; Hipel, K.W.; Madani, K. Strategic analyses of the hydropolitical conflicts surrounding the grand Ethiopian renaissance dam. *Group Decis. Negot.* **2019**, *28*, 305–340. [CrossRef]

122. Seifi, A.; Hipel, K.W. Interior-point method for reservoir operation with stochastic inflows. *J. Water Resour. Plan. Manag.* **2001**, *127*, 48–57. [CrossRef]

123. Wang, L.; Fang, L.; Hipel, K.W. Mathematical programming approaches for modeling water rights allocation. *J. Water Resour. Plan. Manag.* **2007**, *133*, 50–59. [CrossRef]

124. Rheinheimer, D.E.; Null, S.E.; Lund, J.R. Optimizing selective withdrawal from reservoirs to manage downstream temperatures with climate warming. *J. Water Resour. Plan. Manag.* **2015**, *141*, 1–9. [CrossRef]

125. Ahmad, S.; Simonovic, S.P. System dynamics modeling of reservoir operations for flood management. *J. Comput. Civ. Eng.* **2000**, *14*, 190–198. [CrossRef]

126. Akter, T.; Simonovic, S.P. Aggregation of fuzzy views of a large number of stakeholders for multi-objective flood management decision-making. *J. Environ. Manage.* **2005**, *77*, 133–143. [CrossRef]

127. Jiang, H.; Simonovic, S.P.; Yu, Z.; Wang, W. A system dynamics simulation approach for environmentally friendly operation of a reservoir system. *J. Hydrol.* **2020**, *587*, 124971. [CrossRef]

128. King, L.M.; Simonovic, S.P.; Hartford, D.N.D. Using system dynamics simulation for assessment of hydropower system safety. *Water Resour. Res.* **2017**, *53*, 7148–7174. [CrossRef]

129. Mautner, M.R.L.; Foglia, L.; Herrera, G.S.; Galán, R.; Herman, J.D. Urban growth and groundwater sustainability: Evaluating spatially distributed recharge alternatives in the Mexico City Metropolitan Area. *J. Hydrol.* **2020**, *586*, 124909. [CrossRef]

130. Simonovic, S.P.; Ahmad, S. Computer-based model for flood evacuation emergency planning. *Nat. Hazards* **2005**, *34*, 25–51. [CrossRef]

131. Chang, N.B.; Chen, H.W.; Ning, S.K. Identification of river water quality using the fuzzy synthetic evaluation approach. *J. Environ. Manag.* **2001**, *63*, 293–305. [CrossRef] [PubMed]

132. Sharma, A.; Hipel, K.W.; Schweizer, V. Strategic insights into the cauvery river dispute in India. *Sustainability* **2020**, *12*, 1286. [CrossRef]

133. Afshar, A.; Najafi, E. Consequence management of chemical intrusion in water distribution networks under inexact scenarios. *J. Hydroinform.* **2014**, *16*, 178–188. [CrossRef]

134. Hipel, K.W.; Ben-Haim, Y. Decision making in an uncertain world: Information-gap modeling in water resources management. *IEEE Trans. Syst. Man Cybern. Part C Appl. Rev.* **1999**, *29*, 506–517. [CrossRef]

135. Gohari, A.; Bozorgi, A.; Madani, K.; Elledge, J.; Berndtsson, R. Adaptation of surface water supply to climate change in central Iran. *J. Water Clim. Change* **2014**, *5*, 391–407. [CrossRef]

136. Oliveira, R.; Loucks, D.P. Operating rules for multireservoir systems. *Water Resour. Res.* **1997**, *33*, 839–852. [CrossRef]

137. Winz, I.; Brierley, G.; Trowsdale, S. The use of system dynamics simulation in water resources management. *Water Resour. Manag.* **2009**, *23*, 1301–1323. [CrossRef]

138. Moridi, A. A bankruptcy method for pollution load reallocation in river systems. *J. Hydroinf.* **2019**, *21*, 45–55. [CrossRef]

139. Philpot, S.; Hipel, K.; Johnson, P. Strategic analysis of a water rights conflict in the south western United States. *J. Environ. Manag.* **2016**, *180*, 247–256. [CrossRef] [PubMed]

140. Sheikhmohammady, M.; Madani, K. A Descriptive model to analyze Asymmetric multilateral negotiations. In Proceedings of the Universities Council on Water Resources (UCOWR) Conference, Durham, NC, USA, 22–24 July 2008.

141. Read, L.; Madani, K.; Inanloo, B. Optimality versus stability in water resource allocation. *J. Environ. Manag.* **2014**, *133*, 343–354. [CrossRef] [PubMed]

142. Rouhi Rad, M.; Haacker, E.M.K.; Sharda, V.; Nozari, S.; Xiang, Z.; Araya, A.; Uddameri, V.; Suter, J.F.; Gowda, P. MOD\$\$AT: A hydro-economic modeling framework for aquifer management in irrigated agricultural regions. *Agric. Water Manag.* **2020**, *238*, 106194. [CrossRef]

143. Zeff, H.; Kaczan, D.; Characklis, G.W.; Jeuland, M.; Murray, B.; Locklier, K. Potential implications of groundwater trading and reformed water rights in Diamond Valley, Nevada. *J. Water Resour. Plan. Manag.* **2019**, *145*, 1–19. [CrossRef]

144. Housh, M.; Cai, X.; Ng, T.L.; McIsaac, G.F.; Ouyang, Y.; Khanna, M.; Sivapalan, M.; Jain, A.K.; Eckhoff, S.; Gasteyer, S.; et al. System of systems model for analysis of biofuel development. *J. Infrastruct. Syst.* **2015**, *21*, 04014050. [CrossRef]

145. Müller, M.F.; Müller-Itten, M.C.; Gorelick, S.M. How Jordan and Saudi Arabia are avoiding a tragedy of the commons over shared groundwater. *Water Resour. Res.* **2017**, *53*, 5451–5468. [CrossRef]

146. Madani, K.; Zarezadeh, M.; Morid, S. A new framework for resolving conflicts over transboundary rivers using bankruptcy methods. *Hydrol. Earth Syst. Sci.* **2014**, *18*, 3055–3068. [CrossRef]

147. Hassanzadeh, E.; Strickert, G.; Morales-Marin, L.; Noble, B.; Baulch, H.; Shupena-Soulodre, E.; Lindenschmidt, K.E. A framework for engaging stakeholders in water quality modeling and management: Application to the Qu'Appelle river basin, Canada. *J. Environ. Manag.* **2019**, *231*, 1117–1126. [CrossRef]

148. Simonovic, S.P.; Verma, R. A new methodology for water resources multicriteria decision making under uncertainty. *Phys. Chem. Earth* **2008**, *33*, 322–329. [CrossRef]

149. Parsapour-Moghaddam, P.; Abed-Elmdoust, A.; Kerachian, R. A Heuristic evolutionary game theoretic methodology for conjunctive use of surface and groundwater resources. *Water Resour. Manag.* **2015**, *29*, 3905–3918. [CrossRef]

150. Giuliani, M.; Castelletti, A.; Amigoni, F.; Cai, X. Multiagent systems and distributed constraint reasoning for regulatory mechanism design in water management. *J. Water Resour. Plan. Manag.* **2015**, *141*, 1–12. [CrossRef]

151. Hu, Y.; Quinn, C.J.; Cai, X.; Garfinkle, N.W. Combining human and machine intelligence to derive agents' behavioral rules for groundwater irrigation. *Adv. Water Resour.* **2017**, *109*, 29–40. [CrossRef]

152. Noël, P.H.; Cai, X. On the role of individuals in models of coupled human and natural systems: Lessons from a case study in the Republican River Basin. *Environ. Model. Softw.* **2017**, *92*, 1–16. [CrossRef]

153. Zekri, S.; Madani, K.; Bazargan-Lari, M.R.; Kotagama, H.; Kalbus, E. Feasibility of adopting smart water meters in aquifer management: An integrated hydro-economic analysis. *Agric. Water Manag.* **2017**, *181*, 85–93. [CrossRef]

154. Giuliani, M.; Herman, J.D. Modeling the behavior of water reservoir operators via eigenbehavior analysis. *Adv. Water Resour.* **2018**, *122*, 228–237. [CrossRef]

155. Loáiciga, H.A. Analytic game—Theoretic approach to ground-water extraction. *J. Hydrol.* **2004**, *297*, 22–33. [CrossRef]

156. Madani, K.; Dinar, A. Exogenous regulatory institutions for sustainable common pool resource management: Application to groundwater. *Water Resour. Econ.* **2013**, *2*, 57–76. [CrossRef]

157. Kanta, L.; Zechman, E. Complex adaptive systems framework to assess supply-side and demand-side management for urban water resources. *J. Water Resour. Plan. Manag.* **2014**, *140*, 75–85. [CrossRef]

158. Akhbari, M.; Grigg, N.S. Water management trade-offs between agriculture and the environment: A multiobjective approach and application. *J. Irrig. Drain. Eng.* **2014**, *140*, 1–11. [CrossRef]

159. Madani, K. The value of extreme events: What doesn't exterminate your water system makes it more resilient. *J. Hydrol.* **2019**, *575*, 269–272. [CrossRef]

160. Liu, J.; Dietz, T.; Carpenter, S.R.; Alberti, M.; Folke, C.; Moran, E.; Pell, A.N.; Deadman, P.; Kratz, T.; Lubchenco, J.; et al. Complexity of coupled human and natural systems. *Science* **2007**, *317*, 1513–1516. [CrossRef] [PubMed]

161. Müller, M.F.; Levy, M.C. Complementary vantage points: Integrating hydrology and economics for sociohydrologic knowledge generation. *Water Resour. Res.* **2019**, *55*, 2549–2571. [CrossRef]

162. Rahmandad, H.; Sterman, J. Heterogeneity and network structure in the dynamics of diffusion: Comparing agent-based and differential equation models. *Manag. Sci.* **2008**, *54*, 998–1014. [CrossRef]

163. Madani, K. Iran Statement, Statements by Heads of State and Government, High Level Segment Statements of COP23/CMP13/CMA1.2. In Proceedings of the 23rd Conference of the Parties to the United Nations Framework Convention on Climate Change (UNFCCC), Bonn, Germany, 16 November 2017. Available online: http://unfccc.int/files/meetings/bonn_nov_2017/statements/application/pdf/iran_cop23cmp13cma1-2_hls.pdf (accessed on 7 July 2020).

164. Ishtiaque, A.; Sangwan, N.; Yu, D.J. Robust-yet-fragile nature of partly engineered social-ecological systems: A case study of coastal Bangladesh. *Ecol. Soc.* **2017**, *22*. [CrossRef]

165. Rosenberg, D.E.; Madani, K. Water resources systems analysis: A bright past and a challenging but promising future. *J. Water Resour. Plan. Manag.* **2014**, *140*, 407–409. [CrossRef]

166. Brown, C.M.; Lund, J.R.; Cai, X.; Reed, P.M.; Zagona, E.A.; Ostfeld, A.; Hall, J.; Characklis, G.W.; Yu, W.; Brekke, L. The future of water resources systems analysis: Toward a scientific framework for sustainable water management. *Water Resour. Res.* **2015**, *51*, 6110–6124. [CrossRef]

167. Vogel, R.M.; Lall, U.; Cai, X.; Rajagopalan, B.; Weiskel, P.K.; Hooper, R.P.; Matalas, N.C. Hydrology: The interdisciplinary science of water. *Water Resour. Res.* **2015**, *51*, 4409–4430. [CrossRef]

168. Kasprzyk, J.R.; Smith, R.M.; Stillwell, A.S.; Madani, K.; Ford, D.; McKinney, D.; Sorooshian, S. Defining the role of water resources systems analysis in a changing future. *J. Water Resour. Plan. Manag.* **2018**, *144*, 01818003. [CrossRef]

169. Loucks, D.P. From analyses to implementation and innovation. *Water* **2020**, *12*, 974. [CrossRef]

170. Hornberger, G.M.; Perrone, D. *Water Resources: Science and Society*; Johns Hopkins University Press: Baltimore, MD, USA, 2019.

171. Ringler, C.; Cai, X. Valuing fisheries and wetlands using integrated economic-hydrologic modeling—Mekong river basin. *J. Water Resour. Plan. Manag.* **2006**, *132*, 480–487. [CrossRef]

172. Yang, Y.C.E.; Cai, X. Reservoir reoperation for fish ecosystem restoration using daily inflows—Case study of Lake Shelbyville. *J. Water Resour. Plan. Manag.* **2011**, *137*, 470–480. [CrossRef]

173. Escriva-Bou, A.; Lund, J.R.; Pulido-Velazquez, M.; Hui, R.; Medellín-Azuara, J. Developing a water-energy-GHG emissions modeling framework: Insights from an application to California's water system. *Environ. Model. Softw.* **2018**, *109*, 54–65. [CrossRef]

174. Liu, B.; Lund, J.R.; Liao, S.; Jin, X.; Liu, L.; Cheng, C. Optimal power peak shaving using hydropower to complement wind and solar power uncertainty. *Energy Convers. Manag.* **2020**, *209*, 112628. [CrossRef]

175. Shen, J.; Cheng, C.; Cheng, X.; Lund, J.R. Coordinated operations of large-scale UHVDC hydropower and conventional hydro energies about regional power grid. *Energy* **2016**, *95*, 433–446. [CrossRef]

176. Xu, Z.; Yang, Z.; Cai, X.; Yin, X.; Cai, Y. Modeling framework for reservoir capacity planning accounting for fish migration. *J. Water Resour. Plan. Manag.* **2020**, *146*, 04020006. [CrossRef]

177. Shafiee-Jood, M.; Housh, M.; Cai, X. Hierarchical decision-modeling framework to meet environmental objectives in biofuel development. *J. Water Resour. Plan. Manag.* **2018**, *144*, 04018030. [CrossRef]

178. Housh, M.; Yaeger, M.A.; Cai, X.; McIsaac, G.F.; Khanna, M.; Sivapalan, M.; Ouyang, Y.; Al-Qadi, I.; Jain, A.K. Managing multiple mandates: A system of systems model to analyze strategies for producing cellulosic ethanol and reducing riverine nitrate loads in the upper Mississippi river basin. *Environ. Sci. Technol.* **2015**, *49*, 11932–11940. [CrossRef]

179. Cai, X.; Wallington, K.; Shafiee-Jood, M.; Marston, L. Understanding and managing the food-energy-water nexus—Opportunities for water resources research. *Adv. Water Resour.* **2018**, *111*, 259–273. [CrossRef]

180. Su, Y.; Kern, J.D.; Denaro, S.; Hill, J.; Reed, P.; Sun, Y.; Cohen, J.; Characklis, G.W. An open source model for quantifying risks in bulk electric power systems from spatially and temporally correlated hydrometeorological processes. *Environ. Model. Softw.* **2020**, *126*, 104667. [CrossRef]

181. Escriva-Bou, A.; Lund, J.R.; Pulido-Velazquez, M. Optimal residential water conservation strategies considering related energy in California. *Water Resour. Res.* **2015**, *51*, 4482–4498. [CrossRef]

182. Escriva-Bou, A.; Lund, J.R.; Pulido-Velazquez, M. Saving energy from urban water demand management. *Water Resour. Res.* **2018**, *54*, 4265–4276. [CrossRef]

183. Wade, R. The The management of irrigation systems: How to evoke trust and avoid prisoner's dilemma. *World Dev.* **1988**, *16*, 489–500. [CrossRef]

184. Ferdon, H.R. Game Theory Analysis of Intra-District Water Transfers; Case Study of the Berrenda Mesa Water Disctrict. Master's Thesis, California State Polytechnic University, San Luis Obispo, CA, USA, 2016.

185. Ert, E.; Cohen-Amin, S.; Dinar, A. The effect of issue linkage on cooperation in bilateral conflicts: An experimental analysis. *J. Behav. Exp. Econ.* **2019**, *79*, 134–142. [CrossRef]

186. Tayloer, A.C. A Planning Model for a Water Quality Management Agency. *Manag. Sci.* **1973**, *20*, 675–685. [CrossRef]

187. Pittel, K.; Rübbelke, D.T.G. Transitions in the negotiations on climate change: From prisoner's dilemma to chicken and beyond. *Int. Environ. Agreem. Polit. Law Econ.* **2012**, *12*, 23–39. [CrossRef]

188. Soroos, M.S. Global Change, environmental security, and the prisoner's dilemma. *J. Peace Res.* **1994**, *31*, 317–332. [CrossRef]

189. Madani, K. Modeling international climate change negotiations more responsibly: Can highly simplified game theory models provide reliable policy insights? *Ecol. Econ.* **2013**, *90*, 68–76. [CrossRef]

190. Quinn, N.; Blöschl, G.; Bárdossy, A.; Castellarin, A.; Clark, M.; Cudennec, C.; Koutsoyiannis, D.; Lall, U.; Lichner, L.; Parajka, J.; et al. Invigorating hydrological research through journal publications. *Water Resour. Res.* **2020**, *56*, 257–260. [CrossRef]

191. Hjorth, P.; Madani, K. Systems Analysis to Promote Frames and Mental Models for Sustainable Water Management. In Proceedings of the 3rd World Sustainability Forum, Basel, Switzerland, 1–30 November 2013; p. f003. [CrossRef]

192. Ackoff, R.L. Strategies, systems, and organizations: An interview with Russell, L. Ackoff. *Strateg. Leadersh.* **1997**, *25*, 22–27. [CrossRef]

193. Madani, K.; Khatami, S. Water for energy: Inconsistent assessment standards and inability to judge properly. *Curr. Sustain. Energy Rep.* **2015**, *2*, 10–16. [CrossRef]

194. Lund, J.R. Integrating social and physical sciences in water management. *Water Resour. Res.* **2015**, *51*, 5905–5918. [CrossRef]

195. Kurtz, C.F.; Snowden, D.J. The new dynamics of strategy: Sense-making in a complex and complicated world. *IBM Syst. J.* **2003**, *42*, 462–483. [CrossRef]

Review

CHNS Modeling for Study and Management of Human–Water Interactions at Multiple Scales

Kumaraswamy Ponnambalam [1,*] and S. Jamshid Mousavi [2]

[1] Department of Systems Design Engineering, University of Waterloo, Waterloo, ON N2L 5R5, Canada
[2] School of Petroleum, Civil and Mining Engineering, Amirkabir University of Technology (Tehran Polytechnic), Tehran 15875-4413, Iran; jmosavi@aut.ac.ir
* Correspondence: ponnu@uwaterloo.ca

Received: 14 May 2020; Accepted: 11 June 2020; Published: 14 June 2020

Abstract: This paper presents basic definitions and challenges/opportunities from different perspectives to study and control water cycle impacts on society and vice versa. The wider and increased interactions and their consequences such as global warming and climate change, and the role of complex institutional- and governance-related socioeconomic-environmental issues bring forth new challenges. Hydrology and integrated water resources management (IWRM from the viewpoint of an engineering planner) do not exclude in their scopes the study of the impact of changes in global hydrology from societal actions and their feedback effects on the local/global hydrology. However, it is useful to have unique emphasis through specialized fields such as hydrosociology (including the society in planning water projects, from the viewpoint of the humanities) and sociohydrology (recognizing the large-scale impacts society has on hydrology, from the viewpoint of science). Global hydrological models have been developed for large-scale hydrology with few parameters to calibrate at local scale, and integrated assessment models have been developed for multiple sectors including water. It is important not to do these studies with a silo mindset, as problems in water and society require highly interdisciplinary skills, but flexibility and acceptance of diverse views will progress these studies and their usefulness to society. To deal with complexities in water and society, systems modeling is likely the only practical approach and is the viewpoint of researchers using coupled human–natural systems (CHNS) models. The focus and the novelty in this paper is to clarify some of these challenges faced in CHNS modeling, such as spatiotemporal scale variations, scaling issues, institutional issues, and suggestions for appropriate mathematical tools for dealing with these issues.

Keywords: coupled human–natural systems; integrated water resources management; sociohydrology; modeling perspectives; agent-based modeling; differential equations; system dynamics; uncertainty; artificial intelligence; machine learning

1. Introduction

"By the continuance of rain the world is preserved in existence; it is therefore worthy to be called ambrosia", Thirukkural [1]—Couplet 11 and *"Even the wealth of the wide sea will be diminished, if the cloud that has drawn (its waters) up gives them not back again (in rain)"*, Thirukkural [1]—Couplet 17, (From about 2000 years ago).

Hydrology is defined [2] as "the science which deals with the waters of the earth, their occurrence, circulation and distribution on the planet, their physical and chemical properties and their interactions with the physical and biological environment, including their responses to human activity". Integrated water resources management (IWRM) is defined as "a process which promotes the coordinated development and management of water, land and related resources, in order to maximize the resultant economic and social welfare in an equitable manner without compromising the sustainability of vital ecosystems" [3]. By these definitions, hydrology is a descriptive tool and IWRM is a prescriptive tool which, by necessity, depends on descriptive tools and must be interdisciplinary. Choi and Pak [4] clarified that "Interdisciplinarity analyzes, synthesizes and harmonizes links between disciplines into a coordinated and coherent whole". While this is a very difficult requirement, some humble attempts to define interdisciplinary approaches have been made through these proposed fields: hydrosociology was motivated by Falkenmark [5] to encourage developing a field to include, from the humanities viewpoint, the society in the planning of water projects, and Sivapalan et al. [6] introduced sociohydrology (SH) as a new emerging branch of hydrology from a scientific viewpoint focusing on sociohydrologic systems as coupled human–natural (CHNS) systems including the feedback on to both local and global systems, where dynamics of the two systems and their co-evolution should be explicitly accounted for.

These interdisciplinary approaches require tools in order to solve real world problems. A decade before presenting sociohydrology (SH), system dynamics (SD) models, an important tool of systems modelers, was used by Simonovic [7] under the umbrella of integrated modeling to consider coupled human–natural systems in his WorldWater model, although not considering the important feedback of evapotranspirations from land and oceans to precipitation which SH proposes to address. It is noted that SD should not be restricted to any particular mathematical construct but they are only lumped systems models and hence are suitable for global level modeling of a large number of systems. SH has been promoting applications of CHNS, and several works have been contributing to SH from both conceptual frameworks and modeling approaches perspectives (e.g., Di Baldassarre et al. [8], Sivapalan and Blöschl [9], Blair and Buytaert [10], Sivapalan and Blöschl [11], Di Baldassarre et al. [12], Xi-Liu and Qing-Xian [13], and Di Baldassarre et al. [14]).

Xu et al. [15] argued eloquently for including social elements in CHNS models. Wesselink et al. [16] summarized well the differences between SH and hydrosociology (HS), and, although we do not discuss this here in detail, they considered the main differences between SH and HS as descriptive versus critical, objective versus subjective, and nature centric versus society centric, among others. SH aims at addressing dynamic cross-scale interactions and feedbacks between natural and human processes that can cause water sustainability challenges [17]. Additionally, there are arguments on whether or not SH can be considered as a new discipline; for example, Sivakumar [18] and Koutsoyiannis [19] who in spite of Wesselink et al. [16] are not convinced that SH has substantial differences with hydrosociology or is a new science, respectively.

The research cited thus far comes from water specialists, but others such as ecologists and economists [20,21] as well as the USA National Science Foundation have also recognized the importance of studying CHNS. Fu and Wei [22] stated "Humans as a group have learned numerous unpleasant lessons for keeping fit for changes in coupled natural and human systems (CNH). ... Furthermore, ... we have a limited understanding of the dynamic mechanisms of CNH; therefore, we have been unable to provide a manual for humans' ability to keep fit for a more sustainable global environment."

More than two hundred hydrologists all around the world contributed to a specific paper [23], reporting the systematic procedure taken for identifying 23 unsolved questions in hydrology to streamline future research in this field, the same as what David Hilbert did in 1900 for mathematics. The problems identified are mainly about understanding how change propagates across interfaces within the hydrological system and across disciplinary boundaries, and in particular human interactions with nature and water cycle feedbacks [23]. Among the 23 questions introduced, the following questions are directly related to CHNS, its conceptualization, or suitable modeling tools, implying the importance of studying CHNS in the future of hydrology: "Q4. What are the impacts of land cover change and soil disturbances on water and energy fluxes at the land surface, and on the resulting groundwater recharge?" [23], "Q6. What are the hydrologic laws at the catchment scale and how do they change with scale?", "Q7. Why is most flow preferential across multiple scales and how does such behaviour co-evolve with the critical zone?" (Q7 in fact addresses the main issues related to the distribution and nature of flow paths raised in Q5 and Q6), "Q18. How can we extract information from available data on human and water systems in order to inform the building process of socio-hydrological models and conceptualisations?", "Q21. How can the (un)certainty in hydrological predictions be communicated to decision makers and the general public?", "Q22. What are the synergies and tradeoffs between societal goals related to water management (e.g., water–environment–energy–food–health)?", and "Q23. What is the role of water in migration, urbanisation and the dynamics of human civilisations, and what are the implications for contemporary water management?" [23].

What has made CHNS analysis more important than before is what has happened to our societies and the Earth system during the last century, especially in the last seven decades. Humans have been utilizing more natural resources of the Earth such as oil, minerals, soil, and particularly water for different uses, affecting strongly ecosystems services. Therefore, we are currently impacting the Earth system's elements, resources, and processes much more than before. These impacts, which used to be more local, have become more intense (locally) and broader (globally). Consequently, the Earth system's response to the huge amount of human interventions has now returned to impact us and our societies. Global warming and climate change and their impacts are just examples for the mentioned influences. Therefore, what we have done on the Earth has come back to affect us in unplanned and unexpected ways; thus, from a system perspective, we are currently facing a coupled system consisting of two main human and natural systems interacting with each other. As a result, we cannot model each of these two systems without accounting for the feedback it receives from the other one. This high coupling makes interdisciplinarity studies a requirement, and CHNS modeling and the quantitative tools for such modeling are inevitable. Several other works making the above points are related to the idea of "planetary boundaries" in references [24,25] or those referring to water scarcity in ref. [26] and water security in ref. [27].

Figure 1 presents symbolic views of interactions between these two systems in the 18th to 20th centuries and the current 21st century view. The top figure is how most hydrology and water management were studied where the people simply depended on the world's global water cycle as an input and did not consider that people were changing it at global scales, which was a reasonable assumption for most of history. In the view now (the bottom one), this interaction between people and global hydrology is explicitly considered as tightly coupled and indicating huge impacts humans have on Earth entering the Anthropocene, thus bringing enormous challenges to modeling and computational research.

Nikolic and Simonovic [29] suggested a generic multi-method modeling framework for support of IWRM to capture structural complexities of water resources systems and to examine the codependence between these systems and socioeconomic environment. Loucks [30] stressed the need for a new kind of water resources planning and management modeling expertise addressing a wider range of societal concerns that stem from the impact water has on human activities. Given the interconnectedness of water and socioeconomic systems, he reminded systems modelers of the need for viewing a water resource system as a coupled social-economic and ecological system and for developing models capable of estimating the possible social impacts, and capturing the adaptive capacity of these systems to learn and innovate in response to change [30]. However, some of the challenges are due to feedbacks between social and natural (water) systems at different spatiotemporal scales. In the last two decades, there have been several attempts for introducing and dealing with CHNS and the associated frameworks, concepts, and modeling tools. Stevenson [31] presented a framework for CHNS consisting of five elements: human well-being, environmental policy, human activities, stressors (contaminants, pollution loads, etc.), and ecosystem services for environmental management problems. Propagation of thresholds in relationships among the elements through CHNS is the key aspect of the proposed framework. Senf et al. [32] used remote sensing as an information source for modeling the central Europe forest ecosystem dynamics and mapping forest disturbances.

Before Anthropocene

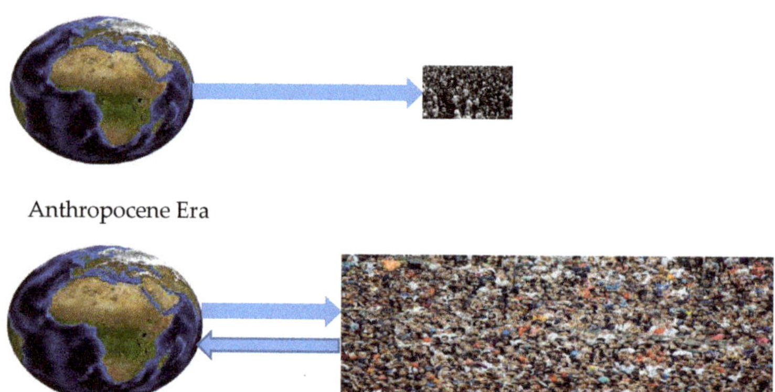

Anthropocene Era

Figure 1. Different perspectives of impacts of people on Earth and vice versa in two different time periods (Small crowd picture: https://unsplash.com/photos/sUXXO3xPBYo Globe: https://www.hiclipart.com/free-transparent-background-png-clipart-dxjin/download Large crowd picture: https://unsplash.com/photos/8I423fRMwjM) [28].

Integrated assessment models (IAMs) and global hydrological models (GHMs) are among other modeling approaches attempting to address feedbacks at a global scale. IAMs as complex socioeconomic models incorporate water resources in terms of both supply (simplified hydrology) and demand (water use) in references [33,34]. IAMs are lumped systems models that use the global change assessment model (GCAM). GCAM [35], being under development for over 35 years, is a global model that represents the behavior of, and interactions between, five systems: the energy system, water, agriculture and land use, the economy, and the climate. These lumped system models have been considered as one of the tools of CHNS in ref. [36]. On the other hand, GHMs incorporate water demand scenarios into complex hydrological models although adding genuine feedbacks is ongoing. GHMs distinguish their approach from other global hydrological models by having a few parameters to calibrate especially at the local scale. Any parameter estimation it does is done at a large scale such as at the ecoregion, large river basins or climatic regions (see, e.g., the review paper by Sood and Smakhtin [37]). GHMs run in a grid format at a spatial resolution of 0.5 degrees (just over 3100 km^2 per grid cell at the Equator) at worse and temporal resolution of a day. Of course, the requirement of data to fit this spatiotemporal scale for the entire globe makes it harder to apply uniformly leading to many different models. The match between various GHMs is normally not that good and hence many uncertainties in their results are yet to be fully understood. Brêda et al. [38] presented a recent GHM application in South America indicating that, without the ability to calibrate, the results are difficult to judge. PCR-GLOBWB 2 is also a GHM that is worth noting as it has 5 arc-min resolution, about 10 km at the equator, and can optionally couple with other well-known models, e.g., MODFLOW [39].

CHNS modeling has been considered in IWRM, complex water resources systems analysis, and water–energy–food–environment nexus approach using different systems analysis simulation and optimization approaches. For instance, Cai et al. [40] formulated a basin-scale integrated hydrologic–agronomic–economic model as a highly nonlinear mathematical program. In particular, the SD method and its advantage in modeling nonlinear feedbacks was employed to simulate interactions between social and natural systems of complex water resources systems. Simonovic's [7] WorldWater SD-based model integrates water resources sector with five driving sectors of industrial growth: population, agriculture, economy, nonrenewable resources, and persistent pollution. The model results demonstrate the strong relationship between the water resources and industrial growth. Simonovic and Davies [41] stated "the current approach to understanding connections between biophysical and socioeconomic systems requires their artificial separation via modelling techniques. Such an approach explicitly excludes those feedbacks critical to understanding the behaviour of climate and socioeconomic systems". Subsequently, they discussed the unreal simplifying assumptions made because of excluding the feedbacks that operate between biophysical and socioeconomic systems. Prodanovic and Simonovic [42] coupled a continuous hydrologic model and a socioeconomic model using SD, where the hydrologic component responds to changes in socioeconomic conditions, and socioeconomic conditions are influenced by hydrologic quantities. Davies and Simonovic [43] pointed out that water resources models have traditionally considered socioeconomic and environmental changes as external drivers, thus they have mainly focused on water systems. Therefore, to provide insight into the nature and structure of connections between water resources and socioeconomic and environmental changes, they presented a SD-based integrated assessment model incorporating dynamic representations of these systems through nonlinear feedbacks. Lam et al. [44] proposed a modeling framework to analyze the sustainability problem in Mississippi River Delta. The framework includes six components from the natural and human systems linked together, and the SD approach is used to model the feedback loops among the components. There are more recent SD applications in water and hydrologic systems in ref. [45], which is a review article, and in ref. [46], which presents a watershed scale application.

Agent-based modeling (ABM) have been utilized as one of the main quantitative techniques for CHNS analysis [47]. Walker et al. [48] offered a conceptual model of a coupled social–water system and proposed analytic approaches to support policymaking for environmental and water resources planning and management. Dziubanski et al. [49] built a sociohydrological model by integrating ABM and a quasi-distributed hydrologic model, in which the impacts of land-cover changes resulting from decisions made by two different agent types are simulated using the curve number method. The model is used to simulate scenarios of crop yields, crop prices, and conservation subsidies considering varied farmer parameters representing the effects of human system variables on peak discharges. Noël and Cai [50] focused on how to quantify the role of individuals in models of CHNS in a basin-scale irrigation management problem, where the human sub-system is a community of farmers. They coupled an agent-based model, simulating farmers' behavior, and a groundwater model and concluded that such behavior can be considered as an additional source of uncertainty in the CHNS model proposed. ABM has also been used for systematically studying interactions among hydrology, climate, and strategic human decision making in a watershed system [51] and among those mentioned elements and landscape-scale forest ecosystems [52] as CHNS.

Pouladi et al. [53] presented a sociohydrological modeling framework for complex water resources systems performance assessment by combining ABM and the theory of planned behavior (TPB) to account for farmers' behavior in the Lake Urmia Basin, Iran. The framework was then extended by Pouladi et al. [54], who integrated ABM and data mining for capturing farmers' sociohydrological interactions and complex behavior in response to drought conditions. They employed the association rule to discover the main patterns from the field data collected, representing the farmers' agricultural decisions. The rules discovered were used then in ABM as the behavioral rules to simulate the agricultural activities. Aghaie et al. [55] presented an agent-based groundwater market model to analyze the economic and hydrologic impacts of different market mechanisms and water buyback programs.

Based on above review of literature, different branches of the Earth system science, IWRM, and SH have been emphasizing the need for CHNS modeling. In the following section, we elaborate further on overlapping and distinctive aspects of IWRM and SH, both of which have contributed to the progress and advancement of CHNS analysis. Then, as part of our aim in this paper, we discuss on how systems analysis approaches, e.g., SD, ABM, stochastic differential equations, and optimization, along with artificial intelligence (AI), machine learning (ML), and data analytics algorithms, are used in modeling and quantifying co-ordination and integration as the heart of IWRM definition and co-evolution stressed in SH, respectively.

2. Materials and Methods

2.1. IWRM and Sociohydrology (SH)

Before presenting our view on similarities and differences between IWRM and SH, we provide some other important views on this matter as mentioned in the 23-unsolved-problems-in-hydrology paper where it is stated "The traditional support that hydrology has provided to water resources management [56] in its dual role of (i) quantifying hydrological extremes and resources relative to societal needs and (ii) quantifying the impact society has on the water cycle, is now broadened in a number of ways. First, more integrated questions of the long-term dynamic feedbacks between the natural, technical and social dimensions of human-water systems. While water resources systems analysis [57] has dealt with such interactions from an optimisation perspective on a case-by-case basis, much is to be learned by developing a general understanding of phenomena that arise from the interactions between water and human systems. Thus, as socio-economic perspectives [58,59] are being integrated in these feedbacks, the interest is not only on decision support but also on the role of society in the hydrological cycle in its own right. Second, ... , the topic of water and health (e.g., Mayer et al. [60] and Dingemans et al. [61]), as well as spatial problems such as the interaction of migration and water issues. Third, ... Also, water is traded globally through the water–energy–food nexus, and it will be interesting to see what role hydrology can play in this nexus [62]." (Blöschl et al. [23] (p. 1152)).

Several important issues are highlighted in GWP's (2000) IWRM definition that it is "a process which promotes ... " [3] such as managing resources in a coordinated way, considering other related resources than just water (soil, ecosystem, etc.), social welfare in addition to economic objectives, equity, sustainability, and environment. However, to apply these important issues in water management practices, it is also necessary to know what kind of science, approaches, and modeling tools we need to help promote the above process. For example, Metz and Glaus [63] mentioned that the integration of water-related policies in IWRM is challenging because policy actors should coordinate their demands and actions across policy sectors, territorial entities, and decision-making levels within a water basin, whereas actors are restricted by the policy framework. Biswas [64] provided a critique of IWRM in practice, but, given the comprehensiveness required by its definition, it is fair that most studies could not satisfy the definition.

Fu and Wei [22] presented the conceptual cascade of "pattern–process–service (function)–sustainability" for understanding of diagnoses and practices for keeping fit in CHNS. They stated "The former refers to understanding the dynamics of CHNS, and the latter refers to management policies and practices for improving sustainability." Therefore, from such a perspective, SH is more related to the diagnostic understanding, whereas IWRM is more concerned with the practices needed for the keeping-fit concept in CHNS. It is worth noting that "keep fit" is considered as the process of matching a socioeconomic system with its biophysical environment across temporal and spatial scales, while bidirectional coupling exists between environmental changes and socioeconomic changes [22].

The concept of IWRM moves away from top-down "water master planning" and toward "comprehensive water policy planning". The IWRM concept already recognizes the role and importance of two other subsectors, socioeconomic and institutional, and then the natural subsector and addresses the interactions between these subsectors as described well in ref. [65].

IWRM has been mainly concerned with management aspects of technical, socioeconomic, and institutional dimensions of water-related decision making. SH has been proposed for studying water–human systems dynamics and emphasizes the consideration of the co-evolution of natural (hydrologic) and social systems, and to consider human systems feedback on global water cycle and vice versa. Therefore, explicit considerations of multiscale feedbacks proposed in SH is useful for a comprehensive application of IWRM, which the CHNS modeling approach facilitates.

The importance of scale-related challenges has been recognized in SH. For example, Blöschl et al. [23] (p. 1152) stated "The challenges lie in linking short-term local processes (what we have mostly studied in the past) to long-term global processes, and vice versa". In terms of temporal scale, both IWRM and SH can address different time scales, especially long-term impacts under notions of sustainability and (time) evolution, respectively. However, the co-evolution of human–natural systems in SH is more relevant to, and compatible with, the need for considering nonstationarity aspects in these complex systems than tools and methods being developed and used in IWRM. Regarding the spatial scale, as a step ahead and moving from a single project-scale approach to a wider spatial-scale modeling framework, IWRM has mainly focused on basin- or regional-scale analyses, thus IWRM considers the impacts of human interventions on water budget components locally at a basin scale. Therefore, human interventions such as urbanization effects, dam constructions, expansion of irrigated areas, etc., and their impacts on local (basin scale) water cycle have already been recognized by current water resources management practices. On the other hand, the impact of global water cycle on local basin-scale water resources systems and societies when considered takes a typical one-way scenario-based approach, with few exceptions (see ref. [66]). Nevertheless, the cumulative impacts of the mentioned human interventions on global water cycle goes far beyond a basin-scale approach. In other words, current hydrologic and water-resource management approaches do not sufficiently account for the impact of basin-scale human-induced water resource-related activities on the large-scale hydrology of Earth, and many long-term hydrologic predictions do not account for global forces that influence local sources and vice versa, something that SH is proposing to address. Van der Ent et al.'s [67] study of moisture tracking on a global scale is a good example of such suggested studies in SH.

In response to the above-mentioned points, SH focuses on some other important issues lacking in well-established science of hydrology and IWRM such as impacts of human interventions on global water cycle directly and explicitly. By "directly and explicitly", we mean SH highlights the importance of considering human (social) and natural (hydrologic) systems as co-evolving coupled systems, whereas their interactions are considered as part of the systems itself, through for example a two-way feedback approach, not just via a one-way scenario-based approach. To do so, SH requires quantifying approaches and modeling frameworks, and CHNS modeling is a possible framework. When studying CHNS, we embed human–natural systems including water. From such a point of view, SH through CHNS becomes a study of complex system of systems, a type of system discussed well for example by Haimes [68].

Let us clarify what we explain above about local- and global-scale impacts through a simple well-known problem in classical hydrology. Urbanization is a good, known example of a human (social) system impact on the hydrologic cycle locally. Local-scale effects of urbanization, especially flooding issues, are typically considered through designing a proper urban drainage system. The impacts are increases in surface runoff; reductions in infiltration, groundwater recharge, and evaporation, degradation of water quality indices, etc. In response to the impacts of such a socioeconomic-driven intervention on local water cycle components, engineers design proper drainage systems and wastewater treatment plants to efficiently manage stormwater resulting from land use change and increased impervious lands. Additionally, comprehensive integrated models are currently available for assessing a longer than normal time horizon (e.g., 100 years) impact of urbanization on a basin, similar to the work of Luo et al. [69], who studied the loss of land from agriculture and forestry to urban areas to increase land runoff to over 60% in China. However, what about the impacts of these changes on the global water cycle components and our responses to them? What is the cumulative impact of millions of small to large developed urbanized areas in different basins, countries, regions, and continents all around the world on the global water cycle? Have we addressed this latter-type impact adequately in our hydrologic and water resources models? We believe this is something SH has correctly attempted to address. In this line, among 23 unsolved problems in hydrology, Question 23 particularly emphasizes migration and urbanization as key topics to focus on in human–water interactions [23].

A point to mention, however, is that, although CHNS has mainly been emphasized in SH, we believe CHNS modeling tools, especially feedback-based socioeconomic–natural systems modeling, has already been used in IWRM. In other words, the authors of this paper think that, while the concept of co-evolving coupled human–natural systems emphasized in SH is essential and of great importance, the SH community may not have adequately recognized several valuable attempts accounting explicitly for coupled human–natural systems modeling done under names and areas other than SH. The attempts have particularly taken place using system dynamics (SD) approach in IWRM, systems approach to complex water resources systems, and water–energy–food–environment nexus approach, some of which were cited in the Introduction (e.g., the works of Simonovic [7], Simonovic and Davies [41], Prodanovic and Simonovic [42], Davies and Simonovic [43], Walker et al. [48], Noël and Cai [50], and Loucks [30]). Additionally, SH does not explicitly talk about the role and importance of institutions and institutional subsystem underlying the co-evolution of human (social) and natural systems, something recognized by IWRM at least conceptually, if not quantitatively. This issue is explained further and discussed later in Section 2.2.4.

The co-evolutionary thinking of SH will be of help and importance in the current and the future of hydrology as a science. However, there is still a significant gap in developing and advancing computational CHNS modeling approaches under any of SH, IWRM, or food–energy–water–environment–health nexus contexts. A tutorial was presented, including as many of the modern computational approaches necessary for considering such challenges in CHNS, by Ponnambalam et al. [28].

It is worth mentioning that, despite the significance of CHNS analysis, there are not as many works that have successfully applied quantitative approaches and models of CHNS as there should be, mainly because of the complexity level of the models required for tackling such a task. In this regard, Loucks [70] stated "Simplification is why we model ... We know that our simplified models will be wrong. But, we develop them because they can be useful. The simpler and hence the more understandable models are the more likely they will be useful, and used, as long as they do the job." Therefore, he raised an important question: " ... what level of model complexity is needed to do a job when the information needs of that job are uncertain and changing?" For instance, we typically assume in IWRM that future water availability and demand values are known or can be estimated (in deterministic or probabilistic sense) without explicitly accounting for the impacts of socioinstitutional systems on their future estimations. However, the point is how challenging the consideration of such impacts would be, and what assumptions, information, and modeling techniques would be needed to overcome the complexities it brings. In this regard, Walker et al. [48] suggested that environmental and water systems decision makers need to consider social responses as well as economic and environmental impacts of their decisions, but predictions of such responses will not be accurate, especially in the future, and hence requiring uncertainty modeling and its inclusion, another thing not well done. To better understand the coupled social and natural components of water resource systems, they then provide some examples of how hard it is to attempt predictions, why, and the consequences if those predictions are wrong.

In the following, we elaborate more on CHNS modeling challenges. In addition, opportunities and potential modeling tools that can address some of these challenges are introduced and discussed.

2.2. CHNS Modeling: Challenges and Opportunities

It is worth noting that SH emphasizes co-evolution to be considered in CHNS modeling. However, from a quantitative and mathematical modeling perspective, it would be very hard and challenging to fully model all behavioral aspects of these very complex coupled systems. Below, we explain and discuss our view on different aspects of such challenges and the complexities involved and provide links to tools:

2.2.1. Mismatch in Temporal Scale and Time Resolution

We first discuss the mismatch in temporal scale and time resolution of models and variables of interest in social systems and those in natural systems. Such a challenge arises while trying to answer questions such as "What is the role of water in migration, urbanisation and the dynamics of human civilisations, and what are the implications for contemporary water management?" as the 23rd question mentioned by Blöschl et al. [23]. To clarify such an inconsistency, suppose a basin-scale water allocation model is the one in which we are going to simulate both natural (physical) processes and social processes and their interrelationships. Governing equations and variables in the natural (physical) system are mass balance equations over time and space or other hydrologic-related equations, reservoir releases, water allocations to demand sites, etc. at a certain temporal scale. On the other hand, governing equations and variables in the socioeconomic system are those related to processes of poverty, migration, income, education, gender equity, etc. at another temporal scale. The type of related equations for the socioeconomic system could be statistical regression equations derived from analysis of corresponding data and information collected from questionnaire or supply and demand curves and functions fitted to data, where societal variables and signals of interest are the number of migrated people from and different sites, average household income, etc.

Traditional natural system-focused water allocation models solve water balance equations over time and space along with operation rules and policies on how to make releases from the reservoirs under predefined demand satisfaction priorities. The corresponding water allocation problem is typically formulated as single- or multi-period network flow programs (NFPs) or linear programs (LPs) solved iteratively by fast out-of-kilter, Lagrangian relaxation, or dual Simplex algorithms. In this approach, socioeconomic considerations are only accounted for indirectly through values of water demands and the priority numbers selected for different demand types and sites and target reservoir levels, which are determined outside the model. What is directly considered inside the model is the mentioned equations related to physical (natural) subsystem. Such a framework is currently being employed in well-known, well-established river basin decision support systems (DSSs) such as ModSim and WEAP.

The question here is: What challenges do we have if we want to develop a new coupled human–natural, sociohydrologic-based water allocation model integrating the mentioned natural system-focused water allocation DSS and a socioeconomic model at varying temporal scales? Such a coupled model aims to simulate both physical and social processes and variables of interest and their interrelations and impacts on each other. This is because we know that more reliable, timely, and adequate water allocations from both water quantity and quality aspects can improve societal and economic signals and indicators in that site or region. On the other hand, a demand area having better welfare indices motivates more development, as well as population and economic growth, resulting in higher levels of water demand, which will therefore put more stress on natural hydrologic system when the quantity of water resources is limited. One rising challenge here is that the typical time resolution considered for (hydrologic) variables of the natural system-focused water allocation model, e.g., water allocation values and reservoir releases, is one day to one month in duration.

Suppose that we are able to develop a quantitative socioeconomic model, simulating social processes and variable of interest, e.g., poverty, income, migration, education quality and level, health, etc., to couple it with the water allocation model. These variables and processes would respond to the changes of water allocation variables on a much longer time interval basis than a month, usually in years of interval. It may not be reasonable to have a monthly-basis socioeconomic model; instead, a model with at least a yearly time step may be more meaningful. Therefore, models of natural systems (physics-based, conceptual, etc.) could be of hourly/daily/monthly basis, while such time resolutions may not be suitable to be selected as the time resolution of societal signals quantifying processes of poverty, unemployment, education level, income, migration, etc. This means that we need to couple two models with two different time resolutions requiring specific considerations from modeling point of view.

To deal with such a challenge generally in CHNS with multiple time resolutions representing different slow to fast natural or social processes concurrently, stiff ordinary differential equations (ODEs) could provide an opportunity and framework. Stiff ODEs are appropriate tools when two or more temporal scales are involved as faced here in coevolution. In addition, these ODEs may have to be stochastic ODEs due to various noisy processes encountered. These problems have been anticipated by many, including Sivapalan and Blöschl [9], but we propose appropriate mathematical solutions for such problems here and in the forthcoming tutorial paper based on Ponnambalam et al.' [28] workshop.

2.2.2. Mismatch in Type of Models and Modeling Approaches

The functions and relationships simulating physical processes in natural systems may form partial deferential equations (PDEs) of conservation of mass, momentum, and energy based on Newtonian physics. Conceptual hydrologic models are also developed on the same basis with different levels of approximations often as lumped systems (ODEs) in modeling the physics of the processes of interest. Then, approximations and uncertainties in input, model structure, and model parameters are accounted for through error-minimizing calibration approaches of these hydrologic models. Of course, applications of data-driven models approximating physics-based and conceptual hydrologic models and processes using model-free artificial intelligence (AI) and machine learning (ML) algorithms have been advancing during the last three decades. This is still an ongoing emerging field in modeling a wide range of analysis and design problems in engineering systems. Similarly, there are physics-based models that are able to simulate physics of macro- or micro-economic processes, e.g., behavior of suppliers and consumers under different market conditions. However, this may not be the case for social processes, i.e., poverty, health, gender equity, migration, education level, etc. In other words, social systems and processes, despite being complex, are being modeled by PDEs or other well established traditional mathematical approaches and equations, but the research is still in early stages for applications in the real world [71].

Traditional mathematics, although very powerful, still is not able to fully capture the physics of these processes and the related disorganized complex social systems. That is why it is said that the social systems are the most complex systems. Future advances in sociology, psychology, and mathematics and interactions and co-operations between sociologists, psychologists, and mathematicians may make it possible in the future to come up with a fuzzy or stochastic PDE solution of which simulates the time evolution of a society. However, up to now, and considering the current available modeling technology, models simulating social processes could be either qualitative or, in the best case, of empirical, statistical type. These empirical models are typically developed based on data and information collected on the variables of interest over time and space using questionnaires encapsulating experts' judgements and knowledge or other socioeconomic data collection and monitoring systems. Consequently, the type of qualitative or quasi-mathematical, empirical models simulating social processes will be different from quantitative mathematical models simulating natural systems. In other words, CHNS modeling requires integration of normative/quantitative, physics/mathematics-based hydro-economic models and subjective, qualitative human mimicking- or data-driven socioeconomic models.

The above-mentioned requirement calls for specific considerations and involves additional complexities when it comes to their calibration and verification and other modeling aspects. For example, model-free AI/ML-based methods with no explicitly defined analytical expressions and functions for quantifying relationships among social variables would restrict putting them among the set of constraints of fast gradient-based network flow programming (NFP) or linear programming (LP) algorithms being used in current basin-scale water allocation DSSs such as ModSim and WEAP. In such a situation, evolutionary optimization algorithms (EOA) that are much slower than NFP and LP would be the only choice to be employed as optimizers in the mentioned DSSs. This is also the case if the model wants to be put into an optimization framework optimizing reservoir releases, water allocations, and other socioeconomic decision variables of interest. Although EOA are very useful, they would slow down the convergence rate of the resulting optimization algorithm when many more function evaluations are required. Therefore, it would be inevitable to think of applying other meta-model-based optimization algorithms. If any single run of the coupled human–natural water allocation model is computationally intensive, then meta-modeling and surrogate optimization will be beneficial to speed up the computation power (Mousavi and Shourian [72], Kamali et al. [73], Mirfenderesgi and Mousavi [74]).

Apart from the above-mentioned challenges and opportunities, there is a great potential for applying fuzzy logic and computations as a branch of AI, introduced systematically by Lofi A. Zadeh in 1964 and less specifically by Luckashewics earlier under multivalued logic context. Fuzzy logic has proved its potential and promise in modeling approximate reasoning as a character of human beings and how they think, behave, and react that can help with social decisions not easily modeled by traditional mathematics. The body of literature on applications of fuzzy logic in water resources systems analysis and hydrology is quite rich and well-established, thus we do not refer to them herein. Instead, we focus on discussing its potential in modeling sociohydrologic processes.

In fuzzy inference systems (FISs), societal variables can be considered as linguistic variables whose values are words such as low, medium, and large as fuzzy numbers, rather than crisp exact numbers. Inexact, nonlinear relationships among these variables are represented by a set of fuzzy if–then rules called a fuzzy rule base. Then, all these linguistic variables and fuzzy rules are put into an inference mechanism, e.g., Mamdani and Takeshi–Sugeno FIS, using compositional inference rules. Such a framework has good potential to be used in modeling social systems and processes since in many instances societal signals and their interrelationships cannot be represented by exact, crisp variables and functions, whereas linguistic variables and fuzzy if–then rules enables us to make use of qualitative information about societal variables and their relationships.

FISs are also beneficial where there is no extensive data available. AI and ML algorithms [75] trained with Big Data are model free, data driven powerful tools with their ability to infer very complex relationships underlying these processes. They can extract and infer governing equations and relationships in complex systems from the data without having any knowledge about physical understanding of the relationships. On the other hand, advances in automated data monitoring, collection, and storage systems have provided a great potential for availability of big databases of socioeconomic variables and their spatiotemporal distributions, e.g., infrastructures locations and characteristics, population growth rates and other related variables, land use, soil type, cropping patterns and areas, crop yields and prices, etc. These data can be stored as different layers of a geographic information system (GIS) that can easily be retrieved whenever required. Therefore, recently-developed deep learning algorithms can be utilized to deal with these big databases of socioeconomic variables. Then, the databases and associated AI-based models trained and validated using a large amount of data can be integrated by other physics-based hydrologic models in the model base of a spatial DSS. Note that with available software and hardware technology, DSSs consisting of a model base, a spatial database, and a graphical user interface connected to each other have already provided powerful computerized framework in which different physics-based, conceptual, and data-driven models communicate different large amounts and types of data and information among each other at different levels of complexity or simplifications (modularity). RiverWare [76], MikeBasin [77], GeoDSS [78], and WEAP [79] are examples for such DSSs among several computerized DSSs developed for water resources and hydrologic systems modeling at different spatiotemporal scales.

Another potential opportunity to deal with the mentioned challenges is agent-based modeling (ABM), which can approximate the system-wide behavior of a complex system from individual level behavior of many elements interacting with each other through simple rules. ABM can simulate the emergent behavior of a system from autonomous individual behaviors necessary for modeling socioeconomic processes. It has already been used in modeling farmers' responses to reduced amounts of water allocations during dry periods, when agriculture systems undergo more severe water scarcity. In this line, there are a number of recently-done or ongoing research works attempting to couple agent-based models, simulating farmers' and food growers' reactions (water consumers), and physics-based water allocation models (see, e.g., Noël and Cai [50], Pouladi et al. [53] Pouladi et al. [54], Aghaie et al. [55]).

As an example problem, we proposed an integrated decentralized reservoir operation optimization model benefiting from ABM for optimal operations of Bukan Dam constructed on Zarineh-Rud River in Lake Urmia (LU), Iran, the second largest salt water lake in the world which is drying out [28]. Considering current critical situation of LU, Bukan Dam operations play a significant role in supplying the environmental water needed for the lake and its restoration plan. That is why planning a concise operational model for water allocations to both agricultural demands and the lake as a vital ecosystem is of crucial importance. In the proposed framework, ABM is one module of a hydro-agro-socioeconomic water allocation scheme embedded in multi-objective optimization to account for farmers' response and behavior against different scenarios of water allocations. A first version of the proposed framework focusing on the supply part of the model was presented by Ponnambalam et al. [28].

These types of coupled models have been developing from about a decade ago as a step towards CHNS modeling in SH. Earlier than that, and apart from the terminology used, the system dynamics (SD) approach has been successfully developing and applied for about two decades in integrated modeling of socioeconomic processes and in lumped hydrologic models. SD-type models in IWRM have paid specific attention to feedback of natural subsystems of water resource systems on their corresponding socioeconomic systems and vice-versa (e.g., Simonovic and Davies [41], Prodanovic and Simonovic [42]). However, as stated above, these works and attempts have used a systems approach for studying complex water resource systems in IWRM or food–energy–water–environment nexus problems.

2.2.3. Mismatch in Spatial Scales and Resolutions of Models

One important issue attracted specific attention in the 23-unsolved-problems-in-hydrology paper is the issue of "Hydrological Change" For example, it is stated in the paper that "the interest no longer resides only in providing scenarios of change (as only a decade ago), but in a rich fabric of experiments, data analysis and modelling approaches geared towards understanding the mechanisms of change." (Blöschl et al. [23] (p. 1152)). However, a challenge in responding to such a requirement is that models quantifying impacts of water cycle on human systems and human systems on local and global water cycle components can have totally different spatial domains and resolutions.

To clarify such a challenge, let us consider the well-established modeling frameworks being used in simulating the impact of climate change on basin-scale water-resource systems, as illustrated in Figure 2, where many important concepts are portrayed. Upscaling (aggregation), downscaling (disaggregation or decomposition), scenario-based analysis, and feedback loops are some of the concepts defined next. Upscaling with a bottom-up approach [80] involves taking an observed or theoretical relationship applicable at the point scale, and altering the relationship so that it is applicable at a larger scale. ABM is perfectly suited for this as individual agents and their actions can be aggregated to provide system level responses. ABM is not only applicable in problems of individual to society level aggregations but also can be equally applied in problems requiring PDEs such as diffusion-advection problems [81]. Thus, ABM is a tool that can be applied to model both lumped and distributed systems.

Downscaling involved taking results from coarse gridded, for example, Global Circulation Models (GCM), to finer resolution at the catchment level both in space and in time. Predictor variables are climatic variables such as specific humidity, total precipitation, convective precipitation, sea level pressure, etc. and soil indices, e.g., total soil moisture, slopes, vegetation indices, etc. Statistical methods including multilinear regression are commonly used (see ref. [82]), as well as the newly proposed combination of physically based statistical models in ref. [83].

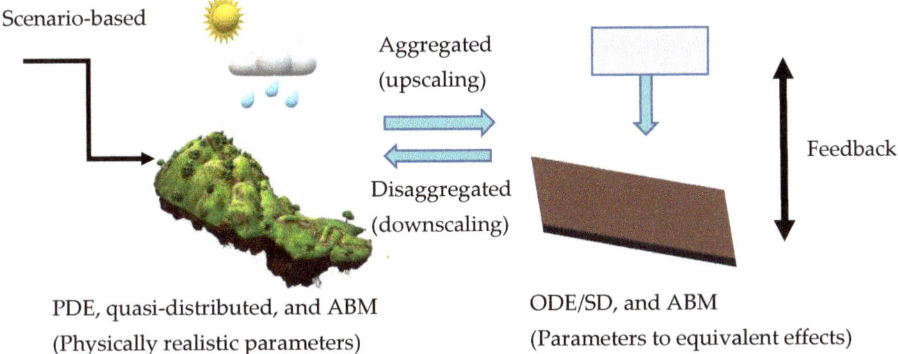

Figure 2. An illustration of frameworks for climate change impact assessment studies.

The typical current climate change impact assessment framework is a one-way, scenario-based modeling approach starting with emission scenarios followed by running global circulation models, downscaling, and employing rainfall–runoff models or other basin-wide, natural system-oriented models (the arrow on the left side of Figure 2 indicates this symbolically). However, the final result in terms of future downscaled climate change-driven fluxes of water and energy calculated at a local scale socioeconomic unit do not provide any feedback on emission scenarios and global water. On the other hand, if we imagine there are a tremendous number of basins or regions (local socioeconomic units) impacted by global water cycle through downscaled outcomes of global circulation models, the cumulative impact of all these units on the global water cycle has not yet been accounted for adequately by appropriate upscaled models. This is an impact which is considered important from the sociohydrologic modeling point of view, emphasizing on the co-evolution of the associated CHNS (the bidirectional arrow on the right side of Figure 2 indicates this feedback). However, estimating such feedback from local socioeconomic units on the global water cycle is not an easy task to tackle quantitatively and mathematically. It is because providing feedback from such units to the global water cycle calls for summing effects of all individual-level responses up, something that requires upscaling. This is despite downscaling currently being used in assessing the impact and consequences of global water cycle on much smaller-size spatial socioeconomic units. Even if we can find proper modeling tools and algorithms capable of tackling upscaling such as the global hydrologic models (GHMs) discussed in Section 1 attempting to deal with this issue [37–39], a coupled model equipped by both global-to-local simulation ability (downscaling) and local-to-global modeling capability (upscaling) would become too complex to solve. In other words, it would not be easy to build a fully coupled, two-way local–global model capable of integration (sum) across individual-level behaviors or processes.

We believe ABM having a bottom-up analysis ability to simulate system-level behavior of the global water cycle system through individual-level interactions of basin- or regional-scale socioeconomic units could be an opportunity to be used for the explained challenge of upscaling effects. Additionally, there are studies trying to estimate the contribution of anthropogenic and natural greenhouse emissions to total global greenhouse gas (GHG) emissions and their time variability. For example, using a statistical analysis, Xi-Liu and Qing-Xian [13] estimated the amount of anthropogenic GHG emissions to be about 55% of the global GHG emissions (2016 value). Such works show that we might be able to build a model-free or model-based mechanism simulating anthropogenic and natural GHG emissions as a function of some influencing factors related to hydroclimatic, hydrologic, and societal variables over different spatial regions (terrestrial or ocean systems). We think such simulation models can be put into an ABM approach, summing regional-scales impacts up, and then use it as a GHG emission-generating module in the above-mentioned climate change impact assessment procedure.

If we can do that, the above one-way scenario-based modeling approach would become a CHNS modeling approach.

Regarding other possible opportunities to deal with local-global impact quantification and upscaling, Budyko modeling approaches in ref. [84] are among other possible good opportunities to be used to fill part of the existing gap and challenges in simulating feedback from small-size sociohydrological units to the global water cycle. Additionally, there have been valuable works recently done in hydrologic modeling of evapotranspiration and precipitation simulating impacts of a change in local evaporation realized in a region (point) on other regions over the globe (points). The authors of this paper propose that the output of works, e.g., by Van der Ent et al. [67] and Roy et al. [85], can be represented as a large-scale response matrix of influential coefficients and used in IWRM management models to trace the changes in impacts due to changes in systems which could have happened from policy changes.

Other challenges could be related to mismatch and inconsistency in computational requirements, model precision degrees, and other issues while coupling a detailed model designed for a spatially small-scale unit and a model developed for a much more spatially coarser global-scale spatial unit. Therefore, there will be serious challenges in joint calibration and verification of such a complex coupled model, considering numerical/computational power limitations, inconsistency in their accuracy levels, their discretization scheme over time and space, extent and type of data required for each of them, etc. It seems that it is even impossible to tackle such a task without making several simplifications and assumptions, undermining the positive role of coupling.

Regarding computational burden difficulties in coupling of local- and global-scale models, having different finer and coarser spatial resolutions, two opportunities of parallel (cloud) computing and ML-assisted meta-modeling can be considered. In the first choice, special software and programming settings are used enabling the required computations to be done by several computers in parallel for building a surrogate or meta-model. The meta-model version is much faster than the original model [73]. However, sufficient sample data are required for training a meta-model, and the training procedure can be done online or offline. Therefore, there will be a tradeoff between the approximation power or accuracy of the built meta-model and the number of training samples, each of which requires running the original computationally-intensive model. Balancing these two aspects is an important challenge for doing meta-modeling successfully [72]. These sorts of meta-models have also been used by integrated assessment modeling community in references [33,36].

Another important point and challenge in applying one-way scenario-based approaches while modeling interacting sub-systems is the issue of interfaces between the sub-systems having different scales. The existing traditional approaches consider the issue via boundary conditions to reduce the complexity. However, such a simplifying approach may not be enough in some cases, as pointed out by Blöschl et al. [23] (p. 1152) stating "There is a broad recognition that we need to learn more about interfaces in hydrology. These have traditionally been imposed as boundary conditions, thereby reducing complexity, but we now need to look at the more typical cases where we can and should not do this, as the interfaces couple rather than constrain system behaviour. These interfaces include those between compartments (e.g., atmosphere–vegetation–soil–bedrock–streamflow–hydraulic structures) in three dimensions, interactions between the hydrological fluxes and the media (e.g., soils, vegetation), and interactions between sub-processes that are usually dealt with by different disciplines. (e.g., water chemistry, ecology, soil science, biogeochemistry). Linking these interfaces conceptually and in a quantitative way is currently considered a real bottleneck".

Overall, given the explanations provided, although there are serious challenges mentioned, we have promising tools and algorithms for shifting from the one-way global-to-local approach towards a two-way feedback-based coupled global-to-local, local-to-global modeling approach. For example, Simonovic and Davies [41] discussed other simplifying assumptions made in traditional climate change impact assessment studies including: (1) predictability of the character of all interactions between biophysical and socioeconomic systems, despite their nonlinear nature; (2) irrelevancy of the interactions between these systems and the behavior of each; and (3) reparability of these two systems so that feedbacks between the systems are external to both. In this line, Davies and Simonovic [66] developed the system dynamic-based ANEMI simulation model for integrated assessment of global change especially the carbon transfers. Additionally, we already referred to a number of numerous sophisticated approaches, including system dynamics and analysis, stochastic simulation and optimization, integrated agro-hydro-socioeconomic modeling, etc., presented by water resources systems analysts and IWRM community before SH's introduction that have gone far beyond scenario based approaches. Koutsoyiannis [19] also stated this fact in his review comments for the first-published SH paper in Hydrological Processes journal.

2.2.4. Institutional and Governance-Related Challenges

The co-evolution of human–natural water systems has not been under the impact of de-regulated social systems. Rather, impacts of people and social systems on the water cycle have been taking place under a complex institutional and governance framework and settings that have dynamically been changing over time with significant differences among social units spread over space at a certain time (provinces, states, countries, etc.). This issue has impacted the past and will impact the future's co-evolution of coupled human–natural systems, making CHNS modeling much more complex and difficult to quantify.

Additionally, there is another type of challenges in terms of participating units, institutions, governance, authorities, and co-operations among socio-administrative units of a coupled human–natural system from an IWRM perspective. For instance, under European Water Framework Directive [86], EU Member States have established Coordination and Participation Boards at the river basin level as multi-agency and multi-actor groups, supporting the development of inclusive and coordinated river basin planning and the inclusion of interested parties in decision-making processes [87]. This challenge is also raised when dealing with local–global impacts of water cycle and societies on each other, requiring specific considerations. Blöschl et al. [23] stated "While water governance is limited to the local and national scales, a global perspective is clearly becoming increasingly more important in the context of the UN Agenda 2030 and Sustainable Development Goals, the societal grand challenge of our time [14]."

Water management policies, institutions, and governance at provincial, state, and national levels directly affect basin and regional-scale water cycle components. On the other hand, international-level policies, agreements, rules, regulations, and protocols (e.g., International Law Association Committee on the Uses of the Waters of International Rivers [88], United Nations [89], and United Nations [90]) and commitments established by international organizations and institutions would affect medium- and long-term state of CO_2 emission, global warming, and climate change conditions that would influence all local socioeconomic units. For example, basin- and national-level water management practices, policies, and governance (local-scale institutions and governance) underlying socioeconomic activities and interventions (e.g., dam constructions, land-use change, expanded irrigation areas, etc.) combined by climate change impacts (global-scale intervention) during last decades have caused significant destroying consequences on the drying Lake Urmia ecosystem in Iran. Brazilian government policies have been influential on recent huge burning of Amazon forests resulting in a more adverse CO_2 emission condition. This means that co-evolution of our future coupled human–natural systems will certainly be under influence of both local (provincial, state, and national) and global (international) water and environment governance-related conditions. By governance we mean all agreements, institutions, rules and regulations, policies, commitments, protocols, etc.

Under the above-mentioned conditions, can we predict the future of the governance underlying future local–global impacts of our water systems and societies? Can we model and quantify the influence of future presidents of countries and their decisions and commitment level to international CO_2 agreements, especially those countries that are more responsible in producing CO_2 and its emission to atmosphere? Therefore, are our future coupled human–natural systems really predictable under such complex governmental and institutional conditions? Do we have better modeling framework and conflict resolution techniques than simple scenario- or feedback-based approaches that can account for deep uncertainties in the future from the aspects just explained? These are some relevant questions in SH and CHNS that are too difficult to answer yet. Loucks [91] stated "What we modellers haven't done yet is to figure out how to make our models suggest planning and management options that we haven't thought of before. This would be an especially important feature for integrated water resources planning and management. Integrated implies that our models have included all the links to all the other major components of our social, economic and, if applicable, ecological environments."

Despite the mentioned difficulties, there are still promising tools and ways to deal with them, which of course are not of mathematical modeling type. Public awareness, NGOs, ease of communications through the Internet and social media, etc. all have promising elements facilitating both national- and international-scale participations and collaborations. Therefore, they provide opportunities for hydrologists to convey their message to policy makers and the society [92].

For example, due to the national-level public awareness and request, the government of Iran established a new organization, Lake Urmia Restoration Program (LURP), coordinating all water management activities for restoring the lake, about six years ago. Under the revised participatory water management policies, help and participation of international parties and collaborators such as FAO and JAICA, and better climatic conditions during last five years, the current situation of the lake has relatively become better, and the continuation of procedures forcing the drying of the lake has fortunately stopped. For the case of Amazon forests, international concerns shown even in the latest summit of the leaders of seven industrialized countries helped the Brazilian government take some actions to control the devastating Amazon forests burning events.

We in this section explained and discussed a number of challenges and complexities system's analysts would encounter while modeling CHNS: (1) mismatches in appropriate time resolutions and space domains of social and physical processes if they want to be included in an integrated coupled model simulating both types of processes; (2) inconsistency among type of models, governing equations, and relationships required to quantitatively simulate societal and physical processes taking place in social and natural systems, hard-to-quantify societal variables of interest, and processes impacting the natural (water) systems both locally and globally; (3) global-to-local (downscaling) and local-to-global (upscaling) challenges and how to model and simulate feedbacks the global water cycle receives from the sum of a huge number of local-scale human-driven impacts; (4) computational CHNS modeling challenges encountered while simulating local and global processes and variables with appropriate spatiotemporal scales and resolutions; and (5) complex and almost impossible-to-predict future governmental and institutional systems, at provincial, basin, national, regional, and international levels, affecting the co-evolution of natural and social systems. For each of these challenges, we also proposed some possible modeling approaches that could help modelers deal with the challenges at least partially and may have to depend on simpler models and meta-models [93]. The new AI/ML techniques provide some new opportunities for promising directions to satisfy both speed and accuracy [94].

3. Final Remarks

Systems analysis and coupled human–natural systems (CHNS) models provide the practical approach needed for applications both in the descriptive science of sociohydrology (SH) and in the prescriptive method of integrated water resources management (IWRM). Although CHNS is of great importance, the extent we can develop a coupled human–natural system model mathematically is limited and ongoing. It is nearly impossible to account for all the mentioned sources of complexity required by SH and IWRM in CHNS modeling, the coupling levels of local-to-global and global-to-local processes would depend on data availability for model calibration and verification in the presence of uncertainty. Such a capacity is also restricted by the level of understanding of social, economic, institutional, and natural processes, their governing equations/relations, and their dynamics and evolution. These difficulties may often lead in practice to simpler models that can be solvable (likely after applying the technique of divide and conquer or decomposition/aggregation to manage different scales in time and space), and verifiable in a limited sense, but adaptive.

Author Contributions: K.P. and S.J.M. both developed and proposed the main ideas behind this paper, and K.P. substantially contributed to the rewriting of the first draft written by S.J.M. Figures 1 and 2 were composed by K.P. All authors have read and agreed to the published version of the manuscript.

Funding: This research received no external funding; however, the first author has received NSERC, Canada funding for much of his multidisciplinary research.

Acknowledgments: Conversations with Slobodan Simonovic, Murugesu Sivapalan, Ximing Cai, Tirupati Bolisetti, James Craig, Tirthankar Roy, and Evan Davies are gratefully acknowledged. We also thank Hamed Hamzekhani and Narjes Ghaderi for their help in formatting references. Constructive comments of two reviewers helped us improve the quality of the final version for which we are very much thankful to them.

Conflicts of Interest: The authors declare no conflict of interest.

References

1. Thirukkural. Chapter 2: The Blessing of Rain. Available online: https://www.ytamizh.com/thirukural/chapter-2 (accessed on 12 June 2020).

2. UNESCO. Intergovernmental Meeting of Experts for the IHD, Paris. 1964. Available online: https://unesdoc.unesco.org/ark:/48223/pf0000153976 (accessed on 12 June 2020).

3. Global Water Partnership Program. 2000. Available online: https://www.gwp.org/ (accessed on 12 June 2020).

4. Choi, B.C.; Pak, A.W. Multidisciplinarity, interdisciplinarity and transdisciplinarity in health research, services, education and policy: 1. Definitions, objectives, and evidence of effectiveness. *Clin. Investig. Med.* **2006**, *29*, 351 364.

5. Falkenmark, M. Main problems of water use and transfer of technology. *GeoJournal* **1979**, *3*, 435–443. [CrossRef]

6. Sivapalan, M.; Savenije, H.H.; Blöschl, G. Socio-hydrology: A new science of people and water. *Hydrol. Process.* **2012**, *26*, 1270–1276. [CrossRef]

7. Simonovic, S.P. World water dynamics: Global modeling of water resources. *J. Environ. Manag.* **2002**, *66*, 249–267. [CrossRef]

8. Di Baldassarre, G.; Viglione, A.; Carr, G.; Kuil, L.; Salinas, J.; Blöschl, G. Socio-hydrology: Conceptualising human-flood interactions. *Hydrol. Earth Syst. Sci.* **2013**, *17*, 3295–3303. [CrossRef]

9. Sivapalan, M.; Blöschl, G. Time scale interactions and the coevolution of humans and water. *Water Resour. Res.* **2015**, *51*, 6988–7022. [CrossRef]

10. Blair, P.; Buytaert, W. Socio-hydrological modelling: A review asking"why, what and how?". *Hydrol. Earth Syst. Sci.* **2016**, *20*, 443–478. [CrossRef]

11. Sivapalan, M.; Blöschl, G. The growth of hydrological understanding: Technologies, ideas, and societal needs shape the field. *Water Resour. Res.* **2017**, *53*, 8137–8146. [CrossRef]

12. Di Baldassarre, G.; Nohrstedt, D.; Mård, J.; Burchardt, S.; Albin, C.; Bondesson, S.; Breinl, K.; Deegan, F.M.; Fuentes, D.; Lopez, M.G. An integrative research framework to unravel the interplay of natural hazards and vulnerabilities. *Earth's Future* **2018**, *6*, 305–310. [CrossRef]

13. Yue, X.-L.; Gao, Q.-X. Contributions of natural systems and human activity to greenhouse gas emissions. *Adv. Clim. Chang. Res.* **2018**, *9*, 243–252. [CrossRef]

14. Di Baldassarre, G.; Sivapalan, M.; Rusca, M.; Cudennec, C.; Garcia, M.; Kreibich, H.; Konar, M.; Mondino, E.; Mård, J.; Pande, S. Socio-hydrology: Scientific challenges in addressing a societal grand challenge. *Water Resour. Res.* **2019**, *55*, 6327–6355. [CrossRef]

15. Xu, L.; Gober, P.; Wheater, H.S.; Kajikawa, Y. Reframing socio-hydrological research to include a social science perspective. *J. Hydrol.* **2018**, *563*, 76–83. [CrossRef]

16. Wesselink, A.; Kooy, M.; Warner, J. Socio-hydrology and hydrosocial analysis: Toward dialogues across disciplines. *Wiley Interdiscip. Rev. Water* **2017**, *4*, e1196. [CrossRef]

17. Srinivasan, V.; Lambin, E.F.; Gorelick, S.M.; Thompson, B.H.; Rozelle, S. The nature and causes of the global water crisis: Syndromes from a meta-analysis of coupled human-water studies. *Water Resour. Res.* **2012**, *48*. [CrossRef]

18. Sivakumar, B. Socio-hydrology: Not a new science, but a recycled and re-worded Hydrosociology. *Hydrol. Process.* **2012**, *26*, 3788–3790. [CrossRef]

19. Koutsoyiannis, D. Review of the First SH Paper Published in J. of Hydrological Processes by Sivapalan et al. in 2012. In Reviewer's Ref.: DK-291, Made Publicly Open by the Reviewer. Available online: http://www.itia.ntua.gr/en/docinfo/1991/ (accessed on 12 June 2020).

20. Turner, B.L.; Matson, P.A.; McCarthy, J.J.; Corell, R.W.; Christensen, L.; Eckley, N.; Hovelsrud-Broda, G.K.; Kasperson, J.X.; Kasperson, R.E.; Luers, A. Illustrating the coupled human–environment system for vulnerability analysis: Three case studies. *Proc. Natl. Acad. Sci. USA* **2003**, *100*, 8080–8085. [CrossRef]

21. Liu, J.; Dietz, T.; Carpenter, S.R.; Folke, C.; Alberti, M.; Redman, C.L.; Schneider, S.H.; Ostrom, E.; Pell, A.N.; Lubchenco, J. Coupled human and natural systems. *AMBIO J. Hum. Environ.* **2007**, *36*, 639–649. [CrossRef]

22. Fu, B.; Wei, Y. Editorial overview: Keeping fit in the dynamics of coupled natural and human systems. *Curr. Opin. Environ. Sustain.* **2018**, *33*, A1–A4. [CrossRef]

23. Blöschl, G.; Bierkens, M.F.; Chambel, A.; Cudennec, C.; Destouni, G.; Fiori, A.; Kirchner, J.W.; McDonnell, J.J.; Savenije, H.H.; Sivapalan, M. Twenty-three unsolved problems in hydrology (UPH)—A community perspective. *Hydrol. Sci. J.* **2019**, *64*, 1141–1158. [CrossRef]

24. Rockström, J.; Steffen, W.; Noone, K.; Persson, Å.; Chapin, F.S., III; Lambin, E.; Lenton, T.M.; Scheffer, M.; Folke, C.; Schellnhuber, H.J. Planetary boundaries: Exploring the safe operating space for humanity. *Ecol. Soc.* **2009**, *14*, 32. [CrossRef]

25. Steffen, W.; Richardson, K.; Rockström, J.; Cornell, S.E.; Fetzer, I.; Bennett, E.M.; Biggs, R.; Carpenter, S.R.; De Vries, W.; De Wit, C.A. Planetary boundaries: Guiding human development on a changing planet. *Science* **2015**, *347*, 1259855. [CrossRef]

26. Hejazi, M.I.; Edmonds, J.; Clarke, L.; Kyle, P.; Davies, E.; Chaturvedi, V.; Wise, M.; Patel, P.; Eom, J.; Calvin, K. Integrated assessment of global water scarcity over the 21st century under multiple climate change mitigation policies. *Hydrol. Earth Syst. Sci.* **2014**, *18*, 2859–2883. [CrossRef]

27. Cook, C.; Bakker, K. Water security: Debating an emerging paradigm. *Glob. Environ. Chang.* **2012**, *22*, 94–102. [CrossRef]

28. Ponnambalam, K.; Mousavi, S.J.; Fletcher, S.G.; Hipel, K.W.; Parker, D. Socio-Hydrology: Opportunities and Challenges—A Department of Systems Design Engineering Workshop. 2019. Available online: http://epoch.uwaterloo.ca (accessed on 12 June 2020).

29. Nikolic, V.V.; Simonovic, S.P. Multi-method modeling framework for support of integrated water resources management. *Environ. Process.* **2015**, *2*, 461–483. [CrossRef]

30. Loucks, D.P. Managing water as a critical component of a changing world. *Water Resour. Manag.* **2017**, *31*, 2905–2916. [CrossRef]

31. Stevenson, R.J. A revised framework for coupled human and natural systems, propagating thresholds, and managing environmental problems. *Phys. Chem. Earthparts A/B/C* **2011**, *36*, 342–351. [CrossRef]

32. Senf, C.; Pflugmacher, D.; Hostert, P.; Seidl, R. Using Landsat time series for characterizing forest disturbance dynamics in the coupled human and natural systems of Central Europe. *ISPRS J. Photogramm. Remote Sens.* **2017**, *130*, 453–463. [CrossRef]

33. Kim, S.H.; Hejazi, M.; Liu, L.; Calvin, K.; Clarke, L.; Edmonds, J.; Kyle, P.; Patel, P.; Wise, M.; Davies, E. Balancing global water availability and use at basin scale in an integrated assessment model. *Clim. Chang.* **2016**, *136*, 217–231. [CrossRef]

34. Ausseil, A.; Daigneault, A.; Frame, B.; Teixeira, E. Towards an integrated assessment of climate and socio-economic change impacts and implications in New Zealand. *Environ. Model. Softw.* **2019**, *119*, 1–20. [CrossRef]

35. Edmonds, J.; Reilly, J. Global Energy and CO2 to the Year 2050. *Energy J.* **1983**, *4*. [CrossRef]

36. Collins, W.D.; Craig, A.P.; Truesdale, J.E.; Di Vittorio, A.; Jones, A.D.; Bond-Lamberty, B.; Calvin, K.V.; Edmonds, J.A.; Kim, S.H.; Thomson, A.M. The integrated Earth system model version 1: Formulation and functionality. *Geosci. Model Dev.* **2015**, *8*, 2203–2219. [CrossRef]

37. Sood, A.; Smakhtin, V. Global hydrological models: A review. *Hydrol. Sci. J.* **2015**, *60*, 549–565. [CrossRef]

38. Brêda, J.P.L.F.; de Paiva, R.C.D.; Collischon, W.; Bravo, J.M.; Siqueira, V.A.; Steinke, E.B. Climate change impacts on South American water balance from a continental-scale hydrological model driven by CMIP5 projections. *Clim. Chang.* **2020**, 1–20. [CrossRef]

39.	Wada, Y.; Wisser, D.; Bierkens, M. Global modeling of withdrawal, allocation and consumptive use of surface water and groundwater resources. *Earth Syst. Dyn.* **2014**, *5*, 15–40. [CrossRef]

40.	Cai, X.; McKinney, D.C.; Lasdon, L.S. Integrated hydrologic-agronomic-economic model for river basin management. *J. Water Resour. Plan. Manag.* **2003**, *129*, 4–17. [CrossRef]

41.	Simonovic, S.P.; Davies, E.G. Are we modelling impacts of climatic change properly? *Hydrol. Process.* **2006**, *20*, 431–433. [CrossRef]

42.	Prodanovic, P.; Simonovic, S.P. An operational model for support of integrated watershed management. *Water Resour. Manag.* **2010**, *24*, 1161–1194. [CrossRef]

43.	Davies, E.G.; Simonovic, S.P. Global water resources modeling with an integrated model of the social–economic–environmental system. *Adv. Water Resour.* **2011**, *34*, 684–700. [CrossRef]

44.	Lam, N.S.-N.; Xu, Y.J.; Liu, K.-B.; Dismukes, D.E.; Reams, M.; Pace, R.K.; Qiang, Y.; Narra, S.; Li, K.; Bianchette, T.A. Understanding the Mississippi River Delta as a coupled natural-human system: Research methods, challenges, and prospects. *Water* **2018**, *10*, 1054. [CrossRef]

45.	Mashaly, A.F.; Fernald, A.G. Identifying Capabilities and Potentials of System Dynamics in Hydrology and Water Resources as a Promising Modeling Approach for Water Management. *Water* **2020**, *12*, 1432. [CrossRef]

46.	Jeong, H.; Adamowski, J. A system dynamics based socio-hydrological model for agricultural wastewater reuse at the watershed scale. *Agric. Water Manag.* **2016**, *171*, 89–107. [CrossRef]

47.	An, L. Modeling human decisions in coupled human and natural systems: Review of agent-based models. *Ecol. Model.* **2012**, *229*, 25–36. [CrossRef]

48.	Walker, W.E.; Loucks, D.P.; Carr, G. Social responses to water management decisions. *Environ. Process.* **2015**, *2*, 485–509. [CrossRef]

49.	Dziubanski, D.; Franz, K.J.; Gutowski, W. Linking economic and social factors to peak flows in an agricultural watershed using socio-hydrologic modeling. *Hydrol. Earth Syst. Sci. Discuss.* **2019**, 1–55. [CrossRef]

50.	Noël, P.H.; Cai, X. On the role of individuals in models of coupled human and natural systems: Lessons from a case study in the Republican River Basin. *Environ. Model. Softw.* **2017**, *92*, 1–16. [CrossRef]

51.	Tesfatsion, L.; Rehmann, C.R.; Cardoso, D.S.; Jie, Y.; Gutowski, W.J. An agent-based platform for the study of watersheds as coupled natural and human systems. *Environ. Model. Softw.* **2017**, *89*, 40–60. [CrossRef]

52.	Rammer, W.; Seidl, R. Coupling human and natural systems: Simulating adaptive management agents in dynamically changing forest landscapes. *Glob. Environ. Chang.* **2015**, *35*, 475–485. [CrossRef]

53.	Pouladi, P.; Afshar, A.; Afshar, M.H.; Molajou, A.; Farahmand, H. Agent-based socio-hydrological modeling for restoration of Urmia Lake: Application of theory of planned behavior. *J. Hydrol.* **2019**, *576*, 736–748. [CrossRef]

54.	Pouladi, P.; Afshar, A.; Molajou, A.; Afshar, M.H. Socio-hydrological framework for investigating farmers' activities affecting the shrinkage of Urmia Lake; hybrid data mining and agent-based modelling. *Hydrol. Sci. J.* **2020**, *65*, 1249–1261. [CrossRef]

55.	Aghaie, V.; Alizadeh, H.; Afshar, A. Agent-Based hydro-economic modelling for analysis of groundwater-based irrigation Water Market mechanisms. *Agric. Water Manag.* **2020**, *234*, 106140. [CrossRef]

56.	Savenije, H.H.; Van der Zaag, P. Integrated water resources management: Concepts and issues. *Phys. Chem. Earthparts A/B/C* **2008**, *33*, 290–297. [CrossRef]

57.	Brown, C.M.; Lund, J.R.; Cai, X.; Reed, P.M.; Zagona, E.A.; Ostfeld, A.; Hall, J.; Characklis, G.W.; Yu, W.; Brekke, L. The future of water resources systems analysis: Toward a scientific framework for sustainable water management. *Water Resour. Res.* **2015**, *51*, 6110–6124. [CrossRef]

58.	Castro, J.E. Water governance in the twentieth-first century. *Ambiente Soc.* **2007**, *10*, 97–118. [CrossRef]

59.	Sanderson, M.R.; Bergtold, J.S.; Heier Stamm, J.L.; Caldas, M.M.; Ramsey, S.M. Bringing the "social" into sociohydrology: Conservation policy support in the C entral G reat P lains of K ansas, USA. *Water Resour. Res.* **2017**, *53*, 6725–6743. [CrossRef]

60.	Mayer, R.E.; Reischer, G.H.; Ixenmaier, S.K.; Derx, J.; Blaschke, A.P.; Ebdon, J.E.; Linke, R.; Egle, L.; Ahmed, W.; Blanch, A.R. Global distribution of human-associated fecal genetic markers in reference samples from six continents. *Environ. Sci. Technol.* **2018**, *52*, 5076–5084. [CrossRef] [PubMed]

61.	Dingemans, M.M.; Baken, K.A.; van der Oost, R.; Schriks, M.; van Wezel, A.P. Risk-based approach in the revised European Union drinking water legislation: Opportunities for bioanalytical tools. *Integr. Environ. Assess. Manag.* **2019**, *15*, 126–134. [CrossRef]

62. Cudennec, C.; Liu, J.; Qi, J.; Yang, H.; Zheng, C.; Gain, A.K.; Lawford, R.; de Strasser, L.; Yillia, P. Epistemological dimensions of the water–energy–food nexus approach: Reply to discussions of "Challenges in operationalizing the water–energy–food nexus". *Hydrol. Sci. J.* **2018**, *63*, 1868–1871. [CrossRef]

63. Metz, F.; Glaus, A. Integrated water resources management and policy integration: Lessons from 169 years of flood policies in Switzerland. *Water* **2019**, *11*, 1173. [CrossRef]

64. Biswas, A.K. Integrated water resources management: Is it working? *Int. J. Water Resour. Dev.* **2008**, *24*, 5–22. [CrossRef]

65. Loucks, D.P.; Van Beek, E. Water Resource Systems Planning and Management: An Introduction to Methods, Models, and Applications. Available online: https://link.springer.com/book/10.1007%2F978-3-319-44234-1 (accessed on 12 June 2020).

66. Davies, E.G.; Simonovic, S.P. ANEMI: A new model for integrated assessment of global change. *Interdiscip. Environ. Rev.* **2010**, *11*, 127–161. [CrossRef]

67. Van der Ent, R.J.; Savenije, H.H.; Schaefli, B.; Steele-Dunne, S.C. Origin and fate of atmospheric moisture over continents. *Water Resour. Res.* **2010**, *46*. [CrossRef]

68. Haimes, Y.Y. *Modeling and Managing Interdependent Complex Systems of Systems*; John Wiley & Sons: New York, NY, USA, 2018.

69. Luo, X.; Li, J.; Zhu, S.; Xu, Z.; Huo, Z. Estimating the Impacts of Urbanization in the Next 100 years on Spatial Hydrological Response. *Water Resour. Manag.* **2020**, 1–20. [CrossRef]

70. Loucks, D.P. Model Complexity—What Is the Right Amount? Integration and Implementation Insights. Available online: https://i2insights.org/2016/11/03/model-complexity/ (accessed on 12 June 2020).

71. Burger, M.; Caffarelli, L.; Markowich, P.A. *Partial Differential Equation Models in the Socio-Economic Sciences*; The Royal Society Publishing: London, U.K, 2014.

72. Mousavi, S.J.; Shourian, M. Adaptive sequentially space-filling metamodeling applied in optimal water quantity allocation at basin scale. *Water Resour. Res.* **2010**, *46*. [CrossRef]

73. Kamali, M.; Ponnambalam, K.; Soulis, E. Comparison of several heuristic approaches to calibration of WATCLASS hydrologic model. *Can. Water Resour. J.* **2013**, *38*, 40–46. [CrossRef]

74. Mirfenderesgi, G.; Mousavi, S.J. Adaptive meta-modeling-based simulation optimization in basin-scale optimum water allocation: A comparative analysis of meta-models. *J. Hydroinform.* **2016**, *18*, 446–465. [CrossRef]

75. Nassar, L.; Okwuchi, I.; Saad, M.; Karray, F.; Ponnambalam, K.; Agrawal, P. Prediction of Strawberry Yield and Farm Price Utilizing Deep Learning. Procceddings of the IEEE World Congress in Computational Intelligence, Glagow, UK, 19–24 July 2020.

76. Zagona, E.A.; Fulp, T.J.; Shane, R.; Magee, T.; Goranflo, H.M. Riverware: A generalized tool for complex reservoir system modeling 1. *J. Am. Water Resour. Assoc.* **2001**, *37*, 913–929. [CrossRef]

77. Water, D. Environment. In *MIKE BASIN: A Versaztile Decision Support Tool for Integrated Water Resources Management and Planning*; DHI: Hørshelm, Denmark, 2006.

78. Triana, E.; Labadie, J.W. GIS-based decision support system for improved operations and efficiency conservation in large-scale irrigation systems. *J. Irrig. Drain. Eng.* **2012**, *138*, 857–867. [CrossRef]

79. Sieber, J.; Purkey, D. *WEAP (Water Evaluation and Planning System), User Guide*; Stockholm Environment Institute (SEI): Stockholm, Sweden, 2016; p. 392.

80. Harvey, L.D. Upscaling in global change research. *Clim. Chang.* **2000**, *44*, 225–263. [CrossRef]

81. Garcia Hernandez, J.A.; Ponnambalam, K. Developing Novel Relations between Differential Equations and Agent-Based Models. In Proceedings of the IEEE Conference on Systems, Man, and Cybernetics, Toronto, ON, Canada, 11–14 October 2020.

82. Najafi, M.R.; Moradkhani, H.; Wherry, S.A. Statistical downscaling of precipitation using machine learning with optimal predictor selection. *J. Hydrol. Eng.* **2011**, *16*, 650–664. [CrossRef]

83. Gaur, A.; Simonovic, S.P. Introduction to physical scaling: A model aimed to bridge the gap between statistical and dynamic downscaling approaches. In *Trends and Changes in Hydroclimatic Variables*; Elsevier: Rotterdam, The Netherlands, 2019; pp. 199–273.

84. Mianabadi, A.; Coenders-Gerrits, M.; Shirazi, P.; Ghahraman, B.; Alizadeh, A. A simple global Budyko model to partition evaporation into interception and transpiration. *Hydrol. Earth Syst. Sci. Discuss.* **2017**, 1–29. [CrossRef]

85. Roy, T.; Martinez, J.A.; Herrera-Estrada, J.E.; Zhang, Y.; Dominguez, F.; Berg, A.; Ek, M.; Wood, E.F. Role of moisture transport and recycling in characterizing droughts: Perspectives from two recent US droughts and the CFSv2 system. *J. Hydrometeorol.* **2019**, *20*, 139–154. [CrossRef]
86. Directive, W.F. *Common Implementation Strategy for the Water Framework Directive (2000/60/EC) Guidance Document No 11 Planning Process*; European Commission: Brussels, Belgium, 2003.
87. Pellegrini, E.; Bortolini, L.; Defrancesco, E. Coordination and Participation Boards under the European Water Framework Directive: Different approaches used in some EU countries. *Water* **2019**, *11*, 833. [CrossRef]
88. International Law Association. Helsinki Rules on the Uses of the Waters of International Rivers. 1966. Available online: https://www.internationalwaterlaw.org/documents/intldocs/ILA/ILA-HelsinkiRules1966-as_amended.pdf (accessed on 12 June 2020).
89. Kyoto Protocol to the United Nations Framework Convention on Climate Change. 1988. Available online: https://en.wikipedia.org/wiki/Kyoto_Protocol (accessed on 12 June 2020).
90. Nations, U. *The Future We Want: Outcome Document Adopted at Rio+ 20*; United Nations: Rio de Janeiro, Brazil, 2012.
91. Loucks, D.P. Moving from Models that Synthesize to Models That Innovate, Integration and Implementation Insights. 2016. Available online: https://i2insights.org/2016/03/24/models-that-innovate/ (accessed on 12 June 2020).
92. Re, V.; Misstear, B. Education and Capacity Development for Groundwater Resources Management. In *Advances in Groundwater Governance*; CRC Press/Taylor & Francis: London, UK, 2018; pp. 212–229, Chapter 11.
93. Fraser, C.; McIntyre, N.; Jackson, B.; Wheater, H. Upscaling hydrological processes and land management change impacts using a metamodeling procedure. *Water Resour. Res.* **2013**, *49*, 5817–5833. [CrossRef]
94. Hong, E.-M.; Pachepsky, Y.A.; Whelan, G.; Nicholson, T. Simpler models in environmental studies and predictions. *Crit. Rev. Environ. Sci. Technol.* **2017**, *47*, 1669–1712. [CrossRef]

Article

Systems Approach to Management of Water Resources—Toward Performance Based Water Resources Engineering

Slobodan P. Simonovic

Department of Civil and Environmental Engineering, The University of Western Ontario, London, ON N6A 5B9, Canada; simonovic@uwo.ca; Tel.: +1-519-661-4075

Received: 29 March 2020; Accepted: 20 April 2020; Published: 24 April 2020

Abstract: Global change, that results from population growth, global warming and land use change (especially rapid urbanization), is directly affecting the complexity of water resources management problems and the uncertainty to which they are exposed. Both, the complexity and the uncertainty, are the result of dynamic interactions between multiple system elements within three major systems: (i) the physical environment; (ii) the social environment; and (iii) the constructed infrastructure environment including pipes, roads, bridges, buildings, and other components. Recent trends in dealing with complex water resources systems include consideration of the whole region being affected, explicit incorporation of all costs and benefits, development of a large number of alternative solutions, and the active (early) involvement of all stakeholders in the decision-making. Systems approaches based on simulation, optimization, and multi-objective analyses, in deterministic, stochastic and fuzzy forms, have demonstrated in the last half of last century, a great success in supporting effective water resources management. This paper explores the future opportunities that will utilize advancements in systems theory that might transform management of water resources on a broader scale. The paper presents performance-based water resources engineering as a methodological framework to extend the role of the systems approach in improved sustainable water resources management under changing conditions (with special consideration given to rapid climate destabilization). An illustrative example of a water supply network management under changing conditions is used to convey the basic principles of performance-based water resources engineering methodology.

Keywords: water resources systems; performance-based engineering; simulation; resilience

1. Introduction

Two paradigms are identified by Simonovic [1] as shaping contemporary water resources management: "The first paradigm focuses on the complexity of the water resources management domain (increases with time), and the complexity of the modeling tools (decreases with time), in an environment characterized by continuous, rapid technological development (sharp increase in development over time). The illustrative presentation of the complexity paradigm is shown in Figure 1a. The extension of temporal and spatial scales characterizing contemporary water resources management problems leads to an increase in the complexity of decision-making processes (which could be measured using a number of state variables on the vertical axis in Figure 1a). The evolution of systems analysis with increasing computational power (expressed for example using computational time required for the solution of a problem on the vertical axis in Figure 1a) results in more complex analytical tools being replaced by simpler and more robust search tools and very often by simple simulation (assessed using a number of mathematical relationships on the vertical axis in Figure 1a).

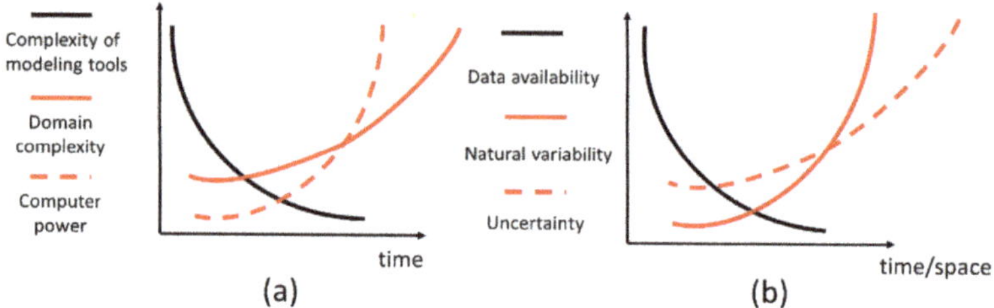

Figure 1. Illustration of the (**a**) complexity and (**b**) uncertainty paradigms (after [1]).

The second paradigm deals with the water resources-related data availability (represented for example by the number of observation stations on the vertical axis in Figure 1b) and the natural variability of the domain variables (for example measured by the range of values that a particular state variable can take on the vertical axis in Figure 1b) in time and space that affect the uncertainty (possibly expressed by the statistical dispersion of the values attributed to a measured quantity on the vertical axis in Figure 1b) of water resources management decision-making (Figure 1b). Data necessary for management of water resources are costly and collected by various agencies. The financial constraints of government agencies that are responsible for the collection of water-related data have resulted in reduction of data collection programs in many countries."

The traditional understanding of water resources management is that it is the management of water resources [2–5]. But the language behind the concept is simpler since there is a set of complex interactions between the water resources, people and the environment that they all share. The two paradigms call for a question: What are we managing? We try to manage environments (water, land, air, etc.). We keep try to manage the behavior of people within environments [6]. It seems that every time we introduce a change at one point, it causes an unexpected response somewhere else—the first fundamental systems principle.

It is argued by Simonovic [7] (based on [6]) that the system in our focus is a social system. It describes the way water resources are interacting with people to clearly define the management problem and determine the best strategies for systems intervention. The water resources system includes four tightly connected subsystems: individuals, organizations, society, and the environment. To sustainably manage water resources, the interactions between the four subsystems must be appropriately mapped.

Individuals are the players in organizations and society and affect the way they behave. As individual decision-makers, they have a direct role in the use and management of water resources. Organizations are used by individuals as an instrument to obtain outcomes that they cannot produce. The structure of the organizations is developed to realize a particular set of goals. Structure of the organizations defines resource and information flows and governs the organizational behavior. Individuals and organizations are subsets of society. The society is a system that encompasses the relationships between people, the rules of behavior and the mechanisms that are used to regulate it. Societies are nested within the environment. The environment includes concrete elements such as water, air, raw materials, natural systems, as well as the universe of ideas including the expectation of future water shortages and future global change impacts that define concern for sustainable water resources management.

Every open system includes inputs of energies—resources—that are transformed into outputs. Systems inputs and outputs include resources, information and values. They link individuals, organizations, society and environment. Information and resource flows link people and organizations. Value systems are attached to information and resource flows. They are generated by the individuals and/or organizations and provide meaning for information and resource flows.

Each subsystem relies on other subsystems and on the environment for its resources. The physical environment applies passive pressure on the subsystems and can limit action by exhausting resources. In that way, the resources can become more valuable (i.e., climate change).

Each of the subsystems utilizes information to make decisions on communicating with other subsystems and the environment. In the case where flows of information from outside of the subsystem are not available, it must rely on its own knowledge that increases the risk that the subsystem may lose connections with the other subsystems.

Since data does not have meaning by itself, interpretation between information and meaning is necessary and provided by values. They provide meaning to flows of information. Flows are then used to determine resource use by each subsystem. Value systems are embedded in the culture of society and organizations. They determine what resources individuals, organizations and societies need. Using value systems, the interpretation of information is provided and behavior of the subsystems is determined.

The decision-making choice is always related to the availability of resources. Feedback information on the availability of resources signals to the decision-maker (individuals, organizations, or society) the subsystem's response to the implemented management procedures. According to [6,7], the most effective options for sustainable water resources management are those that condition access to resources. Each subsystem is using different procedures (combination of options and various interactions) to maximize its access to resources.

The next section of the paper will briefly review the success of the systems approach in management of water resources systems up to now. It is followed by identifying one view of the future that presents the concept of performance-based water resources engineering. The following section illustrates the performance-based concept using an example of a water supply network management under changing conditions. The paper ends with the conclusions.

Systems Approach to Management of Water Resources—A Success Story

During the past five decades, since the introduction of the water resources systems analysis within the Harvard Water Program [8], we have witnessed a great evolution in water resources systems management [9–13]. Three of the characteristics of this evolution are noted in particular [12].

First—*the application of the systems approach to complex water management problems.* It has been recognized as the most important advance in the field of water resources management by providing an improved basis for decision-making.

Second—*transformation of attitude by the water resources management community towards environmental concerns.* The past five decades have brought many examples of initiatives taken for environmental assessment and planning, as well as significant investment in environmental technologies for recovering or removing pollutants.

Third—*introduction of sustainability paradigm.* The publication of the Brundtland Commission's report "Our Common Future" in 1987 started the application of the sustainability principles to water resources decision-making by (a) changing management objectives and (b) obtaining deeper understanding of the complicated inter-relationships between existing ecological, economic and social issues. Brown et al. [13] advocate for water resources systems analysis as a conceptual framework for sustainable management of water resources.

The evolution of water resources systems management is occurring in the context of rapid development of information technology which moved the computer directly into knowledge processing as a partner for more effective decision-making.

Let me repeat the basic definition of a system here. Simonovic [12] defines "a system as a collection of various structural and non-structural elements that are connected and organized in such a way as to achieve some specific objective through the control and distribution of material resources, energy and information". The systems approach is characterized by emergence (the whole is different than the sum of its parts), self-organization (cooperation, interdependence and competition yield stabilizing homeostasis), nonlinearity (small changes in part of the system can have excessively significant effects across the whole), and feedback loops (the outputs of the system affect its inputs).

Let me summarize the current state of the water resources systems approach:

(i)   A very reachable portfolio of applications and the evolution of water resources systems approach today offer a scientific interdisciplinary context for dealing with the complex practical issues of water management and prediction of the water resources future. Together they form the basis for a sustainable water management necessary to address the global water challenges of this century.

(ii)   Systems approach is helping all those who are responsible for water resources management to organize water related information in order to distinguish between the noise and important information and improve the decision-making.

(iii)   The data necessary to understand resource flows and the larger water resources management setting are being identified in close collaboration with the general public to understand the relationships between human behavior and environmental and economic impacts of water resources management decisions [2].

(iv)   The systems approach is helping the improvement of water resources planning and forecasting. Clear articulation of assumptions, use of models, identification of feedback relationships, and monitoring system behavior can help decision-makers better anticipate future conditions and make smarter management decisions.

(v)   The tools of systems analysis (simulation, optimization and multi-objective analysis) are helping to improve the quality of water resources related decision-making [4]. They provide decision-makers with the information for full understanding of the dynamics that direct the interactions between the social (people and economy), natural (water, land and air) and constructed systems (buildings, roads, bridges etc.).

(vi)   The systems approach is contributing to the improvement in human behavior by using systems thinking. It enables everyone involved in water resources management to see themselves as a group of actors in making decisions that involve feedback, developing situations, and advancing the awareness of producing one outcome or another [3].

(vii)   The systems approach today leads to greater practical and safer risk management policies for the simple reason that most water resources systems are nonlinear and therefore hard to predict [5]. Water resources management requires smarter and more adaptable participants, capable of learning and being able to anticipate changing conditions.

A success reached today must contribute to further evolution of the water resources systems approach to successfully address the serious water challenges faced by society. The future activities must continue: to deal with the most difficult complex water problems (that include competing objectives, multidisciplinary cooperation, and changing values); to conduct further practice-based as well as fundamental research (balancing research for basic understanding and providing solutions to current water problems); and provide further capacity building to insure that ranks of water resources systems specialists will not decline (the opposite has been documented by [13]).

2. One View of the Future-Performance-Based Water Eesources Engineering

Performance-based engineering is dealing with the design, evaluation and building of engineered systems that meet—as economically as possible—the uncertain future demands of people and nature in the most economically efficient way. It is an approach to the analysis of any complex system. A system managed in this way should meet quantitative or predictable performance requirements, such as demand load or economic efficiency, without a specific prescribed method for attaining those requirements. This is very different from traditional prescribed standards (code provisions), which mandate specific practices, such as pipe size, levee height, and minimum drinking water quality, for example. Such an approach is very flexible in developing tools and methods to evaluate the entire water resources system management process. The main assumption is that performance levels and objectives can be measured, that performance can be predicted using analytical tools, and that the impact of improved performance can be evaluated to allow rational trade-offs based on lifecycle considerations rather than a single criterion alone, such as construction costs for example.

Much of the current research on performance-based engineering focuses on earthquakes [14,15]. Performance-based engineering offers opportunities for better management of water resource systems faster and more cost effectively. It can be implemented for revitalization of the decaying infrastructure. It can utilize emerging technologies to monitor the strength of existing facilities through sensor technology. It can be deployed in performance control with active control systems and smart materials.

Performance-based engineering also offers great opportunities for research and teaching of the processes involved in the design and construction of engineered water resources systems. Adoption of performance-based engineering requires major changes in practice and education of water resources engineers. Perhaps most important is a shift away from the dependence on empirical and experience-based tools, and toward a design and assessment process based on a scientifically oriented systems approach that emphasizes accurate characterization and prediction of system behavior.

2.1. Challenges

Water infrastructure facilities are designed and managed to withstand demands imposed by their service requirements and by environmental events such as floods, droughts, ice, windstorms and earthquakes. Most of the water resources management decisions are being made according to current prescriptive standards (code provisions) and usually provide adequate levels of safety. However, changing conditions, extreme environmental and human-made events may still result in severe damage and economic losses. In an era of rapid changes in engineering design and construction practices, and heightened public awareness of water infrastructure performance, engineers are now seeking to achieve levels of performance in the built environment beyond what currently is provided by prescriptive standards and to better meet public expectations. This discussion introduces a performance-based engineering approach as the replacement for traditional use of prescriptive standards. Performance-based engineering offers an opportunity for heightening the role of simulation combined with quantitative resilience assessment.

2.2. Need for Performance-Based Water Resources Engineering

Globally changing conditions, including rapid population growth, land use change (especially urbanization) and climate change, are affecting water resources engineering planning, design and operations. Air and surface temperature, and precipitation patterns and intensity are directly linked to climate change [16].

According to IPCC [17] a large proportion (1/6) of the world's population live in snowmelt-fed river basins and will be affected by the seasonal changes in streamflow, a change in the ratio of winter to annual flows, and possibly the reduction in low flows. Sea-level rise will extend areas of salinization of groundwater and estuaries. These changes will result in a decrease in freshwater availability for human consumption and the needs of ecosystems. Increased precipitation intensity and variability is projected to increase the risk of flooding. Higher water temperatures, increased precipitation intensity, and longer periods of low flows exacerbate many forms of water pollution, with impacts on ecosystems, human health, water infrastructure system dependability and operating costs [17].

The presence of global change (especially climate change) complicates the development of risk-informed engineering standards significantly. Current assessments of reliability treat the operational and environmental demands as stationary in nature. This assumption is not defensible when global change effects are considered. Furthermore, the uncertainties in global change effects projected over the 21st century are extremely large. Finally, achieving the necessary consensus on global change effects on the built environment within some standard committees will present challenges.

A number of key questions must be addressed to consider the imperatives of global change in standards development, among them: (i) How should one model the nonstationarity in water-related natural hazard occurrence and intensity that arises as a consequence of global change? (ii) How should these uncertainties be integrated in time-dependent infrastructure performance analysis to estimate future behavior and to demonstrate compliance with performance objectives? (iii) How should we deal with lifecycle cost issues when implementing global change effects in practical design criteria?

One possible answer, proposed in this discussion, is: performance-based engineering based on system simulation modeling and resilience assessment.

2.3. Implementation of Performance-Based Water Resources Engineering

Performance-based engineering has gained traction in earthquake engineering, where the incentives are strongly economic in nature and the shortcomings of traditional prescriptive approaches to design, planning and operations are known [18]. Research is underway to extend the performance-based approach to water resources engineering (including hazards such as flooding, drought, sea level rise and tsunami), and to develop planning, design and operations procedures in which the consequences of competing hazards are properly balanced and investments in damage reduction and recovery can be made appropriately.

Main deficiencies of the prescriptive framework include: (i) checking only a single performance level; (ii) applying only a single system disturbance event; (iii) linear static or dynamic analysis; and (iv) no local acceptance criteria. Current, prescriptive water resources engineering frameworks rely on risk analysis tools for modeling uncertainties associated with water resources decision making related to system loads and responses.

Very different tools will be essential to the successful implementation of performance-based water resources engineering in providing a framework for managing the impacts of external disturbances on the performance of the built environment and for guiding water resources management decisions related to the recovery of existing water infrastructure systems affected by changing conditions. These tools should allow: (i) checking multiple performance levels; (ii) application of multiple system disturbance events; (iii) possible utilization of nonlinear analysis; (iv) implementation of detailed local acceptance criteria; and (v) joint consideration of system structural and nonstructural components.

The performance-based water resources engineering process is illustrated in Figure 2. It starts with the identification of system disturbance as a consequence of global change. System disturbance could be a flood, an extreme precipitation event or a long-term drought event, just to name a few. Selection of performance criteria follows, that should allow for measurement of impacts that system disturbance may have on the system. For example, a performance criterion could be area inundated by flood waters, or the total damage from the drought event, and similar. Each system performance can be measured in its own units. The following step includes identification of alternative options (plans/designs/operations strategies) for responding to the disturbance. Options may include structural solutions (flood protection infrastructure for example) and nonstructural measures (change of regulations for example) alone or combined together. System performance capability is then tested by doing calculation of system performance in response to selected disturbance and alternative response according to a performance criterion. A system simulation approach is recommended for the implementation at this stage. It is a preferable approach because it does not pose any limitations for the complexity of system structure description. Calculated system performance is subject to multiple uncertainties. Risk approach could be one way to assess the system performance. However, the risk approach has many deficiencies. It is static (in time and space). It includes difficulties in assessing probability of extreme events and integrating physical, social, economic and ecological concerns at the same time. Here, it is proposed to integrate system performance into a single measure of dynamic system resilience (in time and space) that can be easily implemented in the broader evaluation of alternative options not limited to the assessment of direct and indirect losses only.

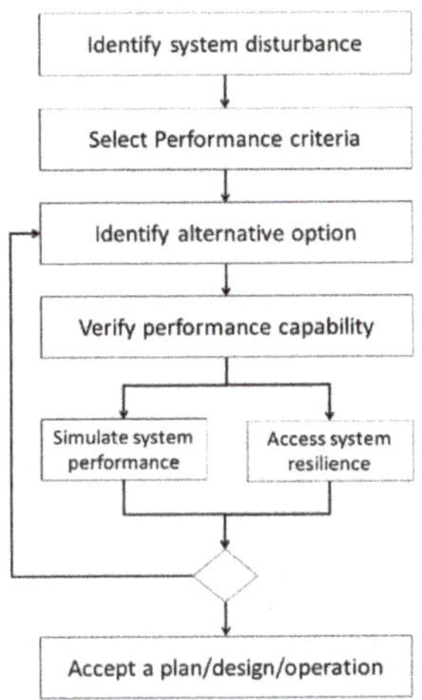

Figure 2. Performance-based engineering process.

The performance-based water resources engineering process in Figure 2 can be implemented (i) in an iterative way by examining alternative options (plans/designs/operational strategies) ahead of system disturbance or (ii) in a real-time by responding to system disturbance and managing recovery from it. Verification of system performance capability is done by combined use of simulation and quantitative resilience assessment (see Figure 2). More details on the tools for supporting the performance-based water resources engineering follow.

2.4. Simulation

The classical simulation approach involves understanding of system structure through decomposition of the problem that helps in the system description. The simulation process starts with identification of elements and their mathematical description. The procedure continues with the development of a computer program based on the mathematical description of the model. In the next step, each model parameter is calibrated, and the model performance is verified using different data. The computer program of the model is then operated using various input data. Detailed analysis of the output is the final step in the simulation process.

The performance-based engineering approach can take advantage of system dynamics simulation, which is defined by Simonovic [12] "as a rigorous method of system description, which facilitates feedback analysis via a simulation model of the effects of alternative system structures and control policies on system behavior. In the context of water resources engineering a system is defined as a collection of elements which continually interact over time to form a unified whole". The underlying map of interactions between the system elements is called the system structure. The term dynamics in the definition refers to change of system behavior over time. A dynamic system is a system in which the variables interact to generate changes over time. The way in which the system elements, or variables, vary over time is referred to as the system behavior. System dynamics simulation is not new to water resources engineering. Multiple applications are documented in the literature (for example see [7]).

System dynamics simulation lends itself well to the assessment of engineering system performance over time. Complex systems can be easily built using object-oriented system dynamics simulation software packages that allow for a high level of detail to be included in the description of system structure. By running deterministic simulations of potential system planning, design and operating conditions, the system dynamics model facilitates investigation of nonlinear behavior in complex water resources infrastructure systems. Outputs from the system dynamics simulation model include the values of variables at each time step in the simulation. Such information gives insight into the system response and recovery, which can be assessed using dynamic resilience.

In order to move away from static estimates of risk towards dynamic estimates of system performance before, during and after the occurrence of an undesirable event, a new approach is necessary that deals with system performance over time. The main recommendation of this discussion is to implement systems dynamics simulation as a foundation for assessing complex water infrastructure system resilience. The methodology involves the utilization of simulation to generate change in infrastructure system performance as a consequence of a wide range of operating conditions. The simulation outputs provide information that can be used to estimate dynamic system resilience by assessing the change in system performance and its adaptive capacity.

2.5. Quantitative Resilience Assessment

The quantitative dynamic resilience measure, first introduced by [19], followed by [20], is defined by Simonovic and Peck [19] as "the ability of a system and its component parts to anticipate, absorb, accommodate or recover from the effects of a system disruption in a timely and efficient manner, including through ensuring the preservation, restoration or improvement of its essential basic structures and functions". Resilience is defined in this way: (a) performs well during periods without system disturbance, and (b) captures a system's adaptation ability to respond during periods when the system is under disturbance. Quantitative resilience is the system characteristic applicable to built and natural physical environments; social and economic systems; and institutions and organizations. Resilience is founded on two basic concepts: system performance level and its adaptive capacity. Figure 3 illustrates generic system performance under a disturbing event. For example, let us consider water supply reservoir release under reduced inflow. System disturbance in this case is the reduced amount of inflow. The performance can be the water supply reservoir release amount expressed in flow units (m^3/s). Generic system performance used for the quantification of dynamic resilience is shown in Figure 3 (after [19] and [21]). Application of numerous adaptation measures results in the change of the performance curve shape (two options presented as (a) and (b) are presented in Figure 3 using dashed lines). For example, proactive measures of water supply demand control may result in curve (a), and reactive measures of ground water supplemental supply may result in curve (b). It should be noted that changing the amount of supplemental supply may place curve (b) at a different location.

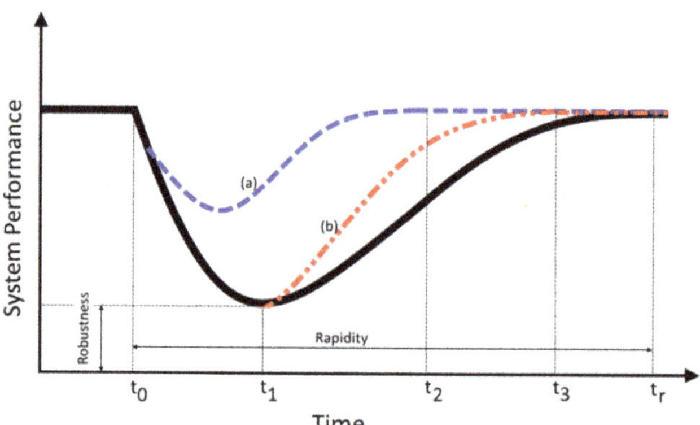

Figure 3. Generic representation of system performance (after [19]).

While traditional risk-based engineering focuses on the reduction of predisturbance vulnerabilities, resilience is realized by considering adaptation options that allow for the system to adapt to changing conditions and increase the ability of the physical, social, economic sectors to maintain some level of performance during the disturbance.

Change of system performance forms the basis for quantification of system resilience. The transformation of system performance into system resilience is captured in Figure 4. Illustration in Figure 4 is not related to the simple example from Figure 3. Notation in Figure 4 includes: t_0—time of the the beginning of the disturbance; t_1—time of the end of system disturbance; t_r—time of the end of the recovery period; $P(t)$—system performance; P_0—initial system performance level; $P_{e'}(t)$—degraded ending system performance level; $P_{e''}(t)$—improved ending system performance level; the area between P_0 and performance line (full black line) $P(t)$ represents the loss of system performance; and the shaded area under the performance line $P(t)$ denotes the system resilience.

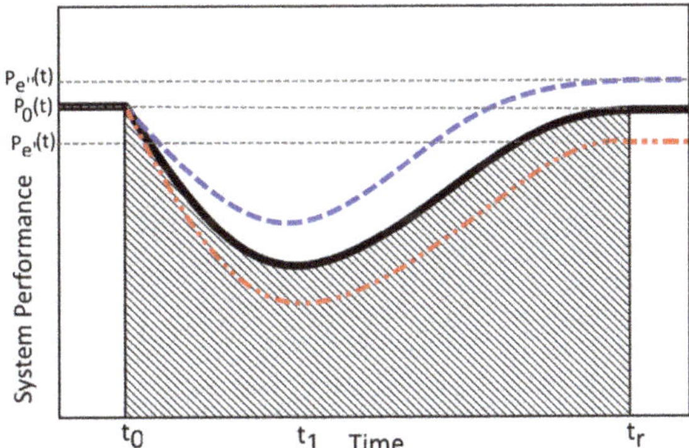

Figure 4. Illustration of transformation of system performance into resilience.

In mathematical form, the loss of performance (ρ) shown in Figure 4 as the area between the start of the system disruption event (t_0) and the end of the disturbance recovery process (t_r):

$$\rho(t) = \int_{t_0}^{t} [P_0 - P(\tau)]d\tau, \; t \in [t_0, t_r] \tag{1}$$

where $P(\tau)$ represents degree of system performance and P_0 is the initial system performance level. The remaining system performance (shaded area in Figure 4) is defined as system resilience $r(t)$, and is obtained by normalizing the value of (ρ):

$$r(t) = 1 - \left(\frac{\rho(t)}{P_0 \times_t (t - t_0)} \right) \tag{2}$$

Normalization eliminates the units of system performance and substitutes them with units of resilience between 0 and 1. Generic presentation of resilience is provided in Figure 5 (this illustration is also not related to the simple example from Figure 3).

The calculation, using system dynamics simulation, of resilience is performed at each point in time by solving the following differential equation:

$$\frac{\partial r(t)}{\partial t} = AC(t) - P(t) \tag{3}$$

where AC stands for adaptive capacity. The solid black line in Figure 5 represents the consequence of integrated system performance (shaded area in Figure 4) under the disturbance with current system adaptation capacity. There are three conceivable outcomes in resilience simulation: (i) return of resilience value to predisturbance level (value of 1), captured by the solid black line in Figure 5; (ii) improved resilience value compared to predisturbance level (ending system performance level $P_{e''}(t)$, resilience value > 1), shown by the blue dashed line in Figure 5; or (iii) declined resilience value compared to predisturbance level (ending system performance level $P_{e'}(t)$, value < 1), shown by the dashed and dotted red line in Figure 5.

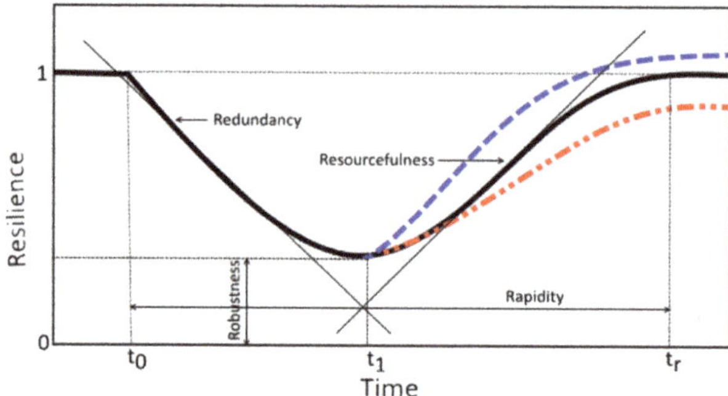

Figure 5. Generic representation of system resilience.

Introduction of dynamic measure of resilience into performance-based water resources engineering offers additional information that can be of value in the decision-making process. The shape of the resilience curve is defined by the system adaptive capacity and it provides additional insights into system robustness, redundancy, resourcefulness and rapidity. They are graphically presented Figure 5. The slope of the declining resilience curve section (time $t_0 < t < t_1$; slope $P_t - SP_{t0}/t-t_0$) defines system redundancy (defined as the inclusion of extra system components which are not firmly necessary to maintain system functioning, in case of failure of other components). The slope of the rising section of the resilience curve (time $t_1 < t < t_r$; slope $P_t-P_{tr}/t-t_r$) offers information about system resourcefulness (defined as the ability to mobilize resources necessary to overcome difficulties caused by system disruption). Robustness of the system (defined as the minimum value of the remaining system performance after the disturbance) and rapidity (duration of system performance under the disturbance) are clearly illustrated with the system resilience level at time t_1 and difference in time between t_0 and t_r, respectively. Implementation of numerous adaptation actions results in the change of resilience curve shape.

The performance-based water resources engineering approach proposed in this paper rests on the power of system simulation and quantitative dynamic resilience. The simulation approach is a tool for the analyses of water resources system performance. Use of resilience as a metric for the assessment of system response to changing conditions provides a much more complete insight into the characteristics of the system structure and system response, allowing for a more meaningful investigation of system vulnerabilities. Various planning/design/operations options including capital upgrades and maintenance could be compared by using resilience to measure the loss of performance due to undesirable events, system response time and level of performance after recovery. Overall system resilience can be assessed by looking at the resilience of individual system components and taking into consideration their interactions.

3. An Illustrative Example

A simplified water supply network problem, modified after Kong et al. [22], is selected only as an illustrative application of the performance-based water resources engineering methodology. Water supply is one of the essential services that provides support for the economic productivity, security, and population quality of life. There are practical links between disaster risk management, global change adaptation and sustainable development leading to the reduction of disaster risk and re-enforcing resilience as a new development paradigm. Both, system disturbance types, natural (such as floods, severe weather, earthquakes, hurricanes, and similar) or human caused (such as terrorist threats, chemical spills, and similar), always affect geographically restricted areas. In this example, a geographical location of interest (where the water supply network components are located) is presented using the cell space method. The water supply network model is founded on the network theory. The model of a system includes two basic components, nodes and edges. Water supply network is represented as a complex system of intakes, reservoirs, pumping stations, pipelines, conduits, and other components by which water is collected, cleaned, stored, and distributed to an urban area. In this network, intakes, reservoirs and pumping stations are denoted as nodes with different characteristics and water distribution pipes, and conduits are denoted as edges. Water supply network is a directed network, as the water flows from an intake to pumping stations and storage facilities through distribution pipes. In the directed networks, the downstream nodes and edges will not be able to operate unless all the upstream nodes and edges function normally. A detailed mathematical simulation model of network structure and dynamic behavior is available in [22].

The network example system includes 16 (4 × 4) cells shown in Figure 6. To simplify the network model simulation, one node is assumed to exist in every cell, as shown in Figure 6. Blue color filled nodes are representing main components of the water supply network, such as intakes, treatment plants, etc. Blue color empty nodes represent storage facilities such as pump stations, reservoirs, etc. The edges are used for representation of water transmission pipes and conduits. The example network includes 16 main components, storage facilities and pump stations, and 17 water transmission pipes and conduits.

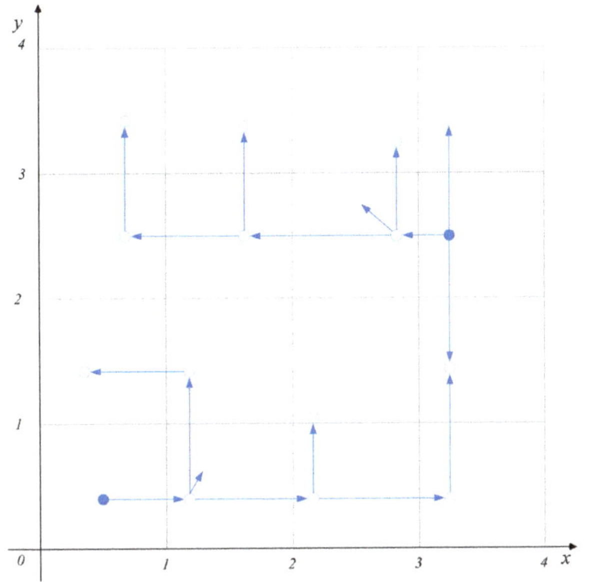

Figure 6. Example water supply network.

The problem to be addressed in this illustrative example is the problem of network recovery after a major flood disaster. In the network theory, disturbance to the infrastructure system is always captured by the removal of nodes and/or edges from the system network. It is assumed that components of the water supply network layer in the same cell are affected simultaneously. Fluvial flooding develops slowly and can last for days and weeks. The water usually spreads over a large area and inundates infrastructure network components located in the floodplains.

Following the performance-based engineering process (see Figure 2) the *first step* is the identification of disturbance. It is assumed that (i) flood occurs once, (ii) affects a large area, and (iii) lasts over a longer time. Many water supply network elements located in the floodplains are affected, due to submergence. In this example, water supply network elements located in the four bottom and four right cells (see Figure 6), with coordinates {$0 \leq x \leq 4, 0 \leq y \leq 1$} $\cup$ {$3 \leq x \leq 4, 1 \leq y \leq 4$}, are assumed to be affected. The selected flood could be a historical event or any statistical flood event. The whole process can be repeated for as many disturbance events as the user would like to investigate.

In the *second* step performance criteria is selected as a simple state of the water supply network. To simplify the simulation, the network is considered to be in one of two states: function and malfunction—denoted with the value of 1 or 0 as shown in Figure 7.

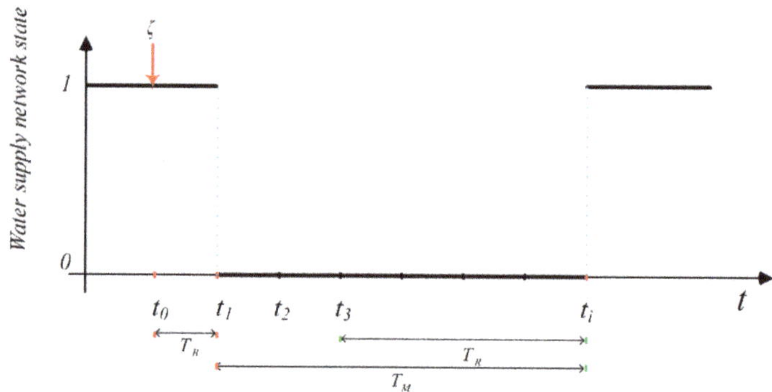

Figure 7. Performance dynamics of the example water supply network after a large flood.

The notation used in Figure 7 includes: t_0—time of occurrence of disturbing event (assumed flood in this case); T_B—buffering time; T_R—repair time; T_M—network malfunction time.

In the *third step* a set of five repair options taken from [22] is being identified[: (i) first repair components that failed first (*RS-FF*)—this approach is usually used during the emergency when time and space may not be available for a more comprehensive response; (ii) first repair components that failed last (*RS-FL*); (iii) first repair important components independently (*RS-IE*)—this strategy is used to maximize the benefits of a water supply sector in an interconnected case (for example when water supply is connected to electricity supply network, information network, etc.); (iv) first repair the obviously dependent components (*RS-OD*)—this approach considers obvious or physical interdependencies of infrastructure elements (for example, node–node, node–edge and node–edge-cluster dependencies); and (v) first repair the hidden dependent elements (*RS-HD*)—the fifth repair approach takes interdependencies between the water supply network and other networks that water may be connected to (usually illustrated as node–edge-path dependencies).

The *fourth step* of performance-based engineering process involves verification of system capability by simulating system performance and performing resilience assessment. The example water supply network system performance simulation is performed following the flow diagram in Figure 8. System performance is assessed for all five response strategies. General water supply system simulation (in Figure 8) is adopted to all five response strategies (details are available in Kong et al., 2019).

Simulation results, presented in Figure 9, clearly show the difference in system performance as a function of the response strategy. The black line (P_0) in Figure 9 shows system performance without any response. The other five lines are describing system performance according to the selected five response strategies (see the Figure 9 legend). Water supply network performance under *RS-OD* and *RS-HD* outperforms performance under other response strategies, and *RS-FL* and *RS-IE* result in the worst performance. Simulation results under all five strategies confirm that in this example case, the water supply system cannot exceed the initial performance level after the flood. The possible explanation for these results is that no water supply network system improvements can be built in a short period of time. Therefore, the additional resilience characteristic of rapidity and the end of recovery time are determined as the time when the system performance recovers to the preflood performance level.

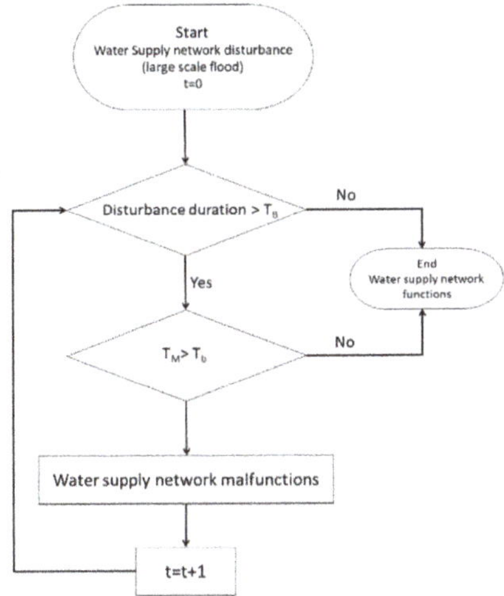

Figure 8. Flow diagram of the water supply network performance simulation under a large flood.

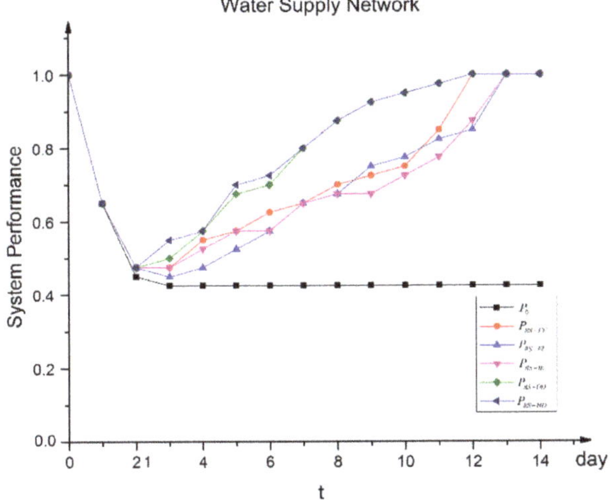

Figure 9. Performance of the example water supply network after a large flood under five response strategies.

The resilience of the example water supply network under various response options is calculated using a modified Equation (2), adjusted for the network systems [22].

The final, *fifth step,* of the performance-based water resources engineering includes the decision-making based on the results of system simulation and resilience assessment. The example water supply network resilience values follow the system performance and are shown in Figure 10. Resilience is the integral of the system adaptive capacity. The higher value, the more resilient the system. As shown by Equations (1) and (2), the adaptive capacity, *AC*, is a function of response option *RS*. Application of different response strategies *RS* results in different system resilience of the same water supply network. In the example case, the resilience value under *RS-HD* is the highest. As the number of destroyed elements is always lower than the number of malfunctioning elements. The rapidity (recovery time) under *RS-HD* is longer than the recovery time under *RS-OD*. If the average resilience during the recovery time is compared, the *RS-HD* approach results in a higher resilience than the *RS-OD*, and the recovery time (rapidity) of the latter is longer. This phenomenon is common for water supply networks that include multiple interdependencies. The results clearly show that the recovery time (rapidity) should be taken into consideration for a more wide-ranging decision making.

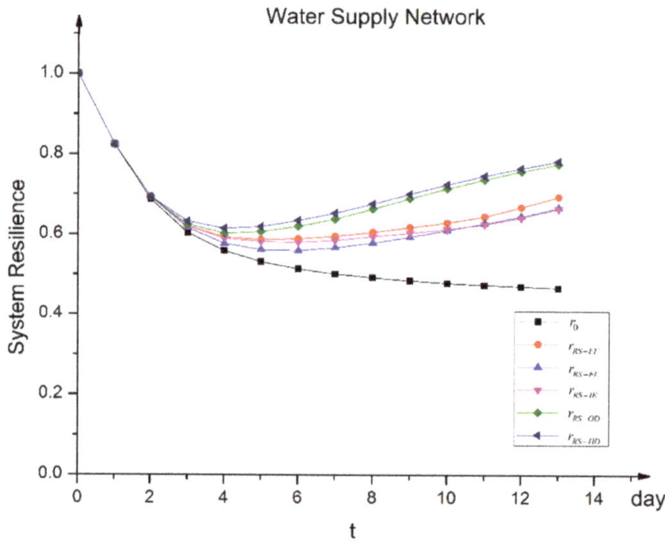

Figure 10. Resilience of the example water supply network after a large flood under five response strategies.

The application of performance-based analysis in the example case shows that from the five proposed repair options, two (*RS-HD* and *RS-OD*) are clearly outperforming the others. Both of them include more interdependencies in the system recovery process. They are recommended for application and further enhancement of the decision-making that can be done by including other characteristics of the quantitative resilience measure, as for example, rapidity.

4. Conclusions

The systems approaches to managing water resources provide proven strategies for more efficient resolution of water resources management challenges imposed by global change. Looking forward from the current practice, this paper explores the future opportunities based on the advances in systems theory that can, on a broader scale, majorly transform management of water resources. The performance-based engineering is proposed as the replacement for the current prescriptive approach based on the risk-informed engineering standards which are very difficult to implement in the presence of global change (especially climate change).

Performance-based engineering is the design, evaluation and construction of engineered systems that meet the uncertain future demands of owner-users and nature. It is an approach to the analysis of any complex system. The performance-based water resources engineering offers an opportunity for heightening the role of systems science, especially simulation, combined with quantitative resilience assessment for addressing various sources of uncertainty. The implementation of the performance-based water resources engineering is presented as a five-step approach that is taking advantage of system simulation and assessment of quantitative resilience. Performance-based engineering approach is suggested for use in system dynamics simulation as defined earlier in the paper. Assessment of system performance obtained by simulation is to be done using the quantitative dynamic resilience measure.

A simple water supply network problem is selected as an illustrative application of the performance-based water resources engineering. The problem addressed in this illustrative example is the problem of water supply network recovery after a major flood disaster. A set of five network repair options is evaluated by using network performance simulation and resilience assessment.

The performance-based water resources engineering can be implemented in solving complex planning, design and operations problems. It is identified as a methodological framework to improve water resources management in the face of rapid climate change so that sustainability becomes the standard, not the infrequent, success story.

Funding: The author is thankful for the financial support provided by the Natural Sciences and Engineering Council of Canada, BC Hydro and Chaucer Re for the work presented in this paper.

Conflicts of Interest: The author declares no conflict of interest.

References

1. Simonovic, S.P. *Systems Approach to Management of Disasters: Methods and Applications*; John Wiley & Sons Inc.: New York, NY, USA, 2011; p. 348. ISBN 978-0-470-52809-9.
2. UN-WWAP (United Nations World Water Assessment Programme). *UN World Water Development Report 1: Water for People, Water for Life*; UNESCO and Berghahn Books: Paris, France; New York, NY, USA; Oxford, UK, 2003.
3. UN-WWAP *UN World Water Development Report 2: Water, a Shared Responsibility*; UNESCO and Berghahn Books: Paris, France; New York, NY, USA; Oxford, UK, 2006.
4. UN-WWAP. *UN World Water Development Report 3: Water in a Changing World*; UNESCO: Paris, France; Earthscan: London, UK, 2009.
5. UN-WWAP. *UN World Water Development Report 4: Managing Water under Uncertainty and Risk*; UNESCO: Paris, France, 2012.
6. Martin, P. *Just What are We Trying to Manage, Anyway?* Profit Foundation Pty Ltd.: Melbourne, Australia, 2000.
7. Simonovic, S.P. Water Resources Management: A Systems View. *Water Front* **2009**, *1*, 12–13.
8. Maass, A.; Hufschmidt, M.; Dorfman, R.; Thomas, H.; Marglin, S.; Fair, G. *Design of Water-Resource Systems: New Techniques for Relating Economic Objectives, Engineering Analysis, and Governmental Planning*; Harvard Univ. Press: Cambridge, MA, USA, 1962.
9. Loucks, D.P.; Stedinger, J.R.; Haith, D.A. *Water Resources Systems Planning and Analysis*; Prentice Hall: Englewood Cliffs, NJ, USA, 1981.
10. Yeh, W.W.G. Reservoir management and operations models: A state-of-the-art review. *Water Resour. Res.* **1985**, *21*, 1797–1818. [CrossRef]
11. Loucks, D.P.; van Beek, E. *Water Resources Systems Planning and Management: An Introduction to Methods, Models and Applications*; UNESCO: Paris, France, 2005; p. 680.
12. Simonovic, S.P. *Managing Water Resources: Methods and Tools for a Systems Approach*; UNESCO: Paris, France; Earthscan James & James: London, UK, 2009; p. 576. ISBN 978-1-84407-554-6.
13. Brown, C.M.; Lund, J.R.; Cai, X.; Reed, P.M.; Zagona, E.A.; Ostfeld, A.; Hall, J.; Characklis, G.W.; Yu, W.; Brekke, L. The future of water resources systems analysis: Toward a scientific framework for sustainable water management. *Water Resour. Res.* **2015**, *51*, 6110–6124. [CrossRef]
14. Porter, K.A. An Overview of PEER's Performance-Based Earthquake Engineering Methodology. In Proceedings of the Ninth International Conference on Applications of Statistics and Probability in Civil Engineering (ICASP9), San Francisco, CA, USA, 6–9 July 2003.
15. FEMA. *Seismic Performance Assessment of Buildings, Volume 1—Methodology*; Federal Emergency Management Agency: Washington, DC, USA, 2012.
16. UNESCO UN-Water. *United Nations World Water Development Report 2020: Water and Climate Change*; UNESCO: Paris, France, 2020; p. 235.

17. IPCC—Intergovernmental Panel on Climate Change. Summary for Policymakers. In *Climate Change, The Physical Science Basis*; Contribution of Working Group I to the Fifth Assessment Report of the Intergovernmental Panel on Climate Change; Stocker, T.F., Qin, D., Plattner, G.-K., Tignor, M., Allen, S.K., Boschung, J., Nauels, A., Xia, Y., Bex, V., Midgley, P.M., Eds.; Cambridge University Press: Cambridge, UK; New York, NY, USA, 2013.

18. Attar, A.; Lounis, Z. (Eds.) NRC—National Research Council Canada. Climate Change & Codes Implementation. In Proceedings of the International Workshop, Ottawa, ON, Canada, 18–19 January 2017; p. 224.

19. Simonovic, S.P.; Peck, A. Dynamic Resilience to Climate Change Caused Natural Disasters in Coastal Megacities—Quantification Framework. *Br. J. Environ. Climate Change* **2013**, *3*, 378–401. [CrossRef] [PubMed]

20. Cutter, S.L.; Barnes, L.; Berry, M.; Burton, C.; Evans, E.; Tate, E.; Webb, J. A place-based model for understanding community resilience to natural disasters. *Global Environ. Change* **2008**, *18*, 598–606. [CrossRef]

21. Simonovic, S.P. From risk management to quantitative disaster resilience: A paradigm shift. *Int. J. Safety Security Eng.* **2016**, *6*, 85–95. [CrossRef]

22. Kong, J.; Simonovic, S.P.; Zhang, C. Resilience Assessment of Interdependent Infrastructure Systems: A Case Study Based on Different Response Strategies. *Sustainability* **2019**, *11*, 6552. [CrossRef]

Book Review

Transboundary Hydro-Governance: From Conflict to Shared Management: Book Review. Written by Jacques Ganoulis and Jean Fried. Springer: Cham, Switzerland, 2018, 222 pages. ISBN 978-3-319-78624-7; eBook ISBN 978-3-319-78625-4

Slobodan P. Simonovic

Department of Civil and Environmental Engineering, University of Western Ontario, London, ON N6A 3K7, Canada; simonovic@uwo.ca

Received: 20 September 2019; Accepted: 25 September 2019; Published: 27 September 2019

This book came as a support of the UNESCO International Hydrological Programme (IHP) activities on the International Shared Aquifer Resources Management (ISARM) project launched in the year 2000, with the goal of developing wise practices and guidance tools for the shared management of groundwater resources and to contribute to the multifaceted efforts required for global water cooperation. A key result of the project was the publication by UNESCO of the first world map of 592 transboundary aquifers. In spite of progress made, UNESCO's work required continuation. The book written by Jack Ganoulis and Jean Fried, with both authors being long-term collaborators of the UNESCO IHP, comes as a significant contribution to the transboundary groundwater education and training, offering a bridge between theory and practice.

The objectives set by the authors for the book included addressing the very important question of "water security" in an era of "major systemic risk" for humanity. The book starts with an old task of balancing water availability and water demand through "hydro-governance", defined as "an interactive process for managing at different levels and with different actors all kinds of water, including not only natural water resources but also new-water from wastewater after recycling and seawater desalination" (page vii). The book is correctly focusing on "transboundary hydro-governance" of water transboundary catchments that cover almost half of the world's land surface, including about 60% of global river flow and an area inhabited by 40% of the world's population. The main innovation introduced in the book is "integrated transboundary hydro-governance", which includes both surface water and groundwater. The book provides methodologies, practical tools, and examples of "effective transboundary hydro-governance".

The book is divided into three parts and eight chapters, starting with the main characteristics of transboundary waters (Part I), followed by the definition of transboundary hydro-governance (Part II), and ending with the discussion of transboundary hydro-governance in practice (Part III).

The first part offers (in three chapters) basic information on transboundary waters, water security, and topics of water conflicts and cooperation. The justification for employing a shared approach is clearly presented, and useful recommendations are provided for collaborative scientific approaches. Water security is also analyzed in this part, starting from the water variability in space and time. The ending chapter of Part I suggests the recognition of water districts at risk of conflict and the assessment and classification of different causes and various types of transboundary conflicts. The ending section focuses on the investigation of conflict prevention strategies and the ways to reverse potential conflicts into cooperation.

The second part of the book (in three chapters) provides clear distinctions between management, policy, and governance and analyzes some basic tools used to implement these concepts. This part is a bit more technical. The main innovation presented in this section of the book is a multi-disciplinary integrated approach and its adoption to international settings. Discussion extends to an examination of what international joint institutions can be established and instruments that are needed to implement transboundary hydro-governance in practice. Tools and instruments (economic, shared, legal, and diplomatic) are reviewed in the last chapter of the Part II.

The third part of the book (in two chapters) is devoted to examples of hydro-governance in practice. Organizations and examples are studied from all around the world. From Danube to Mekong, and from Rhine to Senegal, the authors systematically analyze the pros and cons of approaches that lead to "good" transboundary hydro-governance. The analytical model of good hydro-governance can be derived from the experience presented in the book. However, in agreement with the authors and based on my own personal experience, there is still a long way to go.

This book is timely and pragmatically addresses worldwide problems, like climate change and significant floods and droughts, especially in parts of the world where every drop of water counts. The book can be used by experts and practitioners. Examples from the book could serve university courses on water resources management. The book has the potential to be a guiding manual and a reference tool.

Funding: This research received no external funding.

Conflicts of Interest: The author declares no conflict of interest.

MDPI
St. Alban-Anlage 66
4052 Basel
Switzerland
Tel. +41 61 683 77 34
Fax +41 61 302 89 18
www.mdpi.com

Water Editorial Office
E-mail: water@mdpi.com
www.mdpi.com/journal/water

www.ingramcontent.com/pod-product-compliance
Lightning Source LLC
LaVergne TN
LVHW071530170726
843515LV00008B/2107